T0211054

Lecture Notes in Computer Science　　10503

Commenced Publication in 1973
Founding and Former Series Editors:
Gerhard Goos, Juris Hartmanis, and Jan van Leeuwen

Editorial Board

David Hutchison
  Lancaster University, Lancaster, UK
Takeo Kanade
  Carnegie Mellon University, Pittsburgh, PA, USA
Josef Kittler
  University of Surrey, Guildford, UK
Jon M. Kleinberg
  Cornell University, Ithaca, NY, USA
Friedemann Mattern
  ETH Zurich, Zurich, Switzerland
John C. Mitchell
  Stanford University, Stanford, CA, USA
Moni Naor
  Weizmann Institute of Science, Rehovot, Israel
C. Pandu Rangan
  Indian Institute of Technology, Madras, India
Bernhard Steffen
  TU Dortmund University, Dortmund, Germany
Demetri Terzopoulos
  University of California, Los Angeles, CA, USA
Doug Tygar
  University of California, Berkeley, CA, USA
Gerhard Weikum
  Max Planck Institute for Informatics, Saarbrücken, Germany

More information about this series at http://www.springer.com/series/7407

Nathalie Bertrand · Luca Bortolussi (Eds.)

Quantitative Evaluation of Systems

14th International Conference, QEST 2017
Berlin, Germany, September 5–7, 2017
Proceedings

 Springer

Editors
Nathalie Bertrand
Inria
Rennes
France

Luca Bortolussi
University of Trieste
Trieste
Italy

ISSN 0302-9743   ISSN 1611-3349   (electronic)
Lecture Notes in Computer Science
ISBN 978-3-319-66334-0   ISBN 978-3-319-66335-7   (eBook)
DOI 10.1007/978-3-319-66335-7

Library of Congress Control Number: 2017949515

LNCS Sublibrary: SL1 – Theoretical Computer Science and General Issues

© Springer International Publishing AG 2017
This work is subject to copyright. All rights are reserved by the Publisher, whether the whole or part of the material is concerned, specifically the rights of translation, reprinting, reuse of illustrations, recitation, broadcasting, reproduction on microfilms or in any other physical way, and transmission or information storage and retrieval, electronic adaptation, computer software, or by similar or dissimilar methodology now known or hereafter developed.
The use of general descriptive names, registered names, trademarks, service marks, etc. in this publication does not imply, even in the absence of a specific statement, that such names are exempt from the relevant protective laws and regulations and therefore free for general use.
The publisher, the authors and the editors are safe to assume that the advice and information in this book are believed to be true and accurate at the date of publication. Neither the publisher nor the authors or the editors give a warranty, express or implied, with respect to the material contained herein or for any errors or omissions that may have been made. The publisher remains neutral with regard to jurisdictional claims in published maps and institutional affiliations.

Printed on acid-free paper

This Springer imprint is published by Springer Nature
The registered company is Springer International Publishing AG
The registered company address is: Gewerbestrasse 11, 6330 Cham, Switzerland

Preface

Welcome to the proceedings of QEST 2017, the 14th International Conference on Quantitative Evaluation of Systems. QEST is a leading forum on quantitative evaluation and verification of computer systems and networks, through stochastic models and measurements. This year's QEST was held in Berlin, Germany, and collocated with the 28th Conference on Concurrency Theory (CONCUR 2017), the 15th International Conference on Formal Modeling and Analysis of Timed Systems (FORMATS 2017), and the 14th European Performance Engineering Workshop (EPEW 2017).

As one of the premier fora for research on quantitative system evaluation and verification of computer systems and networks, QEST covers topics including classic measures involving performance and reliability, as well as quantification of properties that are classically qualitative, such as safety, correctness, and security. QEST welcomes measurement-based studies as well as analytic studies, diversity in the model formalisms and methodologies employed, as well as development of new formalisms and methodologies. QEST also has a tradition in presenting case studies, highlighting the role of quantitative evaluation in the design of systems, where the notion of system is broad. Systems of interest include computer hardware and software architectures, communication systems, embedded systems, infrastructural systems, and biological systems. Moreover, tools for supporting the practical application of research results in all of the aforementioned areas are also of interest to QEST. In short, QEST aims to encourage all aspects of work centered around creating a sound methodological basis for assessing and designing systems using quantitative means.

This year's edition of QEST comes with the novelty of special sessions on frontier topics in the current research landscape. The two topics selected this year are Smart Energy Systems over the Cloud and Machine Learning and Formal Methods.

The Program Committee (PC) consisted of 33 experts and we received a total of 58 submissions. Each submission was reviewed by three reviewers, either PC members or external reviewers. Based on the reviews and the PC discussion phase, 20 full papers and 4 tool demonstration papers were selected for the conference program. The two special topics, Smart Energy Systems over the Cloud and Machine Learning and Formal Methods, attracted several submissions, leading to two special sessions of three papers each.

The program was greatly enriched with the QEST keynote talk of Romualdo Pastor-Satorras (University of Catalunya, Spain), a joint keynote talk with Formats 2017 of Morten Bisgaard (GomSpace, Denmark) and the joint keynote talk with Concur 2017 and Formats 2017 of Hongseok Yang (University of Oxford, UK). We believe the overall result is a high-quality conference program of interest to QEST 2017 attendees and other researchers in the field.

We would like to thank a number of people. Firstly, thanks to all the authors who submitted papers, as without them there simply would not be a conference. In addition, we would like to thank the PC members and the additional reviewers for their hard work and for sharing their valued expertise with the rest of the community, as well as EasyChair for supporting the electronic submission and reviewing process. We are also indebted to Alfred Hofmann and Anna Kramer for their help in the preparation of this LNCS volume, and we thank Springer for kindly sponsoring the prize for the best paper award. Also thanks to the Local Organization Chair and General Chair, Katinka Wolter, for her dedication and excellent work. Finally, we would like to thank Jane Hillston, chair of the QEST Steering Committee, for her guidance throughout the past year, as well as the members of the QEST Steering Committee.

We hope that you find the conference proceedings rewarding and will consider submitting papers to QEST 2018.

July 2017

Nathalie Bertrand
Luca Bortolussi

Organization

General Chair

Katinka Wolter   Free University Berlin, Germany

Program Co-chairs

Nathalie Bertrand   Inria Rennes, France
Luca Bortolussi   University of Trieste, Italy

Workshop Chair

William Knottenbelt Imperial College London, UK

Tools Chair

Marco Paolieri   University of Southern California, USA

Tutorials Chair

Susanna Donatelli   University of Turin, Italy

Student Forum Chair

Huaming Wu   Tianjin University, China

Publicity Co-chairs

Ezio Bartocci   TU Wien, Austria
Antonio Filieri   Imperial College London, UK

Local Organization

Katinka Wolter   Free University Berlin, Germany

Steering Committee

Alessandro Abate   University of Oxford, UK
Javier Campos   University of Zaragoza, Spain
Pedro D'Argenio   Universidad Nacional de Cordoba, Argentina
Jane Hillston   University of Edinburgh, UK
András Horváth   University of Turin, Italy

Joost-Pieter Katoen	RWTH Aachen University, Germany
William Knottenbelt	Imperial College London, UK
Andrea Marin	University of Venice, Italy
Gethin Norman	University of Glasgow, UK
Anne Remke	University of Münster, Germany
Enrico Vicario	University of Florence, Italy

Program Committee

Alessandro Abate	University of Oxford, UK
Erika Abraham	RWTH Aachen University, Germany
Gul Agha	University of Illinois at Urbana-Champaign, USA
Nail Akar	Bilkent University, Turkey
Varsha Apte	Indian Institute of Technology Bombay, India
Ezio Bartocci	TU Wien, Austria
Tomas Brazdil	Masaryk University, Czech Republic
Ana Busic	Inria Paris, France
Giuliano Casale	Imperial College London, UK
Florin Ciucu	University of Warwick, UK
Andres Ferragut	Universidad ORT, Uruguay
Antonio Filieri	Imperial College London, UK
Jane Hillston	University of Edinburgh, UK
András Horváth	University of Turin, Italy
Kaustubh Joshi	AT&T Labs Research, USA
William Knottenbelt	Imperial College London, UK
Jan Kretinsky	TU Munich, Germany
Boris Köpf	IMDEA Software Institute, Spain
Fumio Machida	NEC Corporation, Japan
Paulo Maciel	Universidade Federal de Pernambuco, Brazil
Andrea Marin	University of Venice, Italy
Annabelle McIver	Macquarie University, Australia
Sasa Misailovic	University of Illinois at Urbana-Champaign, USA
Sayan Mitra	University of Illinois at Urbana-Champaign, USA
Gethin Norman	University of Glasgow, UK
Pavithra Prabhakar	Kansas State University, USA
Guido Sanguinetti	University of Edinburgh, UK
Miklos Telek	Budapest University of Technology and Economics, Hungary
Benny Van Houdt	University of Antwerp, Belgium
Enrico Vicario	University of Florence, Italy
Carey Williamson	University of Calgary, Canada
Huaming Wu	Tianjin University, China
Lijun Zhang	Chinese Academy of Sciences, China

Additional Reviewers

Ashok, Pranav
Brinkmann, Andre
Caravagna, Giulio
Carnevali, Laura
Chen, Li
Ganty, Pierre
Garcia Soto, Miriam
Genest, Blaise
Hahn, Ernst Moritz
Hasanbeig,
    Mohammadhosein
Horvath, Illes
Howar, Falk
Jansen, David N.
Jansen, Nils
Konnov, Igor
Krämer, Julia
Kucera, Antonin
Kwon, Youngmin
Lal, Ratan
Li, Wen-Juan
Luisa Vissat, Ludovica
Massa Ferreira Lima,
    Ricardo

Meggendorfer, Tobias
Meszaros, Andras
Michaelides, Michalis
Milios, Dimitrios
Nenzi, Laura
Novotný, Petr
Palmskog, Karl
Paolieri, Marco
Peruffo, Andrea
Polgreen, Elizabeth
Rehak, Vojtech
Rizk, Amr
Rossi, Sabina
Silva, Bruno
Spegni, Francesco
Turrini, Andrea
Vandin, Andrea
Wijesuriya, Viraj Brian
Wimmer, Ralf
Xu, Ming
Yadav, Anshu
Zamani, Majid
Zhang, Guanqun
Zhu, Lulai

Contents

Parametric Verification

Machine Learning and Formal Methods (Special Session)

Tools

Statistical Model Checking

Probabilistic Modelling

Extending Parikh's Theorem to Weighted and Probabilistic Context-Free Grammars

Vijay Bhattiprolu[1], Spencer Gordon[2(✉)], and Mahesh Viswanathan[2]

[1] Carnegie Mellon University, Pittsburgh, PA 15213, USA
vpb@cs.cmu.edu
[2] University of Illinois at Urbana-Champaign, Urbana 61801, USA
{slgordo2,vmahesh}@illinois.edu

Abstract. We prove an analog of Parikh's theorem for weighted context-free grammars over commutative, idempotent semirings, and exhibit a stochastic context-free grammar with behavior that cannot be realized by any stochastic right-linear context-free grammar. Finally, we show that every unary stochastic context-free grammar with polynomially-bounded ambiguity has an equivalent stochastic right-linear context-free grammar.

1 Introduction

Two words u, v over an alphabet Σ are said to be Parikh equivalent, if for each $a \in \Sigma$, the number of occurrences of a in u and v are the same. The Parikh image of a language L, is the set of Parikh equivalence classes of words in L. One of the most celebrated results in automata theory, Parikh's theorem [27], states that for any context-free language L, there is a regular language L' such that the Parikh images of L and L' are the same. For example, the context-free language $\{a^n b^n \mid n \geq 0\}$ has the same Parikh image as the regular language $(ab)^*$; both the Parikh images only consist of those equivalence classes where the numbers of as is equal to the number of bs. An important and immediate consequence of this result is that every context-free language over the unary alphabet is in fact regular. Parikh's theorem has found many applications—in automata theory to prove non-context-freeness of languages [13], decision problems for membership, universality and inclusions involving context-free languages and semi-linear sets [11,17–19]; in verification of subclasses and extensions of counter machines [8,11,14,15,20,23,33,35,37]; automata and logics over unranked trees with counting [2,34]; PAC-learning [23].

Weighted automata [10,32] are a generalization of classical automata (finite or otherwise) in which each transition has an associated weight from a semiring. Recall that a semiring is an algebra with two operations $\oplus$ and $\otimes$ such that $\oplus$ is a commutative monoid operation, $\otimes$ is a monoid operation, $\otimes$ distributes

V. Bhattiprolu—This work was started while this author was at the University of Illinois, Urbana-Champaign.

M. Viswanathan—Partially suported by NSF CNS 1314485.

© Springer International Publishing AG 2017
N. Bertrand and L. Bortolussi (Eds.): QEST 2017, LNCS 10503, pp. 3–19, 2017.
DOI: 10.1007/978-3-319-66335-7_1

over $\oplus$, and the identity of $\oplus$ is an anhilator for $\otimes$. Unlike classical automata that compute Boolean-valued functions over words, weighted automata compute more general functions over words—the weight of an accepting computation is the product of the weights of its transitions, and the weight of a word is the sum of the weights of all its accepting computations. Since the seminal paper of Schützerberger [32], weighted automata have inspired a wealth of extensions and further research (see [10] for a recent handbook compilation). Weighted automata have found applications in verification [5,6,9,24], reasoning about competitive ratio of online algorithms [1], digital image compression [7,16,21,22], in speech-to-text processing [4,25,26], and data flow analysis [30,31]. A special case of weighted automata are probabilistic automata [28,29] that model randomized algorithms and stochastic uncertainties in the system environment.

In this paper, we investigate whether Parikh's theorem can be generalized to the weighted case. In particular we investigate if for any weighted context-free grammar G there is a weighted right-linear grammar G' such that for any Parikh equivalence class C, the sum of the weights of words in C under G and G' is the same. It is easy to see that if the weight domain is not commutative (i.e., $\otimes$ is not a commutative operation) then Parikh's theorem does not hold. Thus we focus our attention on commutative weight domains.

Our first result concerns weight domains that are additionally idempotent, which means that $\oplus$ is an idempotent operation. A classical example of such a semiring is the min-plus or tropical semiring over the natural numbers where min is the "addition" operation, and + is the "product" operation. We show that Parikh's theorem does indeed hold for weighted automata over commutative, idempotent semirings.

Next, we show that our assumption about idempotence of semirings is necessary. In particular, we give an example of a stochastic context-free grammar G over the unary alphabet such that the function computed by G cannot be realized by any stochastic right linear grammar.

Our last result concerns unary grammars that are polynomially ambiguous. Recall that a grammar is polynomially ambiguous if there is a polynomial p such that on any word of length n in the language, the number of derivation trees for the word is bounded by $p(n)$. We prove that Parikh's theorem extends for such grammars. Specifically, we show that, over the unary alphabet, any probability function realized by a stochastic context-free grammar can also be realized by a right-linear grammar. Though we present this result in the context of stochastic grammars, the proof applies to any polynomially ambiguous weighted context-free grammar over a semiring that is commutative, but not necessarily idempotent.

The rest of the paper is organized as follows. We introduce the basic models and notation in Sect. 2. The Parikh's theorem for weighted automata over commutative, idempotent semirings is presented in Sect. 3. In Sect. 4, we present an example unary stochastic context-free grammar, and show that there is no stochastic right-linear grammar that is equivalent to it. Section 5 contains our proof for Parikh's theorem for polynomially ambiguous grammars. Eventhough this proof is presented in the context of stochastic grammars, it is easy to see that it extends to any weighted context free grammar over a commutative (but not necessarily idempotent) semiring. Finally, we present our conclusions and directions for future work in Sect. 6.

2   Preliminaries

Strings. Let us fix a finite string/word $w \in \Sigma^*$ over Σ. For a subset $\Gamma \subseteq \Sigma$, $w\!\upharpoonright_\Gamma$ will denote the string over Γ obtained by removing the symbols not in Γ from w. The **Parikh map**, or Parikh image, of $w \in \Sigma^*$, denoted by $\mathrm{Pk}(w)$, is a mapping from Σ to $\mathbb{N}$, such that for $a \in \Sigma$, $\mathrm{Pk}(w)(a)$ is the number of occurrences of a in w. The **Parikh equivalence class** of w, $[w]_{\mathrm{Pk}} = \{w' \mid \mathrm{Pk}(w') = \mathrm{Pk}(w)\}$, is the set of all words with the same Parikh image as w. We can extend the Parikh map to languages $L \subseteq \Sigma^*$, defining $\mathrm{Pk}(L) \triangleq \{\mathrm{Pk}(w) \mid w \in L\}$.

Context Free Grammars. We will consider context free grammars in **Greibach Normal Form**. Formally (in this paper) a **context-free grammar** is G = (V, Σ, P, S), where V and Σ are disjoint sets of variables (or non-terminals) and terminals, respectively; $S \in V$ is the start symbol; and $P \subseteq V \times \Sigma V^*$ is a finite set of productions where each production is of the form $A \rightarrow a\beta$ with $a \in \Sigma$ and $\beta \in V^*$. Without loss of generality, we assume that every production in the grammar is used in some derivation from S to a string in Σ^*. A **sentence** is a string in $(\Sigma \cup V)^*$. A **right-linear grammar** is a context-free grammar where the productions have at most one non-terminal on the right-hand side, i.e., $P \subseteq V \times (\Sigma(\{\epsilon\} \cup V))$. It is well known that a language is generated by a right-linear grammar if and only if it is regular.

We will find it convenient to partition the variables of a grammar into those that have exactly one derivation tree and those that have more than one. Formally, the set of **single-derivation variables** $X \subseteq V$ is the smallest set containing all variables A with exactly one production of the form $A \rightarrow a$ (with $a \in \Sigma$) and having the property that if a variable A has exactly one production of the form $A \rightarrow a\alpha$ where $a \in \Sigma$ and $\alpha \in X^*$ then $A \in X$. The remaining variables, i.e. $Y = V \setminus X$, are **multiple-derivation variables**.

Prioritized leftmost derivations. In this paper we will consider special derivation sequences of a context-free grammar that expand the leftmost variable while giving priority to single-derivation variables. We call these **prioritized leftmost (PLM) derivations**, and we define them precisely next.

Definition 1. *Consider a context-free grammar* G = (V, Σ, P, S), *where the non-terminals* V *have been partitioned into the set of single-derivation variables* X *and multiple-derivation variables* Y. *We say that* $\alpha A \beta$ *rewrites in a single prioritized leftmost derivation step to* $\alpha\gamma\beta$ *(denoted as* $\alpha A \beta \Rightarrow_{\mathrm{plm}} \alpha\gamma\beta$*) iff* $\exists \pi \in P, \pi = (A \rightarrow \gamma)$ *such that either*

1. $A \in X$, $\alpha \in (\Sigma \cup Y)^*$, and $\beta \in V^*$, or
2. $A \in Y$, $\alpha \in \Sigma^*$, and $\beta \in (\Sigma \cup Y)^*$

In other words, either A is the leftmost single-derivation variable in $\alpha A\beta$, or A is the leftmost multiple-derivation variable and $\alpha A\beta$ has no single-derivation variables. If $\alpha \Rightarrow_{\mathrm{plm}} \beta$ by application of π, we'll write $\alpha \overset{\pi}{\Rightarrow}_{\mathrm{plm}} \beta$. Note that if $\alpha \Rightarrow_{\mathrm{plm}} \beta$ there is always a unique π such that $\alpha \overset{\pi}{\Rightarrow}_{\mathrm{plm}} \beta$.

A prioritized leftmost (PLM) derivation *is a sequence $\psi = \alpha_1, \ldots, \alpha_n$ such that $\alpha_1 \Rightarrow_{\mathrm{plm}} \alpha_2 \Rightarrow_{\mathrm{plm}} \cdots \Rightarrow_{\mathrm{plm}} \alpha_n$. The set of all PLM derivations is denoted* $\mathrm{Der}_{\mathrm{plm}}(G)$.

The **language generated by** G is $L(G) \triangleq \left\{ \alpha \in \Sigma^* \ \middle| \ S \overset{*}{\Rightarrow}_{\mathrm{plm}} \alpha \right\}$ where $\overset{*}{\Rightarrow}_{\mathrm{plm}}$ is the reflexive and transitive closure of $\Rightarrow_{\mathrm{plm}}$. Finally, the **parse** of a word $w \in (\Sigma \cup V)^*$, denoted $\mathrm{parse}_G(w)$, is the set of all PLM derivations yielding w:

$$\mathrm{parse}_G(w) \triangleq \{\alpha_1, \ldots, \alpha_n \in \mathrm{Der}_{\mathrm{plm}}(G) \mid \alpha_1 = S \text{ and } \alpha_n = w\}.$$

Example 1. We present a simple example to illustrate the definitions. Consider the grammar $G = (\{S, B\}, \{\mathsf{a}, \mathsf{b}\}, P, S)$ where P consists of the following productions: $\pi_1 = S \rightarrow \mathsf{a}SB$, $\pi_2 = S \rightarrow \mathsf{a}B$, and $\pi_3 = B \rightarrow \mathsf{b}$. The set of single-derivation variables is $\{B\}$ and the set of multiple-derivation variables is $\{S\}$. An example of a prioritized leftmost derivation is

$$S \overset{\pi_1}{\Rightarrow}_{\mathrm{plm}} \mathsf{a}SB \overset{\pi_3}{\Rightarrow}_{\mathrm{plm}} \mathsf{a}Sb \overset{\pi_2}{\Rightarrow}_{\mathrm{plm}} \mathsf{a}\mathsf{a}Bb \overset{\pi_3}{\Rightarrow}_{\mathrm{plm}} \mathsf{a}\mathsf{a}bb$$

The language generated by this grammar is $\{\mathsf{a}^n \mathsf{b}^n \mid n \geq 1\}$.

Derivation trees. The set of all derivation trees for G will be denoted as Δ_G. For a derivation tree τ, a node n in τ, and a path p from the root in τ, $\ell(\tau)$, $\ell(n)$ and $\ell(p)$ will denote the label of the root, the node n, and the node reached by path p in τ, respectively. For any node n in a tree τ and path p from the root, we denote the subtree rooted at n by $\tau(n)$, and the subtree rooted at the node reached by path p by $\tau(p)$. The **frontier of a tree** τ, denoted $\mathrm{Fr}(\tau)$ is the sentence $\ell(n_1)\ell(n_2)\ldots\ell(n_k)$ where $n_1, \ldots, n_k$ are the leaves of τ in left-to-right order.

For any variable $A \in V$, $\Delta_G(A) \triangleq \{\tau \in \Delta_G \mid \ell(\tau) = A\}$ is the subset of derivation trees rooted at A. A tree τ for which $\mathrm{Fr}(\tau) \in \Sigma^*$ is called a **complete derivation tree**, and the set of all complete derivation trees rooted at A is $\Delta_G^\Sigma(A) \triangleq \{\tau \in \Delta_G(A) \mid \mathrm{Fr}(\tau) \in \Sigma^*\}$. The set of all complete derivation trees is $\Delta_G^\Sigma \triangleq \{\tau \in \Delta_G \mid \mathrm{Fr}(\tau) \in \Sigma^*\}$. A tree $\tau \in \Delta_G(A)$ is said to be an A-**pumping tree** if $\mathrm{Fr}(\tau)\!\restriction_V = A$. The set of A-pumping trees is $\Delta_G^p(A) \triangleq \{\tau \in \Delta_G(A) \mid \mathrm{Fr}(\tau)\!\restriction_V = A\}$. The set of all pumping trees is given by $\Delta_G^p = \{\tau \in \Delta_G \mid \mathrm{Fr}(\tau)\!\restriction_V \in V\}$.

Remark 1. In a context-free grammar (where all productions are "useful"), every single-derivation variable is the root of exactly one complete derivation tree, and every multiple-derivation variable is the root of at least two complete derivation trees.

Tree Notation. We will use the following notation on derivation trees. Let $\tau \in \Delta_G$, n be a node in τ, and p be a path from the root in τ. The leftmost child of the node reached by path p, will be the one reached by the path $p \cdot 0$ with the other children corresponding to the paths $p \cdot 1$, $p \cdot 2$, etc. For $\tau_1 \in \Delta_G(\ell(n))$ ($\tau_1 \in \Delta_G(\ell(p))$), $\tau[n \mapsto \tau_1]$ ($\tau[p \mapsto \tau_1]$) denotes the derivation tree obtained by replacing $\tau(n)$ ($\tau(p)$) by the tree τ_1. We denote by $\mathsf{rem}_p(\tau)$ the tree obtained by replacing $\tau(p)$ by the root of $\tau(p)$, i.e., by "removing" all the children of p. Finally, for a rule $A \rightarrow \mathsf{a}\alpha$ with $\alpha = A_1 A_2 \cdots A_k$, and trees $\tau_i \in \Delta_G(A_i)$, $A_{\mathsf{a}\alpha}(\tau_1, \tau_2, \ldots \tau_k)$ denotes the tree with root labeled A and children $\mathsf{a}, \tau_1, \ldots \tau_k$ from left-to-right. Thus, A_a denotes the tree with root labeled A and one child labeled a.

Cuts. Observe that, for any string $\alpha \in (V \cup \Sigma)^*$, there is a bijection between derivation trees τ with $\mathsf{Fr}(\tau) = \alpha$ and PLM derivations in $\mathsf{parse}_G(\alpha)$. A set of nodes separating the root of τ from all of the leaves in τ is a **cut** of τ. Now consider the unique PLM derivation Ψ corresponding to τ. Every sentence in Ψ corresponds to a cut $\mathcal{C}$ in τ. We call any such $\mathcal{C}$ a **prioritized leftmost (PLM) cut** of τ. For a set of trees T and a variable $A \in V$, the **Parikh supremum** of variable A in T, denoted by $\mathsf{sup}_{\mathsf{Pk}}(A, T)$, is the maximum number of occurrences of A in any PLM cut of any tree $\tau \in T$. Observe that any PLM derivation sequence corresponding to a tree τ in T can have at most $\mathsf{sup}_{\mathsf{Pk}}(A, T)$ occurrences of the variable A in any sentence.

Ambiguity. We will say that a set of trees Γ is **ambiguous** if there are two distinct trees τ_1, τ_2 such that $\mathsf{Fr}(\tau_1) = \mathsf{Fr}(\tau_2)$; if Γ is not ambiguous, we say it is **unambiguous**. The **ambiguity function** $\mu_G : \mathbb{N} \rightarrow \mathbb{N}$ for a grammar G is a function mapping every natural number n to the maximal number of PLM derivations which a word of length n may have. Formally, $\mu_G(n) = \max_{w \in L(G), |w| = n} |\mathsf{parse}_G(w)|$. A grammar is said to have **exponential ambiguity** if its ambiguity function is in $2^{\Theta(n)}$, and it is said to have **polynomially-bounded ambiguity**, or to be polynomially ambiguous, if its ambiguity function is in $O(n^d)$ for some $d \in \mathbb{N}_0$. Any grammar G has either exponential ambiguity or polynomially-bounded ambiguity [36]. The following characterization of polynomial ambiguity was proved in [36].

Theorem 1 [36]. *A context-free grammar G has polynomially-bounded ambiguity if and only if Δ_G^p is unambiguous.*

We conclude the preliminaries by recalling a classical result due to Parikh [27].

Theorem 2 (Parikh's Theorem [27]). *For every context-free grammar G, there is a right-linear context-free grammar G' such that $\mathsf{Pk}(L(G)) = \mathsf{Pk}(L(G'))$.*

2.1   Weighted and Stochastic Context-Free Grammars

Weighted context-free grammars define a function that associates a value in a semiring with each string. Stochastic context-free grammars are special weighted context-free grammars that associate probabilities with strings. We recall these classical definitions in this section. We begin by defining a semiring.

Semiring. A ***semiring*** is a structure $\mathbb{D} = (D, \oplus, \otimes, 0_D, 1_D)$ where $(D, \oplus, 0_D)$ is a *commutative monoid* with identity 0_D, $(D \setminus \{0_D\}, \otimes, 1_D)$ is a *monoid* with identity 1_D, $\otimes$ distributes over $\oplus$ (i.e., $(a \oplus b) \otimes c = (a \otimes c) \oplus (b \otimes c)$ and $a \otimes (b \oplus c) = (a \otimes b) \oplus (a \otimes c)$, for every $a, b, c \in D$), and 0_D is an *annihilator* for $\otimes$ (i.e., $a \otimes 0_D = 0_D \otimes a = 0$ for every $a \in D$). We abuse notation and use $\mathbb{D}$ to denote the semiring and the underlying set where the meaning is clear from context. We define $\mathbb{D}_0 = D \setminus \{0_D\}$. When considering an abstract semiring $\mathbb{D}$, we'll write $0_{\mathbb{D}}$ and $1_{\mathbb{D}}$ for 0_D and 1_D respectively. An ***idempotent semiring*** satisfies the additional requirement that for all $a \in \mathbb{D}$, $a \oplus a = a$. A ***commutative semiring*** is one where $\otimes$ is commutative, i.e., $(D \setminus \{0_D\}, \otimes, 1_D)$ is a commutative monoid as well.

Example 2. Classical examples of a semiring are the tropical semiring and the probability semiring. The tropical or min-plus semiring is $(\mathbb{N} \cup \{\infty\}, \min, +, \infty, 0)$, where ∞ is taken to be larger than everything in $\mathbb{N}$. It is commutative and idempotent as $\min(a, a) = a$ for any a. The probability semiring is $([0, 1], +, \times, 0, 1)$, where $[0, 1]$ is the set of reals between 0 and 1. It is commutative as $\times$ is commutative. However, since the addition of two numbers is not idempotent, the probability semiring is not idempotent.

Weighted context-free grammars. A ***weighted context-free grammar*** is a pair (G, W) where $G = (V, \Sigma, P, S)$ is a context-free grammar, and $W : P \to \mathbb{D}$ assigns a weight from $\mathbb{D}$ to each production in P, for some semiring $\mathbb{D}$. (Note that W may assign $0_{\mathbb{D}}$ to some productions in P.) The ***weight*** of a PLM derivation $\psi = \alpha_1 \overset{\pi_1}{\Rightarrow}_{\text{plm}} \alpha_2 \overset{\pi_2}{\Rightarrow}_{\text{plm}} \cdots \overset{\pi_{n-1}}{\Rightarrow}_{\text{plm}} \alpha_n$ of G, is given by $W(\psi) \triangleq \otimes_{i=1}^{n-1} W(\pi_i)$. For $w \in \Sigma^*$, $W(w) \triangleq \oplus_{\psi \in \text{parse}_G(w)} W(\psi)$; we assume that if $\text{parse}_G(w) = \emptyset$ (i.e., $w \notin L(G)$) then $W(w) = 0_{\mathbb{D}}$. The ***semantics*** of a weighted grammar (G, W), denoted $[\![G]\!]_W : \Sigma^* \to \mathbb{D}$, is the function mapping each word to its weight in G, i.e., $[\![G]\!]_W(w) \triangleq W(w)$.

Example 3. Let G be the grammar described in Example 1. Consider a weight function W that assigns weights from the tropical semiring, with the weight of every production $\pi \in P$ being equal to 1. Then the semantics of (G, W) is given as $[\![G]\!]_W(a^n b^n) = 2n$ and $[\![G]\!]_W(w) = 0$, when $w \notin L(G)$.

Definition 2. *The* Parikh image *of a weighted context-free grammar* (G, W), *written as* $\text{Pk}\,[\![G]\!]_W$ *is function defined as*

$$\text{Pk}\,[\![G]\!]_W(w) \triangleq \bigoplus_{w' \in [w]_{\text{Pk}}} [\![G]\!]_W(w')$$

Stochastic Context-free Grammars. A **stochastic context-free grammar** is a weighted context-free grammar $(G = (V, \Sigma, P, S), W)$ where the weight domain is the probability semiring $([0, 1], +, \times, 0, 1)$, and for any $A \in V$ and $\mathsf{a} \in \Sigma$, we have

$$\sum_{\alpha \in V^* : (A \to \mathsf{a}\alpha) \in P} W(A \to \mathsf{a}\alpha) \in [0, 1] \, .$$

3   A Parikh's Theorem for Weighted CFGs

The main result of this section is that for any weighted context-free grammar over an idempotent, commutative semiring (like the tropical semiring), there is a Parikh equivalent weighted right-linear context-free grammar. Thus, this observation extends the classical result to weighted CFGs over idempotent semirings.

Theorem 3 (Weighted Parikh's Theorem). *For every weighted context-free grammar* (G, W) *over an* idempotent, commutative *semiring, there exists a Parikh-equivalent weighted right-linear grammar* (G_*, W_*), *that is, we have*

$$\mathrm{Pk} \, \llbracket G \rrbracket_W = \mathrm{Pk} \, \llbracket G_* \rrbracket_{W_*} \, .$$

Proof. The full proof can be found in [3]. Here we present the broad ideas.

Let $G = (V, \Sigma, P, S)$ be a context-free grammar and let $W : P \to \mathbb{D}$ be a weight function over a commutative, idempotent weight domain $\mathbb{D}$. Consider the following homomorphism $h : P^* \to \Sigma^*$ defined as $h(\pi) = \mathsf{a}$, where $\pi = A \to \mathsf{a}\alpha \in P$.

We begin by first constructing a weighted context-free grammar (G_1, W_1) over the alphabet P, whose image under h gives us G. Formally, $G_1 = (V, P, P_1, S)$ has as productions $P_1 = \{A \to \pi\alpha \mid \exists \mathsf{a} \in \Sigma. \, \pi = A \to \mathsf{a}\alpha \in P\}$. In addition, take W_1 to be $W_1(A \to \pi\alpha) = W(\pi)$. It is easy to see that $h(L(G_1)) = L(G)$ by construction. Moreover, given W_1 and W, we can conclude $\llbracket G \rrbracket_W(w) = \bigoplus_{w \in P^* : h(w) = w} \llbracket G_1 \rrbracket_{W_1}(\omega)$.

By Parikh's theorem (Theorem 2), there is a right-linear grammar $G_2 = (V_2, P, P_2, S_2)$ such that $\mathrm{Pk}(L(G_2)) = \mathrm{Pk}(L(G_1))$. Define the weight function W_2 as $W_2(A \to \pi B) = W(\pi)$ to give us the weighted CFG (G_2, W_2). Using the fact that $\otimes$ is commutative, and $\oplus$ is idempotent, we can prove that $\mathrm{Pk} \, \llbracket G_1 \rrbracket_{W_1} = \mathrm{Pk} \, \llbracket G_2 \rrbracket_{W_2}$.

Finally, we consider the context free grammar G_3 obtained by "applying the homomorphism h" to G_2. Formally, $G_3 = (V_2, \Sigma, P_3, S_2)$, where $P_3 = \{A \to h(\pi)B \mid A \to \pi B \in P_2\}$. The weight function W_3 is defined in such a way that weight of $A \to \mathsf{a}B$ is the sum of the weights of all productions $A \to \pi B$, where $h(\pi) = \mathsf{a}$, i.e.,

$$W_3(A \to \mathsf{a}B) = \bigoplus_{\pi : h(\pi) = \mathsf{a}} W_2(A \to \pi B).$$

(G_3, W_3) and (G_2, W_2) share the same relationship as (G, W) and (G_1, W_1). That is, we have $h(L(G_2)) = L(G_3)$ and $[\![G_3]\!]_{W_3}(w) = \bigoplus_{w \in P^*:h(\omega)=w} [\![G_2]\!]_{W_2}(\omega)$.

(G_3, W_3) is the desired weighted grammar, i.e., $\text{Pk} [\![G_3]\!]_{W_3} = \text{Pk} [\![G]\!]_W$.   □

Corollary 1. *If* (G, W) *is a weighted context-free grammar over an idempotent, commutative weight domain and a unary alphabet, then there exists a weighted right-linear context-free grammar* (G', W') *such that* $[\![G']\!]_{W'} = [\![G]\!]_W$.

Example 4. Starting with the weighted grammar (G, W) from Example 3, the construction used in the proof of Theorem 3 would have P_1 containing the following productions: $S \to \pi_1 SB$, $S \to \pi_2 B$, $B \to \pi_3$. The language of this grammar is $L(G_1) = \{\pi_1^n \pi_2 \pi_3^{n+1} \mid n \geq 0\}$.

One candidate for G_2 would have as productions $S \to \pi_1 J$, $S \to \pi_2 K$, $J \to \pi_3 S$, and $K \to \pi_3$. The language is $L(G_2) = \{(\pi_1 \pi_3)^n \pi_2 \pi_3 \mid n \geq 0\}$.

In that case (G_3, W_3) would have productions and weights as follows:

$$W_3(S \to aJ) = 1 \qquad\qquad W_3(S \to aK) = 1$$
$$W_3(J \to bS) = 1 \qquad\qquad W_3(K \to b) = 1$$

The language of the underlying grammar would be $L(G_3) = \{(ab)^n ab \mid n \geq 0\}$, and

$$[\![G_3]\!]_{W_3}(w) = \begin{cases} 2k & \text{if } w = (ab)^k \text{ for some } k \geq 1 \\ 0 & \text{otherwise} \end{cases}$$

4   A Counterexample to Parikh's Theorem for Stochastic Grammars

Theorem 3 crucially relies on the semiring being idempotent. In this section, we show that Theorem 3 fails to generalize if we drop the requirement of idempotence. We give an example of a stochastic context-free grammar over the unary alphabet that is not equivalent to any stochastic right-linear grammar. Before presenting the example stochastic context-free grammar and proving the inexpressivity result, we recall some classical observations about unary stochastic right linear grammars.

4.1   Properties of Unary Stochastic Right-Linear Grammars

Stochastic right-linear grammars satisfy pumping lemma type properties. Here we recall such an observation for unary stochastic right-linear grammars.

Theorem 4 (Pumping Lemma). *Let* $(G = (V, \{a\}, P, S), W)$ *be a stochastic right-linear grammar over the unary alphabet. There is a number k, and real numbers $c_0, c_1, \ldots c_k$ with $c_0 + c_1 + \cdots + c_k = 1$ such that for every $\ell \in \mathbb{N}$*

$$[\![G]\!]_W(a^{\ell+k+1}) = \sum_{i=0}^{k} c_i [\![G]\!]_W(a^{\ell+i})$$

Proof. The result is a consequence of the Cayley-Hamilton theorem and the fact that 1 is an eigen value of stochastic matrices. We skip the proof as it is a specialization of Theorem 2.8 in Chapter II.C in [28]. □

Let (G, W) be a unary weighted context-free grammar. The generating function of such a grammar is $P(x) = \sum_{k=0}^{\infty} [\![G]\!]_W(a^k)x^k$. We conclude this section by observing that if G is right-linear, then its generating function must be a rational function, i.e., $P(x)$ is an algebraic fraction where both the numerator and denominator are polynomials.

Theorem 5. *Let* (G, W) *be a stochastic right-linear grammar over the unary alphabet. Then the generating function* $P(x) = \sum_{k=0}^{\infty} [\![G]\!]_W(a^k)x^k$ *is a rational function.*

Proof. Observe that Theorem 4 says that the sequence $\langle [\![G]\!]_W(a^n) \rangle_{n \in \mathbb{N}}$ satisfies a linear homogeneous recurrence with constant coefficients. Thus, its generating function must be rational. □

4.2 The Counterexample

We now present a unary weighted CFG and show that there is no weighted right-linear CFG that is equivalent to it. Consider the grammar $G_* = (\{S\}, \{a\}, \{(S \to a), (S \to aSS)\}, S)$. Let p be some number in $(0, 1)$. The weight function W_* is defined as follows: $W_*(S \to a) = 1 - p$, and $W_*(S \to aSS) = p$. Taking c_n to be the nth Catalan number, we can see that $[\![G_*]\!]_{W_*}(a^{2k+1}) = c_k p^k (1 - p)^{k+1}$; this is because the probability of any PLM derivation for a^{2k+1} is $p^k(1-p)^{k+1}$ and there are c_k elements in $\text{parse}_G(a^{2k+1})$. Taking $b_k = [\![G_*]\!]_{W_*}(a^k)$, we have

$$b_k = \begin{cases} c_{(k-1)/2}p^{(k-1)/2}(1-p)^{(k-1)/2+1} & \text{if } k \text{ is odd} \\ 0 & \text{otherwise} \end{cases}$$

Recall that the generating function for the Catalan numbers, $C(z) = \sum_{k \geq 0} c_k z^k$, is given by $C(z) = \frac{1-\sqrt{1-4z}}{2z}$. Based on the above observations, the generating function for the grammar (G_*, W_*), $P(z) = \sum_{k \geq 0} b_k z^k$ can be written as follows.

$$P(z) = \sum_{k \geq 0} b_k z^k = \sum_{k \geq 0} b_{2k+1} z^{2k+1}$$

$$= z \sum_{k \geq 0} b_{2k+1} \left(z^2\right)^k = z \sum_{k \geq 0} c_k p^k (1-p)^{k+1} \left(z^2\right)^k$$

$$= z(1-p) \sum_{k \geq 0} c_k \left(z^2 p(1-p)\right)^k$$

$$= z(1-p)C(z^2 p(1-p))$$

$$= z(1-p)\frac{1-\sqrt{1-4z^2p(1-p)}}{2z^2p(1-p)}$$

$$= \frac{1}{2zp}\left(1-\sqrt{1-4z^2p(1-p)}\right)$$

Having identified an expression for the generating function for (G_*, W_*), we are ready to prove that there is no Parikh equivalent right-linear grammar for (G_*, W_*). First notice that if a weighted grammar (G, W) is over the unary alphabet, then $[\![G]\!]_W = Pk [\![G]\!]_W$. Therefore, to establish the result of this section, it suffices to prove the statement that there is no right-linear grammar that is equivalent to (G_*, W_*); this is the content of the next theorem.

Theorem 6. *There is no stochastic right-linear grammar* (G, W) *such that* $[\![G]\!]_W = [\![G_*]\!]_{W_*}$.

Proof. Given Theorem 5, it suffices to prove that the generating function $P(z)$ for (G_*, W_*) is not rational. Taking $Q(z) = \sqrt{1-4z^2p(1-p)}$, we see that $P(z) = \frac{1}{2zp}(1-Q(z))$. Given this relationship, we can conclude that if $P(z)$ is rational function, then so is $Q(z)$. Thus, our goal will be to prove that $Q(z)$ is not a rational function.

Assume for contradiction that $Q(z)$ is rational. Then $Q(z) = f(z)/g(z)$ where f and g are both polynomials with greatest common divisor having degree 0. Then $Q^2(z) = f^2(z)/g^2(z)$ and $Q^2(z)g^2(z) = (1-4z^2p(1-p))g^2(z) = f^2(z)$. Thus, $Q^2(z)$ must divide $f^2(z)$. We observe that $Q^2(z) = (1-2z\sqrt{p(1-p)})(1+2z\sqrt{p(1-p)})$ is square-free so $Q^2(z)$ must divide $f(z)$, and $f(z) = Q^2(z)h(z)$ for some polynomial h. Substituting for $f(z)$ in $Q^2(z)g^2(z) = f^2(z)$ and rearranging we obtain $Q^2(z)g^2(z) = (Q^2(z))^2h^2(z) \implies g^2(z) = Q^2(z)h^2(z)$, and by the same argument as above, $Q^2(z)$ divides $g(z)$. Thus, $Q^2(z)$ divides both $f(z)$ and $g(z)$. Since $Q^2(z)$ is not a degree 0 polynomial, we contradict the assumption that the greatest common divisor of f and g has degree 0. □

5   Parikh's Theorem for Unary Polynomially Ambiguous Stochastic Grammars

The weighted stochastic context-free grammar (G_*, W_*) in Sect. 4.2 is exponentially ambiguous; the ambiguity function μ_{G_*} is bounded by the Catalan numbers. Exponential ambiguity turns out to be critical to construct such counterexamples. In this section, we prove that any unary stochastic context-free grammar with polynomial ambiguity is equivalent to a unary stochastic right-linear grammar. The proof of this result relies on an observation that in any PLM cut in a complete derivation tree of a unary polynomially ambiguous grammar, the number of occurences of any variable is bounded by a constant dependent on the grammar. The unary alphabet assumption is crucial in obtaining such a bound (Lemma 3). In the next two subsections we present a proof of this observation by first bounding the number of occurrences in cuts of pumping trees and then using it to bound it in complete derivation trees. In Sect. 5.3, we then present the construction of the right-linear grammar. Though we present this result in the context of stochastic grammars, it applies to any weighted CFG over a commutative (but not necessarily idempotent) semiring.

In the rest of this section, let us fix a unary, polynomially ambiguous, context-free grammar $G = (V, \{a\}, P, S)$ and a stochastic weight function $\Pr$ (for probability). We assume that the set of variables V is partitioned into single-derivation variables X and multiple-derivation variables Y. As we have done throughout this paper, we assume that every production in G is "useful", that is, is used in some complete derivation tree whose root is labeled S. Finally we will assume that m is the maximum length of the right-hand side of any production in P.

5.1   Parikh Suprema in Pumping Trees

In this section we will bound the number of times a variable can appear in any PLM cut of a pumping tree in G. We begin by observing some simple properties about single-derivation variables X and multiple-derivation variables Y. Since every production in the grammar is useful, we can conclude that there is a unique complete derivation tree with root A if $A \in X$, and that there are at least two complete derivation trees with root A if $A \in Y$, i.e., $|\Delta_G^\Sigma(A)| = 1$ if $A \in X$, and $|\Delta_G^\Sigma(A)| > 1$ if $A \in Y$. Next, for $A \in X$, the unique $\tau \in \Delta_G^\Sigma(A)$ has the following properties: (a) no node is labeled by a variable in Y; (b) each variable in X labels at most one node along any path in τ. Property (b) holds because if $A \in X$ has a derivation $A \Rightarrow_{\text{plm}}^* \alpha A \beta$, then A cannot have any complete derivation tree and it would be useless. This also means that any pumping tree $\tau \in \Delta_G^p$ must have $\ell(\tau) \in Y$. These properties allow us to bound the size of the unique complete derivation for variable $A \in X$.

Lemma 1. *For any $A \in X$, the unique tree $\tau \in \Delta_G^\Sigma(A)$ has size at most $m^{|X|}$.*

Proof. Since only variables in X can appear as labels in τ and no variable appears more than once in any path, the height of τ is $\leq |X|$. Finally, since any node has at most m children, we get the bound on the size of τ.   □

Next we prove that Lemma 1 allows one to bound the number of times any single-derivation variable appears in any PLM cut of a pumping tree.

Lemma 2. *For any $A \in Y$ and $B \in X$, $\sup_{\text{Pk}}(B, \Delta_G^p(A)) \leq m^{|X|+1}$.*

Proof. Let $\tau \in \Delta_G^p(A)$ be an arbitrary A-pumping tree, where $A \in Y$. Let $\mathcal{C}$ be an arbitrary PLM cut of τ. We will prove a slightly stronger statement; we will show that the total number of single-derivation variables in $\mathcal{C}$ is $\leq m^{|X|+1}$. This will bound the Parikh supremum for any single-derivation variable.

Without loss of generality, assume that $\mathcal{C}$ has at least one node with label in X. Amongst all nodes in $\mathcal{C}$ that are labeled by a variable in X, let n be the node that is closest to the root, and if there are multiple such nodes, take n to be the leftmost one. From the definition of PLM cuts, the following property holds for

$\mathcal{C}$ and n: (a) any node to right of n in $\mathcal{C}$ that is labeled by a variable in X must be a sibling of n, and (b) all nodes to the left of n in $\mathcal{C}$ labeled by variables of X must be descendents of some left sibling (say n_1) of n that is also labeled by a variable in X. Thus, the number of nodes to the right of n (including n) in $\mathcal{C}$ labeled by X is at most m, and, by Lemma 1, the number of nodes to the left of n in $\mathcal{C}$ labeled by X is at most $m^{|X|}$. Putting these together, the total number of nodes in $\mathcal{C}$ labeled by some variable in X is at most $m + m^{|X|} \leq m^{|X|+1}$.   □

Lemma 3. *For any $A \in V$, and $B \in Y$, $\mathrm{supp}_{\mathrm{Pk}}(B, \Delta_G^p(A)) \leq 2$.*

Proof. Let τ be a A-pumping tree, for some variable A. Note that A must be a multiple-derivation variable because of property (b) before Lemma 1. Let $\mathcal{C}$ be any PLM cut of τ. Since τ is an A-pumping tree it must contain A in its frontier. Then there must be some node n in $\mathcal{C}$ such that the subtree $\tau(n)$ contains A in its frontier. Let $C = \ell(n)$. Observe that C is a multiple-derivation variable because a node labeled $A \in Y$ is a descendent. Thus, there are two complete derivation trees τ_1^C, τ_2^C with roots labeled C (Remark 1).

We'll first show that there cannot be more than two occurrences of nodes labeled C in $\mathcal{C}$. Assume towards the contrary that there are at least three nodes n_1, n_2, n_3 in $\mathcal{C}$ with $\ell(n_1) = \ell(n_2) = \ell(n_3) = C$. Without loss of generality, assume n_1, n_2, and n_3 are in left-to-right order in τ and $n \in \{n_1, n_2, n_3\}$. Since n_1, n_2, n_3 belong to a cut, they are not related by the ancestor/descendent relationship.

Let τ_1 be the tree $\tau[n_1 \mapsto \tau_1^C, n_2 \mapsto \tau_2^C, n_3 \mapsto \tau(n)]$, and let τ_2 be the tree $\tau[n_1 \mapsto \tau_2^C, n_2 \mapsto \tau_1^C, n_3 \mapsto \tau(n)]$. By construction, τ_1 and τ_2 are both A-pumping trees with $\mathrm{Fr}(\tau_1) = \mathrm{Fr}(\tau_2)$ and $\tau_1 \neq \tau_2$. However, since G is polynomially ambiguous, by Theorem 1, the set of pumping trees is unambiguous, giving us the desired contradiction.

Next, we show that there cannot be more than two nodes labeled $B \in Y$ in $\mathcal{C}$, where $B \neq C$. Assume that there are at least three nodes n_1, n_2, n_3 in $\mathcal{C}$ with $\ell(n_1) = \ell(n_2) = \ell(n_3) = B$. Again assume n_1, n_2, and n_3 are in left-to-right order in τ. Further, since $B \in Y$, there are two complete derivation trees τ_1^B and τ_2^B with root labeled B (Remark 1).

Observe that at least two nodes of $\{n_1, n_2, n_3\}$ must lie to one side of n in τ. Without loss of generality we may assume that n_1 and n_2 are those nodes. Let τ_1 be the tree $\tau[n_1 \mapsto \tau_1^B, n_2 \mapsto \tau_2^B]$, and let τ_2 be the tree $\tau[n_1 \mapsto \tau_2^B, n_2 \mapsto \tau_1^B]$. Clearly, τ_1, τ_2 are A-pumping trees with $\mathrm{Fr}(\tau_1) = \mathrm{Fr}(\tau_2)$, and $\tau_1 \neq \tau_2$.   □

5.2   Parikh Suprema in Complete Derivation Trees

We will now use the results in Sect. 5.1 to bound the Parikh supremum of any variable in a complete derivation tree of G. The key property we will exploit is the fact that any complete derivation tree can be written as the "composition" of a small number of pumping trees (see Fig. 1) such that any PLM cut is the union of cuts in each of these pumping trees. The bounds on Parikh suprema will then follow from the observations in Sect. 5.1.

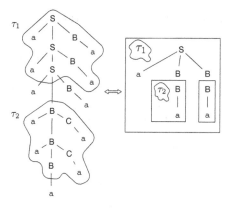

Fig. 1. A complete derivation tree for the grammar in Example 5 on the left and the compressed tree data structure with removed pumping trees on the right.

We begin with some convenient notation. For a $\tau \in \Delta_{\mathrm G}$, let longestpath($\tau$) denote the longest path from the root of τ to a node labeled $\ell(\tau)$. If there are multiple such paths, longestpath(τ) is the lexicographically-first path among them. Note that longestpath(τ) can be ε if the root is the only node with label $\ell(\tau)$ in τ. Let depth(τ) denote the length of the longest path from root to leaf in τ.

We now describe two procedures compress and decompress. Let us fix a complete derivation tree τ. The procedure compress returns a data structure of pumping trees. These pumping trees are small in number and τ is the "composition" of these pumping trees. Let n be the lowest node in τ that has the same label as the root. compress identifies the pumping tree obtained by removing the children of n, and recursively compresses the subtrees rooted at the children of n. Note that if n is the same as the root, then the pumping tree identified by compress will just be the tree with one node.

---

compress(τ):
  If $\tau = A_{\mathrm a}$ for some $A \in V$, **return** τ
  $p \leftarrow$ longestpath(τ)
  Let A, **a** and α be such that $\tau(p) = A_{\mathrm a\alpha}(\tau(p \cdot 1), \ldots, \tau(p \cdot k))$
  $\tau_{\mathrm{pump}} \leftarrow \mathrm{rem}_p(\tau)$
  Return $[\tau_{\mathrm{pump}}, A \rightarrow \mathbf{a}\alpha, \mathrm{compress}(\tau(p \cdot 1)), \ldots, \mathrm{compress}(\tau(p \cdot k))]$

The tree τ is the "composition" of pumping trees in the data structure returned by compress. We describe this "composition operation" itself by an algorithm decompress.

---

decompress(τ_c):
  If $\tau_c = A_{\mathrm a}$ for some $A \in V$, **return** τ_c
  Let τ_c be of the form $[\tau_{\mathrm{pump}}, A \rightarrow \mathbf{a}\alpha, \tau_c^1, \ldots, \tau_c^k]$
  $\tau' \leftarrow A_{\mathrm a\alpha}(\mathrm{decompress}(\tau_c^1), \ldots, \mathrm{decompress}(\tau_c^k))$
  Return $\tau_{\mathrm{pump}}[\mathrm{longestpath}(\tau_{\mathrm{pump}}) \mapsto \tau']$

The following lemma characterizing the relationship between `compress` and `decompress` is easy to see.

Lemma 4. *For any complete derivation tree τ, $\tau = $ decompress(compress(τ)).*

Example 5. Consider a grammar $(\{S, B, C\}, \{a\}, P, S)$ with productions

$$S \rightarrow \mathtt{a}SB | \mathtt{a}BB | \mathtt{a}B, \quad B \rightarrow \mathtt{a}BC | \mathtt{a}, \quad C \rightarrow \mathtt{a}.$$

Consider the complete derivation tree shown on the left in Fig. 1. The output of `compress` will be

$$[\tau_1, S \rightarrow \mathtt{a}BB, [\tau_2, B \rightarrow \mathtt{a}], B_\mathtt{a}]$$

We will now show that the data structure returned by `compress` has a constant number of pumping trees. Consider a call of compress(τ), where p is the longestpath(τ). The key property that we exploit is the fact that the label $\ell(\tau)$ does not appear in the subtrees rooted at the children of p.

Lemma 5. *For any complete derivation tree τ, the number of trees in the data structure returned by* compress(τ) *is at most $m^{|V|}$.*

Proof. Let $p = $ longestpath(τ), and let $\tau(p \cdot 0), \tau(p \cdot 1), \ldots, \tau(p \cdot k)$ be the children of $\tau(p)$. As observed before, the label $\ell(\tau)$ does not appear in the subtrees $\tau(p \cdot i)$. Thus, the depth of the recursion in `compress` is bounded by $|V|$. Finally, observing that $k \leq m$, we get the desired bound.   □

We are now ready to prove the main result of this section.

Lemma 6. *For any variable A, $\sup_{\mathrm{Pk}}(A, \Delta_G^\Sigma) \leq m^{|X|+|V|+1}$*

Proof. By Lemma 5, we know that the number of trees in compress(τ) is at most $m^{|V|}$. Consider any PLM cut $\mathcal{C}$ of τ. Any node in $\mathcal{C}$ belongs to at most one tree in compress(τ). Further for any $\tau_1 \in $ compress(τ), $\mathcal{C}$ restricted to τ_1 is a PLM cut of τ_1. Thus, $\mathcal{C}$ can be seen as the union of at most $m^{|V|}$ PLM cuts in pumping trees. By Lemmas 2 and 3, the Parikh supremum of any variable in any of these pumping trees is at most $m^{|X|+1}$. These observations together establish the bound.   □

5.3   Right-Linear SCFG for Polynomially Ambiguous SCFGs

For this section, let us fix $k = m^{|V|+|X|+1}$. By Lemma 6, in any PLM derivation of G, any variable appears at most k times at any step. Since k is a constant, the right-linear grammar can simulate every PLM derivation of G by explicitly keeping track of only k copies of any variable. This idea is very similar to the one used in [12]. We now give the formal definition of the right-linear grammar based on this intuition.

For a sentence $\alpha \in (\Sigma \cup V)^*$, we define $\boldsymbol{lf}(\alpha) \triangleq \alpha\lceil_\Sigma \alpha\lceil_X \alpha\lceil_Y$. The stochastic right-linear grammar $(\mathrm{G}_1 = (V_1, \{a\}, P_1, S_1), \mathrm{Pr}_1)$ is formally defined as follows.

1. $V_1 = \{\langle \alpha \rangle \mid \alpha \in X^*Y^* \text{such that for each } A \in V. \ Pk(\alpha)(A) \leq k\}$. Thus, the variables of G_1 are sequences of single-derivation variables followed by multiple-derivation variables from G in which each variable appears at most k times.
2. $S_1 = \langle S \rangle$
3. For any production $\pi = (A \rightarrow a\beta) \in P$ and sentence $\alpha \in V^*$ we define a production

$$\pi^\alpha = (\langle A\alpha \rangle \rightarrow a \langle \boldsymbol{lf}(\beta\alpha) \rangle)$$

corresponding to applying the production π as a PLM step from the sentence $A\alpha$. The set P_1 is defined as

$$P_1 = \{\pi^\alpha \mid \pi = (A \rightarrow a\beta) \in P \wedge \langle A\alpha \rangle, \langle \beta\alpha \rangle \in V_1\}$$

4. Finally Pr_1 is defined as $Pr_1(\pi^\alpha) = Pr(\pi)$ for all $\pi^\alpha \in P_1$.

We first observe that (G_1, Pr_1) is a stochastic CFG. The proof is in [3]

Proposition 1. (G_1, Pr_1) *is a stochastic CFG.*

(G_1, Pr_1) is equivalent to (G, Pr). Its proof is in [3].

Theorem 7. *For any unary, stochastic grammar* (G, Pr) *of polynomial ambiguity, there is a stochastic right-linear grammar* (G_1, Pr_1) *such that* $[\![G]\!]_{Pr} = [\![G_1]\!]_{Pr_1}$.

6   Conclusions

In this paper we investigated whether Parikh's theorem generalizes to weighted automata. We proved that it does indeed when the weighted context-free grammar is over a commutative, idempotent semiring. We showed that idempotence of the weight domain is necessary by demonstrating that Parikh's theorem does not extend to unary, stochastic grammars. However, we proved that if the context-free grammar is polynomially ambiguous, then idempotence of the weight domain is not required for Parikh's theorem to hold.

Our proof for Parikh's theorem for commutative and idempotent semirings extends (as is) to pushdown automata (as opposed to context-free grammars). However, the same does not apply to our result for unary, polynomially ambiguous grammars over non-idempotent rings. Our current proof subtly relies on the "one state" property of context-free grammars. It would be interesting to see how to generalize these ideas to the case of pushdown automata. Finally, stochastic context-free grammars have a (weaker) semantics as language acceptors — the grammar accepts a word if its weight is $> \frac{1}{2}$. Our results imply that every unary language accepted by a polynomially ambiguous, stochastic context-free grammar is also accepted by a probabilistic automata (with probability $> \frac{1}{2}$). It is open if this also holds when the grammar is exponentially ambiguous; our counterexample in this paper only shows that there is no probabilistic automaton that satisfies the stronger requirement that words are accepted with the *same* probability.

References

1. Aminof, B., Kupferman, O., Lampert, R.: Reasoning about online algorithms with weighted automata. ACM Trans. Algorithms **6**(2), 28 (2010)
2. Beeri, C., Milo, T.: Schemas for integration and translation of structured and semi-structured data. In: Beeri, C., Buneman, P. (eds.) ICDT 1999. LNCS, vol. 1540, pp. 296–313. Springer, Heidelberg (1999). doi:10.1007/3-540-49257-7_19
3. Bhattiprolu, V., Gordon, S., Viswanathan, M.: Extending Parikh's theorem to weighted and probabilistic context-free grammars. Technical report, University of Illinois, Urbana-Champaign (2017). http://hdl.handle.net/2142/96262
4. Buchsbaum, A., Giancarlo, R., Westbrook, J.: On the determinization of weighted finite automata. SIAM J. Comput. **30**(5), 1502–1531 (2000)
5. Chatterjee, K., Doyen, L., Henzinger, T.A.: Quantitative languages. In: Kaminski, M., Martini, S. (eds.) CSL 2008. LNCS, vol. 5213, pp. 385–400. Springer, Heidelberg (2008). doi:10.1007/978-3-540-87531-4_28
6. Chatterjee, K., Doyen, L., Henzinger, T.A.: Alternating weighted automata. In: Kutyłowski, M., Charatonik, W., Gębala, M. (eds.) FCT 2009. LNCS, vol. 5699, pp. 3–13. Springer, Heidelberg (2009). doi:10.1007/978-3-642-03409-1_2
7. Culik, K., Kari, J.: Image compression using weighted finite automata. Comput. Graph. **17**, 305–313 (1993)
8. Dang, Z., Ibarra, O.H., Bultan, T., Kemmerer, R.A., Su, J.: Binary reachability analysis of discrete pushdown timed automata. In: Emerson, E.A., Sistla, A.P. (eds.) CAV 2000. LNCS, vol. 1855, pp. 69–84. Springer, Heidelberg (2000). doi:10.1007/10722167_9
9. Droste, M., Gastin, P.: Weighted automata and weighted logics. In: Caires, L., Italiano, G.F., Monteiro, L., Palamidessi, C., Yung, M. (eds.) ICALP 2005. LNCS, vol. 3580, pp. 513–525. Springer, Heidelberg (2005). doi:10.1007/11523468_42
10. Droste, M., Kuich, W., Vogler, H. (eds.): Handbook of Weighted Automata. Springer, Heidelberg (2009)
11. Esparza, J.: Petri nets, commutative context-free grammars, and basic parallel processes. Fundam. Inform. **31**(1), 13–25 (1997)
12. Esparza, J., Ganty, P., Kiefer, S., Luttenberger, M.: Parikh's theorem: a simple and direct automaton construction. Inf. Process. Lett. **111**(12), 614–619 (2011)
13. Goldstine, J.: A simplified proof of Parikh's theorem. Discret. Math. **19**, 235–239 (1977)
14. Göller, S., Mayr, R., To, A.: On the computational complexity of verifying one-counter processes. In: Proceedings of the IEEE Symposium on Logic in Computer Science, pp. 235–244 (2009)
15. Gurari, E., Ibarra, O.: The complexity of decision problems for finite-turn multi-counter machines. J. Comput. Syst. Sci. **22**, 220–229 (1981)
16. Hafner, U.: Low bit-rate image and video coding with weighted finite automata. Ph.D. thesis, Universität Würzburg (1999)
17. Huynh, T.-D.: The complexity of semilinear sets. In: Bakker, J., Leeuwen, J. (eds.) ICALP 1980. LNCS, vol. 85, pp. 324–337. Springer, Heidelberg (1980). doi:10.1007/3-540-10003-2_81
18. Huynh, D.: Deciding the ineuivalence of context-free grammars with 1-letter terminal alphabet is σ_2^p-complete. Theor. Comput. Sci. **33**, 305–326 (1984)
19. Huynh, D.: Complexity of equivalence problems for commutative grammars. Inf. Control **66**(1), 103–121 (1985)

20. Ibarra, O.: Reversal-bounded multi-counter machines and their decision problems. J. ACM **25**, 116–133 (1978)
21. Jiang, Z., Litow, B., Vel, O.: Similarity enrichment in image compression through weighted finite automata. In: Du, D.-Z.-Z., Eades, P., Estivill-Castro, V., Lin, X., Sharma, A. (eds.) COCOON 2000. LNCS, vol. 1858, pp. 447–456. Springer, Heidelberg (2000). doi:10.1007/3-540-44968-X_44
22. Katritzke, F.: Refinements of data compression using weighted finite automata. Ph.D. thesis, Universität Siegen (2001)
23. Kopczyński, E., To, A.: Parikh images of grammars: complexity and applications. In: Proceedings of the IEEE Symposium on Logic in Computer Science, pp. 80–89 (2010)
24. Kuperberg, D.: Linear temporal logic for regular cost functions. In: Proceedings of the Symposium on Theoretical Aspects of Computer Science, pp. 627–636 (2011)
25. Mohri, M.: Finite-state transducers in language and speech processing. Comput. Linguist. **23**, 269–311 (1997)
26. Mohri, M., Pereira, F., Riley, M.: The design principles of weighted finite-state transducer library. Theor. Comput. Sci. **231**, 17–32 (1997)
27. Parikh, R.J.: On context-free languages. J. ACM **13**(4), 570–581 (1966)
28. Paz, A.: Introduction to Probabilistic Automata. Academic Press, Cambridge (1971)
29. Rabin, M.: Probabilistic automata. Inf. Control **6**(3), 230–245 (1963)
30. Reps, T., Lal, A., Kidd, N.: Program analysis using weighted pushdown systems. In: Arvind, V., Prasad, S. (eds.) FSTTCS 2007. LNCS, vol. 4855, pp. 23–51. Springer, Heidelberg (2007). doi:10.1007/978-3-540-77050-3_4
31. Reps, T., Schwoon, S., Jha, S., Melski, D.: Weighted pushdown systems and their application to interprocedural dataflow analysis. Sci. Comput. Program. **58**(1–2), 206–263 (2005)
32. Schützenberger, M.: On the definition of a family of automata. Inf. Control **4**, 245–270 (1961)
33. Sen, K., Viswanathan, M.: Model checking multithreaded programs with asynchronous atomic methods. In: Ball, T., Jones, R.B. (eds.) CAV 2006. LNCS, vol. 4144, pp. 300–314. Springer, Heidelberg (2006). doi:10.1007/11817963_29
34. Siedl, H., Schwentick, T., Muscholl, A.: Counting in trees. Texts Log. Games **2**, 575–612 (2007)
35. To, A.W., Libkin, L.: Algorithmic metatheorems for decidable LTL model checking over infinite systems. In: Ong, L. (ed.) FoSSaCS 2010. LNCS, vol. 6014, pp. 221–236. Springer, Heidelberg (2010). doi:10.1007/978-3-642-12032-9_16
36. Wich, K.: Exponential ambiguity of context-free grammars. In: Developments in Language Theory, Foundations, Applications, and Perspectives, Aachen, Germany, 6–9 July 1999, pp. 125–138 (1999)
37. Yen, H.: On reachability equivalence for BPP-nets. Theor. Comput. Sci. **179**, 301–317 (1996)

Exploiting Non-deterministic Analysis in the Integration of Transient Solution Techniques for Markov Regenerative Processes

Marco Biagi[1], Laura Carnevali[1(✉)], Marco Paolieri[2], Tommaso Papini[1], and Enrico Vicario[1]

[1] Department of Information Engineering, University of Florence, Florence, Italy
{marco.biagi,laura.carnevali,tommaso.papini,enrico.vicario}@unifi.it
[2] Department of Computer Science, University of Southern California,
Los Angeles, USA
paolieri@usc.edu

Abstract. Transient analysis of Markov Regenerative Processes (MRPs) can be performed through the solution of Markov renewal equations defined by *global* and *local kernels*, which respectively characterize the occurrence of regenerations and transient probabilities between them. To derive kernels from stochastic models (e.g., stochastic Petri nets), existing methods exclusively address the case where at most one generally-distributed timer is enabled in each state, or where regenerations occur in a bounded number of events. In this work, we analyze the state space of the underlying timed model to identify epochs between regenerations and apply distinct methods to each epoch depending on the satisfied conditions. For epochs not amenable to existing methods, we propose an adaptive approximation of kernel entries based on partial exploration of the state space, leveraging heuristics that permit to reduce the error on transient probabilities. The case study of a polling system with generally-distributed service times illustrates the effect of these heuristics and how the approach extends the class of models that can be analyzed.

Keywords: Non-markovian Petri Nets · Markov Regenerative Process · Enabling restriction · Stochastic state class · Non-deterministic analysis

1 Introduction

In quantitative evaluation of concurrent models, generally distributed (GEN) durations support modeling validity but break the Markov property and rule out efficient solution techniques for Continuous Time Markov Chains (CTMCs). If the model guarantees that, always, with probability 1 (w.p.1), the Markov property will be eventually satisfied at some *regeneration* point, then the underlying stochastic process belongs to the class of Markov Regenerative Processes (MRPs) [12].

© Springer International Publishing AG 2017
N. Bertrand and L. Bortolussi (Eds.): QEST 2017, LNCS 10503, pp. 20–35, 2017.
DOI: 10.1007/978-3-319-66335-7_2

MRPs attain a fortunate trade-off between expressivity of models and feasibility of numerical solution, which is reduced to the evaluation of a *global kernel* and a *local kernel* that characterize behavior in the sequencing of regeneration points and in the *epochs* between them. However, numerical derivation of the kernels has been solved only for some isolated sub-classes of MRP models [7].

Most works address the subclass where at most a single GEN timer is enabled in each state (*enabling restriction*), so that each kernel component can be computed by analyzing the CTMC subordinated to the activity interval of the active GEN [1,6,9]. The method of supplementary variables [8,17] might in principle encompass the case of multiple concurrently enabled GEN timers, but practical feasibility restrains applicability under the enabling restriction. Sampling at equidistant time points [15,19] permits evaluation for models where all timers have either deterministic (DET) or exponentially distributed (EXP) durations.

The method of stochastic state classes [18] enables quantitative evaluation of stochastic processes with multiple concurrent GEN timers, possibly with bounded support; in particular, for models that always reach a regeneration within a bounded number of discrete events, which we call the *bounded regeneration restriction*, exact evaluation of kernels is performed enumerating stochastic transient trees that cover the states between two subsequent regenerations [10].

For models that break both the enabling and the bounded regeneration restriction, kernel components may be still defectively approximated by truncation of stochastic transient trees [10], which may also serve to reduce complexity for models under bounded regeneration. However, this faces an inherent contrast. On the one hand, state space truncation has a different impact on the final evaluation, depending on the probability of reaching truncation points. On the other hand, when the analysis exploits regenerations to decompose state space coverage, each epoch starts from a memoryless condition, which is not able to distinguish whether the probability mass under analysis is relevant or negligible.

In this paper, we exploit non-deterministic analysis to drive integration of different solution techniques, exact and approximate, that are applicable to different regenerative epochs. To this end, we characterize the structure of the state space through terminating and efficient non-deterministic analysis based on the representation of timing domains with Difference Bounds Matrices (DBMs), identifying regenerative epochs and solution techniques that can be applied for kernel components corresponding to each regeneration (Sect. 3.1). This permits integration of the consolidated technique of enabling restriction with exact and approximate solution based on stochastic state classes (Sect. 3.2). Moreover, we also introduce a novel technique that iteratively adapts the approximation of each kernel component so as to optimize the impact of truncation on the defect in the evaluation of transient probabilities (Sects. 3.3 and 3.4). The approach permits to accurately evaluate transient probabilities of markings, and it is open to further adaptation strategies and to integration of other solution techniques, both numerical and simulative. Application is illustrated with reference to an instance of the polling system problem [11,13] with generally distributed service times and exhaustive service subordinated to a deterministic timeout (Sect. 4).

To make the paper self-contained, we recall the formalism of Stochastic Time Petri Nets (STPNs) and transient analysis of MRPs (Sect. 2). Finally, we draw our conclusions and discuss future steps enabled (Sect. 5).

2   Preliminaries

2.1   Stochastic Time Petri Nets

Definition 1. *An STPN is a tuple* $\langle P, T, A^-, A^+, A^{\bullet}, m_0, U, EFT, LFT, F, W \rangle$: *P is the set of places; T is the set of transitions;* $A^- \subseteq P \times T$, $A^+ \subseteq T \times P$, $A^{\bullet} \subseteq P \times T$ *are the sets of precondition, postcondition, inhibitor arcs, respectively;* $m_0 \in \mathbb{N}^P$ *is the initial marking; U associates each transition t with an update function* $U(t) : \mathbb{N}^P \to \mathbb{N}^P$ *which, in turn, associates each marking with a new marking;* $EFT : T \to \mathbb{Q}_{\geq 0}$ *and* $LFT : T \to \mathbb{Q}_{\geq 0} \cup \{\infty\}$ *associate each transition with an earliest and a latest firing time, respectively; F associates each transition t with a Cumulative Distribution Function (CDF)* $F(t)$ *over* $[EFT(t), LFT(t)]$; *and,* $W : T \to \mathbb{R}_{>0}$ *associates each transition with a weight.*

A place p is an *input, output, inhibitor* place for a transition t if $\langle p, t \rangle \in A^-$, $\langle t, p \rangle \in A^+$, $\langle p, t \rangle \in A^{\bullet}$, respectively; precondition and postcondition arcs are represented by arrows, while inhibitor arcs by dotted arrows. A transition t is *immediate* (IMM) if $EFT_t = LFT_t = 0$ and *timed* otherwise; a timed transition t is *exponential* (EXP) if $F_t(x) = 1 - e^{-\lambda x}$ over $[0, \infty]$ with $\lambda \in \mathbb{R}_{>0}$, and *general* (GEN) if it has a non-exponential CDF; a GEN transition t is *deterministic* (DET) if $EFT_t = LFT_t > 0$ and *distributed* otherwise; for each distributed transition t, we assume that F_t is absolutely continuous and thus expressed as the integral function of a Probability Density Function (PDF) f_t, ruling out mixed (continuous and discrete) distributions. IMM, EXP, GEN, DET transitions are represented by thin black, thick white, thick black, thick gray bars, respectively. Update functions and weights are annotated next to transitions as "place ← *expression*" and "*weight = value*", respectively.

The state of an STPN is a pair $\langle m, \phi \rangle$, where m is a marking and $\phi : T \to \mathbb{R}_{\geq 0}$ associates each transition with a time-to-fire. A transition is *enabled* by a marking if each of its input places contains at least one token and none of its inhibitor places contains any token; an enabled transition t is *firable* in a state if its time-to-fire is equal to zero. The next transition t to fire in a state $s = \langle m, \phi \rangle$ is selected among the set $T_{f,s}$ of firable transitions in s with probability equal to $W(t) / \sum_{t_i \in T_{f,s}} W(t_i)$. When t fires, s is replaced by $s' = \langle m', \phi' \rangle$, where: m' is derived from m by (*i*) removing a token from each input place of t (yielding marking $\tilde{m}$), (*ii*) adding a token to each output place of t (yielding marking $\hat{m}$), and (*iii*) applying the update function $U(t)$ to $\hat{m}$; ϕ' is derived from ϕ by (*i*) reducing the time-to-fire of *persistent* transitions (i.e., enabled by m, $\tilde{m}$, $\hat{m}$, m') by the time elapsed in s; (*ii*) sampling the time-to-fire of each *newly-enabled* transition t_n (i.e., enabled by m' but not by $\tilde{m}$) according to F_{t_n}; and, (*iii*) removing the time-to-fire of *disabled* transitions (i.e., enabled by m but not by m').

Given an initial marking m_0 and an initial PDF f_{τ_0} for the vector τ of the times-to-fire of the enabled transitions, the STPN semantics induces a probability space $\langle \Omega_{m_0}, \mathbb{F}_{\tau_0}, \mathbb{P}_{m_0, f_{\tau_0}} \rangle$, where Ω_{m_0} is the set of outcomes (i.e., feasible timed firing sequences of the model) and $\mathbb{P}_{m_0, f_{\tau_0}}$ is a probability measure over them [16]. Note that $\mathbb{P}_{m_0, f_{\tau_0}}$ is zero for outcomes that are not feasible under f_{τ_0}.

Figure 1 shows a running example. The firing of restart enables gen1 and makes reg, enab, and approx firable: the firing of reg enables gen2, which fires w.p.1; the firing of enab enables the cycle exp1–exp2, which can fire an unbounded number of times; the firing of approx enables the cycle gen3–gen4, which can fire an unbounded number of times. In all three cases, gen1 is persistent and will eventually fire w.p.1, bringing the STPN to the initial marking Restart (note that the update function of gen1 flushes places E1, E2, G3, G4).

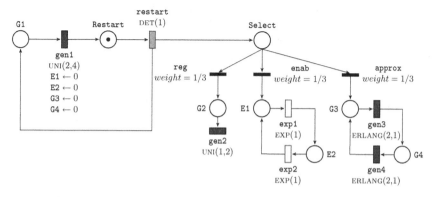

Fig. 1. A simple STPN with multiple concurrent GEN, DET, and EXP transitions: gen1 and gen2 have a uniform distribution over $[2,4]$ and $[1,2]$, respectively; gen3 and gen4 have an Erlang distribution with shape 2 and rate 1; restart has firing time equal to 1; exp1 and exp2 have an EXP distribution with rate 1.

2.2   Transient Analysis of Markov Regenerative Processes

The marking process $\{M(t), t \geq 0\}$, where $M(t)$ is the marking at time t, specifies the logic state of an STPN at each time instant. If the marking process is an MRP [7], its transient evolution is completely characterized by: (i) the initial probabilities of markings; (ii) a local kernel $L_{ij}(t) := P\{M(t) = j, T_1 > t \mid M(0) = i\}$, where T_1 is the time of the first regeneration after regeneration i, characterizing the evolution between two subsequent regenerations (i.e., $L_{ij}(t)$ is the probability that, starting from regeneration i at time 0, no regeneration is reached within time t and the marking at time t is j); and, (iii) a global kernel $G_{ik}(t) := P\{M(T_1) = k, T_1 \leq t \mid M(0) = i\}$ characterizing the occurrence of regenerations (i.e., $G_{ik}(t)$ is the probability that, starting from regeneration i at time 0, the first regeneration is reached on marking k within time t).

Transient probabilities of markings $\pi_{ij}(t) := P\{M(t) = j \mid M(0) = i\}$ are the solution of a set of Markov renewal equations defined by the kernels [5,12]:

$$\pi_{ij}(t) := P\{M(t) = j \mid M(0) = i\} = L_{ij}(t) + \sum_{k \in \Theta} \int_0^t g_{ik}(x)\,\pi_{kj}(t-x)\,dx \quad (1)$$

where $g_{ik}(x) := d\,G_{ik}(x)/dx$. While Eq. 1 can be solved numerically by discretization, kernels can be computed only for some sub-classes of MRP models.

The marking process of the STPN of Fig. 1 is an MRP since the firing of gen1, which always occurs w.p.1. (possibly after an unbounded number of firings), brings the process to the initial regeneration where restart is newly-enabled.

Analysis Under the Enabling Restriction. The *enabling restriction* [6,8] assumes that at most a single GEN time-to-fire is enabled in each state, which in turn implies that it is never the case that a GEN transition *continues* (persists) at the firing of another GEN transition. If an MRP complies with the enabling restriction, then in each regenerative epoch the process behaves either as a CTMC, if only EXP transitions are enabled in the initial regeneration, or as a CTMC subordinated to the activity interval of a GEN transition (i.e., the time interval during which the transition is enabled), if a GEN transition is enabled in the initial regeneration. In this case, the kernels can be computed from CTMC transient probabilities through the method of [6,9].

The marking process of the STPN of Fig. 1 does not satisfy the enabling restriction, since gen2, gen3, and gen4 may be enabled concurrently with gen1.

Analysis Under the Bounded Regeneration Restriction. The method of stochastic state classes [10] permits computation of kernels for models with multiple GEN times-to-fire concurrently enabled, also with overlapping activity intervals, but for exact evaluation requires that: always, a regeneration is eventually reached within a bounded number of discrete events. We term this case as the *bounded regeneration restriction*. The marking process of the STPN of Fig. 1 does not satisfy the bounded regeneration restriction, since both cycles exp1–exp2 and gen3–gen4 may fire an unbounded number of times while gen1 is persistent, without reaching a regeneration.

A *stochastic state class* samples the state of the MRP immediately after a firing, encoding a marking and a joint domain and PDF for the absolute time and for the times-to-fire of the enabled transitions.

Definition 2. *A stochastic state class is a tuple $\Sigma = \langle m, D, f \rangle$ where: m is a marking; D is the support of the random vector $\langle \tau_{\mathrm{age}}, \boldsymbol{\tau} \rangle$, where τ_{age} is the absolute time and $\boldsymbol{\tau}$ is the vector of the remaining times-to-fire of the enabled transitions; and, f is the PDF of $\langle \tau_{\mathrm{age}}, \boldsymbol{\tau} \rangle$, which we term state density function.*

Starting from an initial stochastic state class with $\tau_{\mathrm{age}} = 0$ and independently distributed times-to-fire for the enabled transitions, enumeration of a reachability relation among stochastic state classes yields a *stochastic transient tree*, where the support of the vector τ in each class is a Difference Bounds Matrix (DBM), i.e., a linear convex polyhedron that represents the solution of a set of linear inequalities constraining the difference between pairs of times-to-fire.

Definition 3. *A stochastic state class $\Sigma' = \langle m', D', f' \rangle$ is the successor of a stochastic state class $\Sigma = \langle m, D, f \rangle$ through a transition t with probability μ, which we write $\Sigma \overset{t,\mu}{\Rightarrow} \Sigma'$, iff, given that the marking is m and the random vector $\langle \tau_{age}, \tau \rangle$ is distributed over D according to f, t fires with probability μ, yielding a marking m' and a random vector $\langle \tau'_{age}, \tau' \rangle$ distributed over D' according to f'.*

A stochastic state class is said to be *regenerative* if the Markov property is satisfied immediately after the class is entered, which occurs iff all active GEN times-to-fires have been enabled for a deterministic time [16]:

Definition 4. *A stochastic state class Σ is termed regenerative if the time elapsed from the enabling of each enabled GEN transition t_i until the firing that led to Σ is a deterministic value $d_i \in \mathbb{R}_{\geq 0}$, termed the enabling time of t_i in Σ.*

In *exact regenerative transient analysis* [10], stochastic state classes are enumerated from each regeneration until any regeneration is reached, yielding a set of *stochastic transient trees* that are rooted in a regenerative stochastic state class and contain non-regenerative successors reached before any regeneration. Under the bounded regeneration restriction, each tree is finite, collecting all stochastic state classes that capture the MRP behavior during a regenerative epoch, with (regenerative) leaf nodes characterizing the global kernel and (non-regenerative) inner nodes characterizing the local kernel. For any regenerative stochastic state classes i, integration of the PDF of $\langle \tau_{age}, \tau \rangle$ in the stochastic state classes belonging to the tree rooted in i permits to compute the kernel entries $L_{ij}(t)$ and $g_{ik}(t)$ by summing up the measure of probability of states in the classes of the transient stochastic tree rooted in i, for any non-regenerative stochastic state class j, for any regenerative stochastic state class k, and for any time t.

2.3   Non-deterministic Analysis

An STPN identifies a Time Petri Net (TPN) [3,14] with same outcomes Ω_{m_0}.

Definition 5. *A state class $S = \langle m, D \rangle$ is made of a marking m and a support D for the vector τ of the remaining times-to-fire of the enabled transitions.*

Starting from an initial marking m_0 and an initial domain D_0 for τ, enumeration of the reachability relation among state classes yields a *State Class Graph* (SCG), which represents the continuous set of executions Ω_{m_0} and supports correctness verification of the TPN model (*non-deterministic analysis*).

Definition 6. *$S' = \langle m', D' \rangle$ is the successor of $S = \langle m, D \rangle$ through transition t, i.e., $S \overset{t}{\rightarrow} S'$, iff, given that the marking is m and τ is supported over D, t fires in S, yielding marking m' and a new vector τ' supported over D'.*

If $EFT(t) \in \mathbb{Q}_{\geq 0}$ and $LFT(t) \in \mathbb{Q}_{\geq 0} \cup \{\infty\}$ for every transition t, then the SCG is finite provided that the model generates a finite number of markings [10], which does not comprise a modeling limitation for most applicative scenarios.

3   Integration of Transient Solution Techniques for MRPs

Non-deterministic state space analysis of the underlying TPN of an STPN model permits identification of regeneration epochs and verification of whether each of them satisfies the enabling or bounded regeneration restrictions (Sect. 3.1), driving integration of different solution techniques for the evaluation of kernels (Sect. 3.2). For epochs that satisfy neither of the two restrictions, partial enumeration of stochastic state classes supports approximated evaluation of the kernels, resulting in a safe defective approximation of transient probabilities (Sect. 3.3).

3.1   Analysis of Regenerative Epochs

The set of states collected in a *stochastic* state class identifies a unique underlying *non-deterministic* state class [18] that represents the marking and the support of the vector of the remaining times-to-fire of the enabled transitions when the class is entered. The association between non-deterministic and stochastic state classes is one-to-many (possibly one-to-infinite) and preserves qualitative properties referred to the set of feasible outcomes Ω_{m_0}, while abstracting from quantitative properties depending on the probability measure $\mathcal{P}_{m_0,f_{\tau_0}}$. Given that a stochastic state class is regenerative if it satisfies Definition 4, which depends on Ω_{m_0} but not on $\mathcal{P}_{m_0,f_{\tau_0}}$, state classes can be used to identify regenerations.

To this end, the state space of the underlying TPN is covered by a set of SCGs, which we call *First-Epoch State Class Graphs* (FESCGs), each rooted in a regenerative state class and containing all non-regenerative successors reached before any regeneration (which is also included in the graph). Enumeration of FESCGs can suppress successor relations that correspond to null probability events, i.e., firings that in any associated stochastic state class would be possible in a null measure subset of the support.

Lemma 1. *Let u be an STPN, v be its underlying TPN, R be the set of successor relations $\Sigma = \langle m, D, f \rangle \xrightarrow{t,\mu} \Sigma' = \langle m', D', f' \rangle$ in the stochastic transient tree of u enumerated from a regenerative stochastic state class $\Sigma_0 = \langle m_0, D_0 \rangle$, and $S = \langle m, \bar{D} \rangle \xrightarrow{t} S' = \langle m', \bar{D}' \rangle$ be a succession relation in the SCG of v enumerated from a regenerative state class $S_0 = \langle m_0, \bar{D}_0 \rangle$, such that $\bar{D}$, $\bar{D}'$, and $\bar{D}_0$ are the projections of D, D', and D_0 that eliminate τ_{age}, respectively. A succession relation $\Sigma \xrightarrow{t,\mu} \Sigma' \in R$ has probability $\mu = 0$ iff the projection of D that eliminates DET and IMM timers, conditioned to the firing of transition t, has a null measure in $\mathbb{R}^N$, where N is the number of distributed times-to-fire in Σ and S.*

Proof. Let D_t be D conditioned to the firing of t, i.e., $D_t = D \cap \{\tau_t \leq \tau_{t_i} \forall t_i \in E(m)\}$, where τ_t is the time-to-fire of t and $E(m)$ is the set of transitions enabled by m. Let $\hat{D}_t$ be the projection of D_t that eliminates DET and IMM timers.

(If) If $\hat{D}_t$ has null measure in $\mathbb{R}^N$, either (i) the STPN includes some transition associated with a mixed distribution, or (ii) $\mu = 0$. By Definition 1, the CDF of each GEN transition is absolutely continuous over its support, thus $\mu = 0$.

(Only if) If, ab absurdo, $\hat{D}_t$ had non-null measure in $\mathbb{R}^N$, then the integral over $\hat{D}_t$ of the marginal distribution of distributed times-to-fire in Σ conditioned to the firing of t would not be zero, yielding $\mu \neq 0$.   □

It is straightforward to show that a regenerative epoch complies with the enabling restriction iff at most one GEN transition is enabled in each state class of its FESCG. Conversely, compliance with the bounded regeneration restriction depends on the presence of cycles in the FESCG.

Lemma 2. *A regenerative epoch complies with the bounded regeneration restriction iff its FESCG does not include any cycle.*

Proof. (If) If, ab absurdo, a regenerative epoch did not satisfy the bounded regeneration restriction, the STPN would allow a timed firing sequence made of an unbounded number of firings that never visits a regeneration; given that an STPN and its underlying TPN have the same set of timed firing sequences Ω_{m_0}, also the TPN would allow that behavior. Given that each state class is associated with one or more stochastic state classes having the same marking and time domain, there would exist a state class associated with an unbounded number of stochastic state classes. As a consequence, the FESCG would include a cycle.

(Only if) If, ab absurdo, the FESCG of a regeneration included a cycle, then, by construction, that cycle would not visit any regenerative state class. Hence, there would exist a timed firing sequence that would allow an unbounded number of firings without visiting a regeneration, and the corresponding regenerative epoch would not comply with the bounded regeneration restriction.   □

Figure 2 shows the SCG of the TPN underlying the STPN of Fig. 1, consisting of 5 regenerative and 5 non-regenerative state classes. In particular: the FESCG rooted in S_3 includes S_6 and S_1, satisfying the bounded regeneration restriction (it is cycle free) but not the enabling restriction (two GEN transitions are enabled in S_6); the FESCG rooted in S_5 includes S_8, S_{10}, and S_1, complying with the enabling restriction but not with the bounded regeneration restriction (due to the cycle S_8–S_{10}); and, the FESCG rooted in S_4 includes S_7, S_9, and S_1, satisfying neither the bounded regeneration restriction (due to the cycle S_7–S_9) nor the enabling restriction (two GEN transitions are enabled in S_4, S_7, and S_9). Note that the firing of transition **gen1** in state class S_3 would have probability zero in any associated stochastic state class and thus it is suppressed.

3.2   An Algorithm for Transient Analysis of MRPs

Given an STPN with underlying MRP, the kernel entries of each regenerative epoch can be derived through a different solution technique depending on whether the epoch satisfies the bounded regeneration restriction, or the enabling

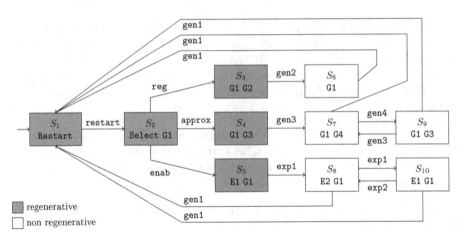

Fig. 2. The SCG of the TPN underlying the STPN of Fig. 1: state classes are represented by rectangles labeled with the marking; successor relations between state classes are represented by arrows labeled with the fired transition.

restriction, or neither of the two conditions. The applicable solution strategy can be efficiently selected through non-deterministic analysis of the underlying TPN of the model, by enumerating the SCG so as to identify the set Θ of regenerative state classes, the set Ψ of reachable markings, and the FESCG of each regenerative state class $i \in \Theta$:

- if the FESCG of i complies with the bounded regeneration restriction (e.g., the FESCG rooted in S_3 in Fig. 2), $L_{ij}(t)$ and $g_{ik}(t)$ are computed through the exact regenerative transient analysis of [10], for any marking $j \in \Psi$, for any regenerative state class $k \in \Theta$, and for any time point t;
- if the FESCG of i satisfies the enabling restriction (e.g., the FESCG rooted in S_5 in Fig. 2), $L_{ij}(t)$ and $g_{ik}(t)$ are derived through the method of [6,9];
- if the FESCG of i breaks both the enabling and the bounded regeneration restrictions (e.g., the FESCG rooted in S_4 in Fig. 2), $L_{ij}(t)$ and $g_{ik}(t)$ can still be estimated by stochastic simulation of the STPN model or they can be approximated by numerical solution as developed in Sect. 3.3.

Note that in so doing the derivation of kernel entries always terminates (even for models with an underlying marking process beyond the class of MRPs), provided that the FESCG of each regenerative state class is finite, which in turn is guaranteed under the fairly general conditions mentioned in Sect. 2.3. Also note that, in the present implementation, regenerative epochs that satisfy both the restrictions are analyzed through exact regenerative transient analysis, but analysis under the enabling restriction could be applied as well; moreover, approximated analysis or simulation might be applied also to regenerative epochs that satisfy one or both the restrictions, as a way to reduce complexity of solution. In-depth comparison and experimentation of the impact of different choices on accuracy and complexity deserves further study.

When kernel entries have been evaluated, transient probabilities of reachable markings are finally derived by numerical integration of the Markov renewal equations of Eq. 1.

3.3   Approximate Evaluation of the Kernels of an MRP

In general, and in particular for regenerative epochs that do not satisfy either the bounded regeneration or the enabling restrictions, an approximation of kernel entries can be derived by truncating the enumeration of the stochastic transient tree computed in the exact regenerative transient analysis [10]. In this case, following the steps of Sect. 2.2, the approximated kernel entries $\tilde{L}_{hj}(t)$ and $\tilde{g}_{ik}(x)$ are computed on a subset of the classes in the stochastic transient tree of the regenerative state class i, and they thus comprise an under-approximation of the exact values $L_{hj}(t)$ and $g_{ik}(x)$. Specifically, denoting $\Delta_{ij} := L_{hj}(t) - \tilde{L}_{hj}(t)$ and $\delta_{ik} := g_{ik}(x) - \tilde{g}_{ik}(x)$, we have $\Delta_{ij} \geq 0$ and $\delta_{ik} \geq 0$.

To characterize the impact of the approximation, the following Lemma provides a bound on $\epsilon_{ij}(t) := \pi_{ij}(t) - \tilde{\pi}_{ij}(t)$, with $\tilde{\pi}_{ij}(t)$ denoting the solution of Eq. 1 obtained with approximated kernel entries:

$$\tilde{\pi}_{ij}(t) = \tilde{L}_{ij}(t) + \sum_{k \in \Theta} \int_0^t \tilde{g}_{ik}(x)\, \tilde{\pi}_{kj}(t-x)\, dx \qquad (2)$$

Lemma 3. *For each regenerative state class $i \in \Theta$, marking $j \in \Psi$, and time t, the error $\epsilon_{ij}(t)$ is non-negative and upper-bounded:*

$$0 \leq \epsilon_{ij}(t) \leq \phi_i(t) + \sum_{k \in \Theta} \int_0^t (\tilde{g}_{ik}(x)\epsilon_{kj}(t-x) + \phi_i(x)(\epsilon_{kj}(t-x) + \tilde{\pi}_{kj}(t-x)))\, dx$$

$$\qquad (3)$$

where $\phi_i(t) := \sum_{j \in \Psi}(L_{ij}(t) - \tilde{L}_{ij}(t)) + \sum_{k \in \Theta}(g_{ik}(t) - \tilde{g}_{ik}(t))$.

Proof. By combining Eqs. 1 and 2, we obtain: $\epsilon_{ij}(t) = \Delta_{ij}(t) + \sum_{k \in \Theta} \int_0^t (\tilde{g}_{ik}(t) + \delta_{ik}(x)) \cdot \epsilon_{kj}(t-x) + \delta_{ik}(x)) \cdot \tilde{\pi}_{kj}(t-x)dx$ Since $\Delta_{ij}(t) \geq 0$ and $\delta_{ik}(t) \geq 0$, $\phi_i(t) \geq \Delta_{ij}(t) \; \forall j \in \Psi$ and $\phi_i(t) \geq \delta_{ik}(t) \; \forall k \in \Theta$. The upper bound of Eq. 3 can thus be obtained by replacing $\Delta_{ij}(t)$ and $\delta_{ik}(t)$ with $\phi_i(t)$.

To prove that $\epsilon_{ij}(t) \geq 0$, $\epsilon_{ij}(t)$ is rewritten as $\epsilon_{ij}(t) = A_{ij}(t) + \sum_{k \in \Theta} \int_0^t (\tilde{g}_{ik}(x) \cdot \epsilon_{kj}(t-x)dx$ where $A_{ij}(t) := \Delta_{ij}(t) + \sum_{k \in \Theta} \int_0^t \delta_{ik}(x)\pi_{kj}(t-x)dx$ Note that $A_{ij} \geq 0$, being $\Delta_{ij}(t) \geq 0$, $\delta_{ik}(x) \geq 0$, and being $\pi_{kj}(t-x)$ a probability. For any discretization step $\tau \in \mathbb{R}_{>0}$, the expression of $\epsilon_{ij}(t)$ can be rewritten by replacing $t = M \cdot \tau$ and $x = m \cdot \tau$, with $m \in [0, M]$. By induction on M, it is easily proven that $\epsilon(t)$ is monotonic non-decreasing with t. Moreover $\epsilon_{ij}(0) = A_{ij}(0) \geq 0$, which proves that $\epsilon(t) \geq 0$. $\qquad \square$

Note that, since $0 \leq \tilde{\pi}_{ij}(t)$ for every markings i, j and time t, summation of probabilities over all reachable markings provides a defective (i.e., lower than 1) evaluation of the total probability mass properly allocated; the complement to 1 of this quantity thus comprises a safe upper bound on the maximum value of each computed probability or summation over them.

3.4   Heuristic Driven Approximation

The quantity $\phi_i(t)$ in Eq. 3 can be safely estimated as the sum of probabilities to reach a truncation point in the partial enumeration of the stochastic transient tree of regenerative class i. According to this, the bounds of Eq. 3 can be used to define a truncation policy in the partial enumeration of regenerative epochs that break both the enabling and the bounded regeneration restrictions (*unrestricted epochs*) with a twofold aim: adapt the error accumulated on kernel entries of each regeneration i to the impact that this epoch takes on the final error $\epsilon_{ij}(t)$; and drive the selection of truncation points within each stochastic transient tree so as to control the trade-off between complexity of enumeration and accuracy of approximation. However, exact implementation of this policy would require repeated evaluation of approximated probabilities $\tilde{\pi}_{ij}(t)$, which in turn implies a major numerical complexity for the solution of Volterra integral equations. Lemma 3 can thus be more conveniently exploited as a ground for the definition of efficient heuristics driving truncation within each regenerative epoch. Note that, while this work emphasizes the use of approximation as a way to make feasible the evaluation of kernel entries, approximation driven by efficient heuristics may be applied also to reduce complexity in epochs that fit the bounded regeneration or the enabling restrictions.

Partial exploration of unrestricted epochs is performed by initially enumerating at most ν_{start} nodes in each tree, and then by iteratively identifying a non-regenerative leaf node and by enumerating at most ν_{iter} of its successors, until the number of classes enumerated in unrestricted epochs is larger than a threshold ν_{max} (*heuristic-based approximate analysis*). Given that the upper-bound of Eq. 3 suggests that the approximation error affects more those regenerative epochs that are visited more often, at each iteration we enumerate the successors of the non-regenerative leaf node with the largest *estimated* probability to be reached. Such estimate is evaluated by analyzing a Discrete Time Markov Chain (DTMC) $\mathbb{D}$ specified as follows:

- $\mathbb{D}$ has a state for each regenerative state class $i \in \Theta$ and for each leaf node j (either regenerative or non-regenerative) belonging to any tree $\mathcal{T}_i \in \mathcal{T}$ (regenerative and non-regenerative leaf nodes are absorbing in every tree);
- $\mathbb{D}$ has an arc from each state representing a regenerative state class $i \in \Theta$ to each state representing a leaf node j in $\mathcal{T}_i$, associated with probability μ_{ij};
- if the epoch rooted in i is analyzed exactly, μ_{ij} is equal to $G_{ij}(\infty)$ under the bounded regeneration restriction and to $G_{ij}(t_n)$ under the enabling restriction; otherwise, if the epoch rooted in i is analyzed in approximate manner, μ_{ij} is equal to $\tilde{G}_{ij}(\infty)$ or to $\tilde{L}_{ij}(\infty)$ depending on whether j corresponds to a regenerative or non-regenerative stochastic state class, respectively.

Steady-state analysis of $\mathbb{D}$ yields the vector of state probabilities P: solution relies on a basic implementation of the evaluation of absorption probabilities, which is not optimized with reference to either general techniques [2] or special techniques that might exploit warm restart in the repeated solution of DTMCs that are each a minor perturbation of the one solved at the previous iteration. Then, the steady-state probability of the states that correspond to non-regenerative leaf nodes are normalized, obtaining the vector of state probabilities $\bar{P}$, i.e., for each state l of the DTMC $\mathbb{D}$ that corresponds to a non-regenerative leaf node in a tree $T_i \in \mathcal{T}$, $\bar{P}_l = P_l / \sum_{h \in \mathcal{S}_L} P_h$, where $\mathcal{S}_L$ is the set of states that correspond to non-regenerative leaf nodes in any tree $T_i \in \mathcal{T}$. Finally, the non-regenerative leaf node that corresponds to the state w with the largest probability $\bar{P}_w$ is selected as the node to be expanded.

4   A Case Study

The approach was implemented on top of the Sirio API of the ORIS Tool [4]. Due to the minimal state space, with a single epoch requiring approximation of kernel entries, the STPN of Fig. 1 does not permit to best illustrate the potential of the approach. Hence, experiments were performed on the STPN of Fig. 3, a variant of a 3-station exhaustive-service polling system [11], where service sojourn is bounded by a DET timeout, polling times have a GEN distribution, and service times have an EXP or GEN distribution. For each station $s \in \{1, 2, 3\}$: place Waiting$s$ encodes the number of pending service requests; places AtService$s$ and Vacant$s$ encode whether the station is being served or not, respectively; and, place Polling$s$ encodes the state where the server is polling station s. In Fig. 3, all stations have no pending requests and the server is polling station 1.

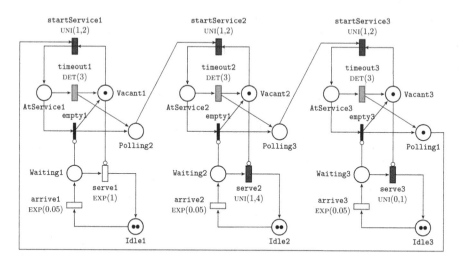

Fig. 3. STPN of a 3-station exhaustive-service polling system with server timeout.

The service at station s begins with the firing of transition startService$s$, with uniform distribution over $[1, 2]$, and it may terminate either when the queue of pending requests (Waiting$s$) is empty or when timeout$s$ fires after a DET maximum duration of value 3. During the service interval, place Vacant$s$ is empty and transition `serves` is enabled, so that any number of requests can be served. Transition `arrives` models the arrival of a new request as an EXP distribution with mean 20. Since the EXP distribution has a null minimum value, the maximum number of requests served during a service interval is limited only by the relation between the timeout value and the minimum duration of each service. Specifically, the number of requests served during a service interval sojourn is unbounded for stations 1 and 3, and it is bounded to 3 for station 2 where each service requires at least 1 time unit.

The underlying marking process regenerates whenever the server arrives to any station (i.e., at firing of `emptys` or `startServices`) or leaves it (i.e., at firing of `emptys` or `timeouts`), which directly implies that starting from any reachable state, w.p.1, a regeneration will be eventually reached, i.e. the process is an MRP. The process behavior falls in different subclasses of MRP during service sojourns at different stations. When the server is at station 1: the process satisfies the *enabling restriction*, given that `timeout1` is the only non-EXP transition enabled in each state; but it does not satisfy the bounded regeneration restriction, as for any natural number n, there exists a non-null probability that `serve1` and `arrive1` are fired more than n times before the expiration of `timeout1`. When the server is at station 2: the process satisfies the *bounded regeneration restriction*, given that `serve2` cannot be fired more than 3 times before the firing of `timeout2`; but the enabling restriction is not satisfied as `timeout2` and `serve2` can be concurrently enabled. When the server is at station 3: the process falls in the *unrestricted case* as `timeout3` and `serve3` are concurrently enabled, and `serve3` may fire an unbounded number of times before the firing of `timeout3`.

Transient analysis is performed through the approach of Sect. 3 with the following parameters: time limit $t_n = 30$ (each station is served at least twice), time step 0.1, $\nu_{start} = 20$ (number of stochastic state classes initially enumerated in each unrestricted epoch), $\nu_{iter} = 20$ (number of stochastic state classes enumerated in each unrestricted epoch at each iteration), and $\nu_{max} = 500$ (threshold on the total number of stochastic state classes enumerated in unrestricted epochs). Overall, the analysis evaluates the kernel entries of 135 regenerative epochs: 99 through the analysis under the bounded regeneration restriction, 18 through the analysis under the enabling restriction, and 18 through the heuristic-based approximate analysis. On a machine equipped with an Intel i5-5200U 2.20 GHz and 8 GB RAM, the evaluation takes nearly 40 min, spending less than 0.1 s to perform non-deterministic analysis and classification of regenerative epochs; nearly 40 s, 0.3 s, and 0.4 s to analyze the state space of regenerative epochs under the bounded regeneration restriction, under the enabling restriction, and beyond both restrictions, respectively; approximately 100 s, 180 s, and 2.5 s to evaluate the kernel entries of regenerative epochs under the bounded regeneration restriction, under the enabling restriction, and beyond both restrictions, respectively; nearly 23 s to evaluate the heuristic criterion; and, approximately 34 min to solve the Markov renewal equations. Numbers show that non-deterministic analysis has relatively negligible computational complexity, and thus it can be efficiently used to select the solution technique applied to each regenerative epoch. Notably,

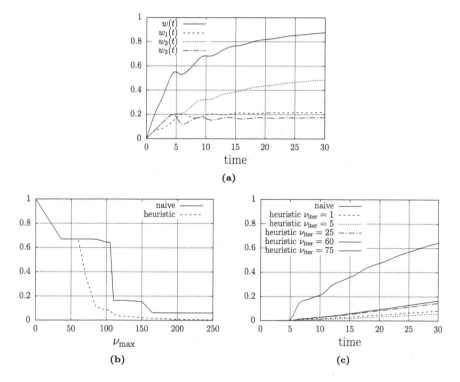

Fig. 4. (a) Average number of messages waiting to be served at time t at station s, i.e., $w_s(t)$ $\forall s \in \{1,2,3\}$, and in the overall system, i.e., $w(t)$; (b) total error $\epsilon(t_n)$ (committed in the evaluation of transient probabilities of markings at the time limit t_n) as a function of the number of classes enumerated in unrestricted epochs; (c) total error $\epsilon(t)$ obtained with 70 stochastic state classes enumerated in unrestricted epochs.

the heuristic criterion has a significantly lower cost with respect to the evaluation of the kernel entries of restricted epochs, which much depends on the number of encountered regenerations. Overall, results suggest that approximate analysis could be applied also to epochs under enabling or bounded regeneration restrictions to limit state space exploration and reduce evaluation complexity.

To illustrate possible rewards of interest, Fig. 4a plots the average number of messages waiting to be served at time t in each station and in the overall system, i.e., $w_n(t) = \sum_{j \in \Psi} \pi_{ij}(t) \cdot j(\texttt{Waiting}n)$ $\forall n \in \{1,2,3\}$ and $w(t) = \sum_{j \in \Psi} \pi_{ij}(t) \cdot \sum_{n=1}^{3} j(\texttt{Waiting}n)$, respectively, where i is the initial regeneration (i.e., a stochastic state class with the marking of Fig. 3, where all enabled transitions are newly-enabled) and Ψ is the set of markings reached within t_n.

To evaluate the impact of different heuristics in approximate analysis, we evaluate the total defect in the evaluation of transient probabilities of markings, i.e., $\epsilon(t) := \sum_{j \in \Psi} \epsilon_{ij}(t)$ where i is the initial regeneration and Ψ the set of markings, which can be easily computed a posteriori as $\epsilon(t) = 1 - \sum_{j \in \Psi} \pi_{ij}(t)$. Figure 4b plots the total error at the time limit $t_n = 30$ as a function of the threshold ν_{max}, comparing results with those obtained with a *naive approximate analysis* that explores all stochastic transient trees of unrestricted epochs, enumerating ν_{max}/U stochastic state classes in each tree, where U is the number of unrestricted epochs. As expected, $\epsilon(t_n)$ decreases as ν_{max} increases, and the two approaches achieve approximately the same values of $\epsilon(t_n)$ for very small values of ν_{max}. Conversely, when ν_{max} becomes larger than 60, the heuristic-based analysis achieves significantly lower values of $\epsilon(t_n)$, in the order of $8 \cdot 10^{-2}$ for $\nu_{max} = 100$ and $7 \cdot 10^{-3}$ from $\nu_{max} = 200$ on, with respect to values in the order of 0.65 and $6 \cdot 10^{-2}$ attained by naive analysis, respectively. Overall, these results could be used to select a convenient value of ν_{max} in a trade-off between the result accuracy and the computational complexity.

Figure 4c plots the total error attained by the two approaches as a function of time, with $\nu_{max} = 100$, $\nu_{start} = 2$, and increasing values of ν_{iter}. All curves are around zero until time 5, due to the very low probability that the server has reached station 3 by that time. From time 5 on, the error attained by naive analysis rapidly increases, being nearly 0.21, 0.48, and 0.64 at $t = 10$, $t = 20$, and $t = 30$, respectively. Conversely, $\epsilon(t)$ increases with a much smaller slope for heuristic-based analysis. As expected, the cases with lower values of ν_{iter} achieves better results; for instance, for $\nu_{iter} = 1$, $\epsilon(t)$ is approximately equal to 0.016, 0.049, and 0.083 at $t = 10$, $t = 20$, and $t = 30$, respectively. Values of $\epsilon(t)$ slightly increase with ν_{iter}, though remaining nearly in the same order of magnitude, showing that heuristic-based analysis yields sufficiently accurate results while permitting to limit the computational cost.

5   Conclusions

We leverage the low computational cost of non-deterministic analysis to drive the integration of different solution techniques in the evaluation of the kernels of an MRP, distinguishing regenerative epochs that can be analyzed through exact approaches from those that need approximate evaluation, due to infinite sequences of discrete events that never visit a regeneration. For the latter epochs, we present a novel approach based on the partial enumeration of stochastic state classes, which are iteratively explored according to a heuristic criterion based on the probability that a regeneration is reached. In so doing, the approximation is limited to the kernel entries of a subset of regenerative epochs, and transient probabilities of markings can be safely and accurately approximated.

Notably, the approximate analysis algorithm is designed to permit the integration of other solution techniques, which can be equivalently analytical or simulative. Other heuristic criteria could be used as well to select the next node to visit in partial enumeration of stochastic state classes, possibly taking into account an estimate of the mean time until when a regeneration is reached. Experimental results show that the heuristic-based approximate analysis provides accurate results while maintaining a moderate computational cost, suggesting that approximation could be used also for regenerative epochs characterized by finite stochastic transient trees, in order to reduce the number of stochastic state classes needed to compute the kernels.

References

1. Amparore, E.G., Buchholz, P., Donatelli, S.: A structured solution approach for Markov regenerative processes. In: Norman, G., Sanders, W. (eds.) QEST 2014. LNCS, vol. 8657, pp. 9–24. Springer, Cham (2014). doi:10.1007/978-3-319-10696-0_3

2. Barrett, R., Berry, M., Chan, T.F., Demmel, J., Donato, J., Dongarra, J., Eijkhout, V., Pozo, R., Romine, C., Van der Vorst, H.: Templates for the Solutions of Linear Systems: Building Blocks for Iterative Methods. SIAM, Philadelphia (1994)

3. Berthomieu, B., Diaz, M.: Modeling and verification of time dependent systems using time Petri nets. IEEE Trans. Softw. Eng. **17**(3), 259–273 (1991)

4. Bucci, G., Carnevali, L., Ridi, L., Vicario, E.: Oris: a tool for modeling, verification and evaluation of real-time systems. STTT **12**(5), 391–403 (2010)

5. Çinlar, E.: Markov renewal theory: a survey. Manag. Sci. **21**(7), 727–752 (1975)

6. Choi, H., Kulkarni, V.G., Trivedi, K.S.: Markov regenerative stochastic Petri nets. Perform. Eval. **20**(1–3), 337–357 (1994)

7. Ciardo, G., German, R., Lindemann, C.: A characterization of the stochastic process underlying a stochastic Petri net. IEEE Trans. Softw. Eng. **20**(7), 506–515 (1994)

8. German, R., Lindemann, C.: Analysis of stochastic Petri nets by the method of supplementary variables. Perform. Eval. **20**(1), 317–335 (1994)

9. German, R., Logothetis, D., Trivedi, K.S.: Transient analysis of Markov regenerative stochastic Petri nets: a comparison of approaches. In: International Workshop on Petri Nets and Performance Models, pp. 103–112. IEEE (1995)

10. Horváth, A., Paolieri, M., Ridi, L., Vicario, E.: Transient analysis of non-Markovian models using stochastic state classes. Perform. Eval. **69**(7–8), 315–335 (2012)

11. Ibe, O.C., Trivedi, K.S.: Stochastic Petri net models of polling systems. IEEE J. Sel. Areas Commun. **8**(9), 1649–1657 (1990)

12. Kulkarni, V.: Modeling and Analysis of Stochastic Systems. Chapman & Hall, London (1995)

13. Kwiatkowska, M., Norman, G., Parker, D.: PRISM: probabilistic symbolic model checker. In: Field, T., Harrison, P.G., Bradley, J., Harder, U. (eds.) TOOLS 2002. LNCS, vol. 2324, pp. 200–204. Springer, Heidelberg (2002). doi:10.1007/3-540-46029-2_13

14. Lime, D., Roux, O.H.: Expressiveness and analysis of scheduling extended time Petri nets. In: IFAC Conference on Fieldbus and their Applications. Elsevier Science (2003)

15. Lindemann, C., Thümmler, A.: Transient analysis of deterministic and stochastic Petri nets with concurrent deterministic transitions. Perform. Eval. **36–37**(1–4), 35–54 (1999)

16. Paolieri, M., Horváth, A., Vicario, E.: Probabilistic model checking of regenerative concurrent systems. IEEE Trans. Softw. Eng. **42**(2), 153–169 (2016)

17. Telek, M., Horváth, A.: Transient analysis of Age-MRSPNs by the method of supplementary variables. Perform. Eval. **45**(4), 205–221 (2001)

18. Vicario, E., Sassoli, L., Carnevali, L.: Using stochastic state classes in quantitative evaluation of dense-time reactive systems. IEEE Trans. Softw. Eng. **35**(5), 703–719 (2009)

19. Zimmermann, A: Modeling and evaluation of stochastic Petri nets with TimeNET 4.1. In: International ICST Conference on Performance Evaluation Methodologies and Tools, pp. 54–63 (2012)

alphaFactory: A Tool for Generating the Alpha Factors of General Distributions
(Tool Paper)

Elvio Gilberto Amparore[✉] and Susanna Donatelli

Dipartimento di Informatica, Università di Torino, Turin, Italy
{amparore,susi}@di.unito.it

Abstract. The Uniformization method computes the probability distribution of a CTMC of maximum rate μ at the time a general event with PDF $f(x)$ fires. Usually, $f(x)$ is taken as the deterministic distribution, leading to the computation of the CTMC probability at time t, but Uniformization may be extended to use other distributions. The extended Uniformization does not manipulate directly the distribution, as the whole computation is based on the *alpha-factors* of $f(x)$, and the maximum CTMC rate μ. This tool paper describes alphaFactory, a tool that computes the series of alpha-factors of a general distribution function starting from $f(x)$. The main goal of alphaFactory is to provide a freely available implementation for the computation of alpha-factors, to be used inside any extended Uniformization method implementation. Truncation of the infinite series of alpha-factors is determined by a novel error bound, which provides a reliable truncation point also in case of defective PDFs. alphaFactory can be easily integrated into other existing tools, and we show its integration inside the GreatSPN framework, to solve Markov Regenerative Stochastic Petri Nets.

Keywords: Alpha-factors · General distributions · Markov Regenerative Processes · Markov Regenerative Stochastic Petri Nets · GreatSPN · Extended Uniformization

1 Introduction

The Uniformization method in its basic form [16,21] computes the probability distribution of a CTMC of maximum rate μ at a fixed time t. This method has been widely applied to the computation of the transient solutions of systems in many domains, systems expressed using a variety of formalisms (queuing networks, stochastic Petri nets, stochastic process algebra, ...). The Uniformization method in its extended form [15] computes the probability distribution at time t, with t distributed according to a random variable. A typical application of this method is found in the steady-state solution of *Markov Regenerative Stochastic Petri Nets* (MRSPN) [11], i.e. stochastic Petri nets where transitions have exponential and generally-distributed delays, subject to the constraint that at most

© Springer International Publishing AG 2017
N. Bertrand and L. Bortolussi (Eds.): QEST 2017, LNCS 10503, pp. 36–51, 2017.
DOI: 10.1007/978-3-319-66335-7_3

one general event is enabled in any state (so called *enabling restriction*). MRSPN are a generalization of Deterministic and Stochastic Petri Nets (DSPN) [1]. Furthermore, the computation of CTMC probabilities at time t, where t is generally distributed, is a central step in the computation of the *subordinated Markov chains* of Markov Regenerative Processes (MRgP) [18].

The general event firing distribution can be described by its probability distribution function (PDF) $f(x)$. Extended Uniformization does not manipulate directly $f(x)$, but instead the whole computation is based on the *alpha-factors* of $f(x)$ and the maximum CTMC rate μ. Intuitively, the alpha-factor $\alpha(m, \mu)$ of $f(x)$ for a CTMC $\mathcal{M}$ of maximum rate μ is the probability of taking m steps in the *uniformized DTMC* of $\mathcal{M}$ before the generally distributed event fires. Alpha-factors form an infinite sequence for $m \in \mathbb{N}_{\geq 0}$, usually truncated according to a certain error bound ϵ. For the deterministic case, the alpha-factors reduce to a sequence of Poisson probabilities, which are usually computed with the Fox-Glynn method [13,17] due to its numerical stability. For arbitrary general distributions $f(x)$, the problem of computing alpha-factors reduces to the computation of an integral of the product of $f(x)$ with a Poisson probability. To allow a variety of general functions, the computation should do symbolic integration of $f(x)$.

Some applications also need a reliable support for *defective PDFs*, i.e. distribution functions $f(x)$ with $0 < \left(\int_0^\infty f(x)\, dx \right) < 1$. This makes the tool more robust when the PDF definition has small numerical error which may happen, for instance, when using fitted expolynomial distributions.

Contribution. This tool paper describes alphaFactory, a program that computes the alpha-factors of general distribution functions from their probability distribution function $f(x)$ and the maximum CTMC rate μ. alphaFactory is written in ISO C++ and has only the Boost-C++ library[1] as a dependency.

alphaFactory provides a freely available implementation for the computation of alpha-factors. The tool is designed to be used both as a standalone command line program, or linked as a component into another program, using a simple API. At its core the tool implements the definition and derivation of alpha-factors provided by German in [15], with a new truncation point of the sequence that is correct for both defective and non-defective PDFs. A proof of the correctness of the new error bound is given in Sect. 5.

The paper also describes how the use of alphaFactory has allowed the extension of the GreatSPN framework [2] to include the solution of MRSPN. Indeed, thanks to the use of alphaFactory the DSPN solver of GreatSPN [3] was easily transformed into an MRSPN solver.

Existing Tools. There is a single tool that we know of which offers an implementation of Extended Uniformization and of the alpha-factors: SPNica [14]. SPNica is a Mathematica package written by R. German for solving MRSPN and, according to our experience, it is a very reliable solver, which unfortunately

---

[1] http://www.boost.org/.

does not scale up to even moderately sized models (few hundreds of states): indeed, by definition of the author himself [14], SPNica is a prototype, a proof of concept. SPNica includes an implementation of the functions for alpha-factors computation, but their use requires the availability of the proprietary Mathematica framework. The software structure of alphaFactory is heavily influenced by the design choices of SPNica, but alphaFactory does not have the dependency on Mathematica as it is all implemented in ISO C++.

A second software that includes Extended Uniformization, again based on the technique proposed by German in [15], is TimeNET [7,23], which supports eDSPN Petri net models with general transitions (basically MRSPN nets). However, the alpha-factors module is not a separate independent component, and no clear analysis of its characteristics were possible, The expression language for general distributions supported by TimeNET is similar to that of SPNica, and also to that of alphaFactory, because of the common SPNica source.

Extended Uniformization is an efficient technique for MRSPN, but there are other tools that can solve MRSPN as a particular case of Non Markovian Stochastic Petri Nets, where typically the enabling restriction of only one general transition enabled in any state, is lifted. The tool WebSPN [6] represents non-Markovian transitions using state-space expansion [19], either discrete (which catches well the behaviour of low-variance distributions like the deterministic or the uniform) or continuous (which catches well high-variance distributions, like hyper-exponentials). In that case, the general distribution behaviour is approximated using a larger state space, represented as a Kronecker product. Another tool that supports general distributions is Oris [8], which again follows a different approach than the one considered in this paper, based on representation and manipulation of mathematical expressions and functions supported over polyhedral and Difference Bound Matrix (DBM) domains [10]. The Oris approach is particularly well suited for expolynomials distributions. Both WebSPN and Oris have a more expensive solution than the one based on Extended Uniformization: WebSPN in terms of larger state spaces, and Oris in computation time due to the need of performing symbolic manipulation of functions. This is certainly not surprising considering that these tools offer a solution for Petri nets with more than one general transitions enabled in a state. Another approach that targets MRSPN analysis is based on Laplace transform inversion, as described in [12].

Paper Outline. Section 2 recalls the Uniformization method and its extended form, along with the alpha-factors definition, whose properties are recalled in Sect. 3, extended to the defective distribution case. Section 4 describes the architecture of the tool, in particular the structure of the alpha-factors evaluation. Sections 5 and 6 describe the computation of the error bound and the alpha-factor algorithm. Section 7 describes how alphaFactory can be integrated into existing tools, a possibility that is illustrated by the integration in GreatSPN; An example of application of the tool to a real model is also shown. Finally, Sect. 8 concludes the paper by identifying new possible research development based on the availability of alphaFactory.

2 Problem Definition

The Uniformization method [16,21] is used to compute the *instantaneous* and *accumulated* transient probabilities of CTMCs. In its extended form [15] it is defined as follows. Let $\mathbf{Q}$ be the infinitesimal generator of the CTMC, and let $\gamma = \max_i (-\mathbf{Q}_{i,i})$ be the maximum rate in any CTMC state. The *uniformized DTMC* $\mathbf{U}$ of $\mathbf{Q}$ is then defined as $\frac{1}{\mu}\mathbf{Q} + \mathbf{I}$, for an arbitrary $\mu \geq \gamma$.

Let g be the event that ends the transient evaluation of the CTMC. g can be seen as an event that is concurrently enabled with the other events represented by the CTMC $\mathbf{Q}$. Let $f(x)$ be the PDF of g. The PDF $f(x)$ is required to be integrable. Let $F(x)$ be the CDF of g. Given an initial probability distribution π_0 over the CTMC states, the *instantaneous* and *accumulated* transient probability distribution at the time g fires are given by:

$$\pi_g^{\text{inst}} = \pi_0 \cdot \sum_{m=0}^{\infty} \mathbf{U}^m \cdot \alpha_f(m, \mu), \qquad \pi_g^{\text{acc}} = \pi_0 \cdot \sum_{m=0}^{\infty} \mathbf{U}^m \cdot \alpha_{\bar{F}}(m, \mu) \qquad (1)$$

The scalar term $\alpha_f(m, \mu)$ is the *alpha-factor* of the PDF $f(x)$ for rate μ. The scalar term $\alpha_{\bar{F}}(m, \mu)$ is the alpha-factor of the *complementary CDF* (CCDF) for rate μ, where the complement CDF $\bar{F}(x)$ is defined as $(1 - F(x))$. The term π_g^{acc} is also commonly referred to as the *cumulative sojourn time* distribution.

The alpha-factors are defined as:

$$\alpha_f(m, \mu) = \int_0^{\infty} e^{-\mu x} \frac{(\mu x)^m}{m!} \cdot f(x)\, dx = \int_0^{\infty} \beta(m, \mu x) \cdot f(x)\, dx$$

$$\alpha_{\bar{F}}(m, \mu) = \int_0^{\infty} e^{-\mu x} \frac{(\mu x)^m}{m!} \cdot \bar{F}(x)\, dx = \int_0^{\infty} \beta(m, \mu x) \cdot \bar{F}(x)\, dx \qquad (2)$$

for $m \in \mathbb{N}_{\geq 0}, \mu > 0$, and with $\beta(m, \lambda) = \dfrac{\lambda^m e^{-\lambda}}{m!}$ the m-th Poisson probability.

In many applications, $f(x)$ is chosen to be distributed as a deterministic event that happens at time t. Hence, its PDF is a Dirac impulse $f_{\text{det}}(x) = \delta(x - t)$, and its CDF is the discontinuous function $F_{\text{det}}(x) = 1$ if $x \leq t$ and 0 otherwise. In this case, the integral can be simplified [22], leading to:

$$\alpha_{f_{\text{det}}}(m, \mu) = e^{-\mu t} \frac{(\mu t)^m}{m!}, \qquad \alpha_{\bar{F}_{\text{det}}}(m, \mu) = \frac{1}{\mu}\left(1 - \sum_{k=0}^{m} e^{-\mu t} \frac{(\mu t)^k}{k!}\right) \qquad (3)$$

However, we are interested in the computation of the alpha-factors as in Eq. (2), which is more general. From an implementation point-of-view, the two most relevant problems of computing Eq. (2) are the necessity of a symbolic integrator, and the numerical stability of the formulas. Both will be treated in Sect. 4.

3 Properties of Alpha Factors

The work in [15, ch. 8] has derived some of the following properties of alpha-factors, that we report. Let $c = \int_0^{\infty} f(x)\, dx$. Then:

Property 1. The sum $\sum_{m=0}^{\infty} \alpha_f(m, \mu) = c$, for any $\mu > 0$.

Property 2. The sequence of $\alpha_f(m, \mu)$ is 0: $\lim_{m \to \infty} \alpha_f(m, \mu) = 0$.

Property 3. If c is finite and $f(x) > 0, \forall x \geq 0$, then it holds that: $0 < \alpha_f(m, \mu) < c$ for all $m \geq 0$.

Property 1 is important because it gives the expected values of the entire sequence of $\alpha_f(m, \mu)$. A non-defective PDF will generate a sequence of alpha-factors $\alpha_f(m, \mu)$ that sums to 1. Alpha factors are upper bounded (Property 3) and converge to 0 (Property 2). This allows to establish a truncation point M to approximate the infinite sequence.

Accumulated alpha-factors are subject to these properties:

Property 4. The accumulated alpha factor $\alpha_{\bar{F}}(m, \mu)$ is given by:

$$\alpha_{\bar{F}}(m, \mu) = \frac{1}{\mu}\left(1 - \sum_{n=0}^{m} \alpha_f(n, \mu)\right) = \frac{1}{\mu} \sum_{n=m+1}^{\infty} \alpha_f(n, \mu)$$

It follows that the sequence of $\alpha_{\bar{F}}(m, \mu)$ converges to 0 when $c = 1$.

Property 4 is useful since the computation of the accumulated alpha-factors can be derived from the sole sequence of $\alpha_f(m, \mu)$. Since we want to consider also defective PDFs, we extend the previous statements (established in [15]) with the following properties:

Property 5. The limit of $\alpha_{\bar{F}}(m, \mu)$ is $\dfrac{1-c}{\mu}$. Therefore, when $c = 1$ the sequence of $\alpha_{\bar{F}}(m, \mu)$ converges to 0.

Proof. A proof of the limit is:

$$\lim_{m \to \infty} \alpha_{\bar{F}}(m, \mu) = \frac{1}{\mu}\left(1 - \lim_{m \to \infty} \sum_{k=0}^{m} \alpha_f(k, \mu)\right) = \frac{1-c}{\mu}$$

Derivation uses Properties 4 and 2. □

Property 6. The sum of the sequence of $\alpha_{\bar{F}}(m, \mu)$ is:

$$\sum_{m=0}^{\infty}\left(\alpha_{\bar{F}}(m, \mu) - \frac{1-c}{\mu}\right) = \int_0^{\infty} x \cdot f(x)\,dx = \mathbb{E}[X]$$

where X is the nonnegative random variable whose distribution is described by $f(x)$. Therefore, the sum does not depend on the value of μ.

Proof. The equivalence can be derived in this way:

$$\sum_{m=0}^{\infty}\left(\alpha_{\bar{F}}(m,\mu)-\frac{1-c}{\mu}\right)=\frac{1}{\mu}\sum_{m=0}^{\infty}\left(\left(1-\sum_{k=0}^{m}\alpha_f(k,\mu)\right)-\left(1-\sum_{k=0}^{\infty}\alpha_f(k,\mu)\right)\right)$$

$$=\frac{1}{\mu}\sum_{m=0}^{\infty}\sum_{k=m+1}^{\infty}\alpha_f(k,\mu)=\frac{1}{\mu}\sum_{m=1}^{\infty}m\cdot\alpha_f(m,\mu)$$

$$=\sum_{m=1}^{\infty}\int_{0}^{\infty}\frac{m}{\mu}\cdot e^{-\mu x}\frac{(\mu x)^m}{m!}\cdot f(x)\,\mathrm{d}x$$

$$=\sum_{m=0}^{\infty}\int_{0}^{\infty}\beta(m,\mu x)\cdot x\cdot f(x)\,\mathrm{d}x$$

$$=\int_{0}^{\infty}x\cdot f(x)\,\mathrm{d}x=\mathbb{E}[X]$$

Derivation uses Properties 4 and 1, and the trivial relation $\sum_{m=0}^{\infty}\beta(m,\mu x)=1$. □

Property 7. If c is finite and $f(x)>0, \forall x\geq 0$, then it holds that:

$$\alpha_{\bar{F}}(m,\mu)\geq\frac{1-c}{\mu}, \qquad \forall m\geq 0$$

Proof. Assuming $f(x)\geq 0$ for $x\geq 0$, it holds that: $\alpha_f(m,\mu)\geq 0$, for any $m\geq 0$. Therefore, Property 7 is a direct consequence of Property 4, which ensures that $\alpha_{\bar{F}}(m,\mu)$ values are monotonically non-increasing, and Property 5, which gives the limiting behaviour of the series. □

Property 5 shows that the sequence $\alpha_{\bar{F}}(m,\mu)$ may converge to a value that is different from 0 for defective PDFs. Property 7 establishes a lower bound for the sequence. A single truncation point for both the $\alpha_f(m,\mu)$ and the $\alpha_{\bar{F}}(m,\mu)$ sequences can then be established based on the convergent behaviours of both.

4 Architecture of alphaFactory

alphaFactory is a small tool written in ISO C++ whose sole purpose is the computation of the alpha-factors $\alpha_f(m,\mu)$ and $\alpha_{\bar{F}}(m,\mu)$, given the textual representation of function $f(x)$ and the rate μ. The tool is made of a single C++ compilation unit, plus a header file. A compile-time macro ALPHAFACTORSLIB controls whether the tool is compiled as a standalone command-line program, or linked inside another program. The main goal of alphaFactory is that of being used inside numerical solvers that use Uniformization for the computation of instantaneous/accumulated transient probabilities. The tool follows the formula derivations found in [15, pp. 394–398]. For the sake of completeness, we report the formulas of the transformation rules derived in that book.

The language of the functions ϕ accepted by alphaFactory is the following:

$$\phi ::= number \mid x \mid \phi \circ \phi \mid \text{Pow}(\phi, \phi) \mid \text{Exp}(\phi) \mid \text{Log}(\phi) \mid \text{I}(\psi) \mid \text{R}(\psi, \psi) \mid$$
$$\text{Uniform}(\psi, \psi) \mid \text{Triangular}(\psi, \psi) \mid \text{Erlang}(\psi, \psi) \mid$$
$$\text{TruncatedExp}(\psi, \psi) \mid \text{Pareto}(\psi, \psi)$$
$$\psi ::= number \mid \psi \circ \psi \mid \text{Pow}(\psi, \psi) \mid \text{Exp}(\psi) \mid \text{Log}(\psi)$$

where $\circ \in \{+, -, *, /\}$. Number literals are floating point real numbers. The term x is the integral variable. The functions Pow, Exp and Log are the power, the exponential and the natural logarithm, respectively. The function $\text{I}(\phi)$ is a Dirac delta unit impulse $\delta(x - \phi)$. It represents the concentration of the probability mass at single point ϕ, and it is interpreted as if the probability of a firing at time ϕ is 1. The function $\text{R}(a, b)$ is a rectangular signal that assumes value 1 over the range $[a, b]$, and 0 outside that range. The language ψ is just a simplified language for algebraic expressions over constant terms.

The remaining elements of ϕ are non-primitive functions:

- $\text{Uniform}(a, b)$ is the uniform distribution, defined as $1/(b - a) * \text{R}(a, b)$.
- $\text{Triangular}(a, b)$ is the triangular distribution, defined as:

$$\frac{4 * (x - a)}{(a - b)^2} * \text{R}\left(a, \frac{a + b}{2}\right) - \frac{4 * (x - b)}{(a - b)^2} * \text{R}\left(\frac{a + b}{2}, b\right)$$

- $\text{Erlang}(\lambda, r)$ is the Erlang function with rate λ and r phases, defined as the standard PDF of the Erlang distribution: $\dfrac{\lambda^r}{(r - 1)!} * x^{r-1} * e^{-\lambda * x}$
- $\text{TruncatedExp}(\lambda, t)$ is the exponential distribution of rate λ truncated at time t, defined as: $\lambda * e^{-\lambda * x} * \text{R}(0, t) + e^{-\lambda * t} * \text{I}(t)$. It is obtained by multiplying the exponential distribution with a rectangular signal $\text{R}(0, t)$, so that after t the distribution is truncated. To compensate the truncation, an impulse of probability $e^{-\lambda * t}$ happens at time t, so that the overall truncated exponential is not a defective PDF.
- $\text{Pareto}(k, s)$ is the Pareto distribution of real scale parameter k and shape parameter s, $s \in \mathbb{N}_{>0}$, defined as: $\begin{cases} \frac{s * k^s}{x^{s+1}} & \text{if } x > k \\ 0 & \text{if } x \le k \end{cases}$

These simple building blocks allow to define common distribution functions, like expolynomials distributions.

The evaluation of a function $f(x)$ starts by building the Abstract Syntax Tree (AST) of the formula. The tool uses a recursive descent parser for this task. ASTs are made by just three node types:

1. *Term* leaf nodes that contain real values.
2. *Symbol* leaf nodes that contain the integration variable x.
3. *Function* nodes, that are n-ary operators for a single arithmetic operand. The function operand is one among $\{+, -, *, /, \text{Pow}, \text{Exp}, \text{Log}, \text{I}, \text{R}\}$, or a non-primitive operand among $\{\text{Uniform}, \text{Triangular}, \text{Erlang}, \text{TruncatedExp}, \text{Pareto}\}$.

Once the parser has finished, it is possible to manipulate the expression of $f(x)$ at the AST level. The tool has four main AST-manipulation functions: `evaluate(e)`, `simplify(e)`, `integrate(e)` and `moment(e, k)`.

- `evaluate(e)` does the numerical evaluation of e. The expression e must have only constant terms or function, i.e. it cannot have the integration variable x.
- `simplify(e)` implements polynomial simplification and rearrangement of the expression argument e into a canonical form. It works by applying a fixed set of transformation rules. Rules use pattern matching and node substitution, and are encoded inside the function. For instance, the function $x * x$ is canonicalized as x^2, or the function x^0 is simplified as 1. The transformation rules are:

```
simplify(ψ)  →  evaluate(ψ)
simplify(φ * 1 or φ + 0)  →  φ
simplify(φ * φ * ... * φ)  →  φⁿ
simplify(φ₁ * (φ₂ + φ₃))  →  φ₁ * φ₂ + φ₁ * φ₃
simplify(φ⁰)  →  1
simplify(φ¹)  →  φ
simplify(any function φ)  →  recursively apply simplify on φ operands
```

The function is also responsible for the expansion of the non-primitive functions $\text{Uniform}(a, b)$, $\text{Triangular}(a, b)$, $\text{Erlang}(\lambda, r)$, $\text{TruncatedExp}(\lambda, t)$ $\text{Pareto}(k, s)$, and operand reordering (terms in a product are always arranged following a specified canonical order).

- `integrate(e)` computes the symbolic integral $\int_0^\infty e(x)\,\mathrm{d}x$ using AST manipulation, assuming that the expression is in canonical form (obtained by `simplify`). As before, it implements a set of transformation rules to compute the symbolic result. The implemented rules are:

```
integrate(x)  →  ½ * x²
integrate(t)  →  t * x
integrate(eˣ)  →  eˣ
integrate(e^{k*x})  →  1/k * e^{k*x}
integrate(xᵐ)  →  1/(m+1) * x^{m+1}
integrate(c * I(t))  →  c
integrate(c * x * I(t))  →  c * t
integrate(c * e^{l*x} * xʰ * R(0, b))  →  c * (-l)^{-h-1} * (Γ(h+1) - Γ(h+1, -b*l))
integrate(c * R(a, b))  →  integrate(c * R(0, b)) - integrate(c * R(0, a))
integrate(c * φ)  →  c * integrate(φ)
integrate(φ₁ - φ₂)  →  integrate(φ₁) - integrate(φ₂)
integrate(sum of φᵢ)  →  sum of integrate(φᵢ)
```

where c, t are constant terms, $\Gamma(z)$ and $\Gamma(s, z)$ are the complete and the upper incomplete gamma functions, respectively. The rules are actually standard integration rules. A product with a rectangular signal $R(a, b)$ is equivalent to computing the integral over the $[a, b]$ range instead of the $[0, \infty)$ range.

- `moment(e, k)` is the moment generating function. It computes the k-th moment of the random variable with PDF e as: `evaluate(integrate(e*x`^{k}`))`.

Once `alphaFactory` has built and simplified the AST e of the expression of $f(x)$, it starts by computing the 0 and 1 moments of $f(x)$. The 0 moment, being the area of $f(x)$, is checked to be 1. If it is not, $f(x)$ is a defective PDF, and a warning message is printed. In some applications, defective PDFs are allowed, so the tool does not stop for this condition. Property 1 also tells that the 0 moment gives the sum of the sequence of $\alpha_f(m, \mu)$, which is used for error bound. The first moment, being $\mathbb{E}[X]$, can be used to bound the sum of $\alpha_{\bar{F}}(m, \mu)$ for non-defective PDFs (Property 6).

At this point, alpha-factors can be computed for AST e using the recursive function $\texttt{alpha}(m, \mu, e)$. This function computes the m-th factor for a CTMC of rate μ. The function $\texttt{alpha}(m, \mu, \phi)$ applies these transformation rules to ϕ:

$$\texttt{alpha}(\phi_1 + \phi_2) \;\rightarrow\; \texttt{alpha}(m, \mu, \phi_1) + \texttt{alpha}(m, \mu, \phi_2)$$
$$\texttt{alpha}(c * \texttt{I}(a)) \;\rightarrow\; c * \beta(m, \mu * a)$$
$$\texttt{alpha}(\texttt{I}(a)) \;\rightarrow\; \beta(m, \mu * a)$$
$$\texttt{alpha}(c * \texttt{R}(a, b)) \;\rightarrow\; \texttt{alpha}(m, \mu, c * \texttt{R}(0, b)) - \texttt{alpha}(m, \mu, c * \texttt{R}(0, a))$$
$$\texttt{alpha}(\texttt{R}(a, b)) \;\rightarrow\; \texttt{alpha}(m, \mu, \texttt{R}(0, b)) - \texttt{alpha}(m, \mu, \texttt{R}(0, a))$$
$$\texttt{alpha}(c * e^{l * x} * x^h) \;\rightarrow\; c * \gamma_a(m, \mu, h, -l, a)$$
$$\texttt{alpha}(c * e^{l * x} * x^h * \texttt{R}(0, a)) \;\rightarrow\; c * \gamma_\infty(m, \mu, h, -l)$$
$$\texttt{alpha}(\texttt{Pareto}(k, s)) \;\rightarrow\; -\frac{e^{-\mu} * k^m * s * \mu^m}{(m-s) * m!}$$

These formulas are directly derived by solving the integral of Eq. 2 over the function argument, and are taken from [15, p. 396]. The rest of the evaluation of the alpha-factors relies on four recursive functions β, γ_a, γ_∞ and η, that are needed to evaluate the integral terms symbolically. Memoization of the partially evaluated results is employed to speed up the computation. The m-th Poisson probability $\beta(m, \lambda)$ function is implemented using the usual recursive relation:

$$\beta(m, \lambda) = \begin{cases} e^\lambda & \text{if } m = 0 \\ \dfrac{\lambda * \beta(m - 1, \lambda)}{k} & \text{otherwise} \end{cases}$$

The two γ factors are derived by integrating the expolynomial equations (6th and 7th rules of `alpha`) with the Poisson function, expanding Eq. (2). The full derivation can be found in [15, pp. 154–155]. This results in a recursive relation for γ_∞ and γ_a, defined as:

$$\gamma_\infty(m, \mu, h, l) = \begin{cases} \dfrac{h!}{(\mu + l)^{h+1}} & \text{if } m = 0 \\ \dfrac{m + h}{m} * \dfrac{\mu}{\mu + l} * \gamma_\infty(m - 1, \mu, h, l) & \text{otherwise} \end{cases}$$

and:

$$
\gamma_a(m, \mu, h, l, a) = \begin{cases} \dfrac{h!}{(\mu + l)^{h+1}} \left(1 - \displaystyle\sum_{i=0}^{m} \beta(i, (\mu + l)a) \right) & \text{if } m = 0 \\[2ex] \dfrac{m+h}{m} * \dfrac{\mu}{\mu + l} * \gamma_a(m - 1, \mu, h, l, a) \\ \quad - \eta(m, \mu, h, l, a) * \beta(m, (\mu + l) * a) & \text{otherwise} \end{cases}
$$

where $\gamma_\infty(m, \mu, h, l)$ is equivalent to $\gamma_a(m, \mu, h, l, \infty)$. The η factor is defined as:

$$
\eta(m, \mu, h, l, a) = \begin{cases} \dfrac{a^h}{\mu + l} & \text{if } m = 0 \\[2ex] \eta(m - 1, \mu, h, l, a) * \dfrac{\mu}{\mu + l} & \text{otherwise} \end{cases}
$$

5   Bounds of α-Tails

Since the series of alpha-factors $\alpha_f(m, \mu)$ and $\alpha_{\bar{F}}(m, \mu)$ are infinite, defined for all $m \in \mathbb{N}_{\geq 0}$, it is necessary to approximate the sequence up to a right truncation point M. Using an accuracy parameter ϵ, the sequence of instantaneous alpha-factors $\alpha_f(m, \mu)$ can be truncated at M:

$$
\sum_{m=M+1}^{\infty} \alpha_f(m, \mu) < \epsilon \implies \sum_{m=0}^{M} \alpha_f(m, \mu) = c - \epsilon
$$

The relation, derived in [15, p. 152], is directly derived from Property 1, which ensures that the sum of the entire sequence is c, and Property 2 which ensures that the sequence converges to 0.

The right truncation point of the sequence of accumulated alpha-factors $\alpha_{\bar{F}}(m, \mu)$ is slightly different, since they converge to 0 only for non-defective PDFs (i.e. $c = 1$). In the general case, the sequence converges to $\dfrac{1 - c}{\mu}$, as stated in Property 5. Therefore, a truncation point M' can be set such that:

$$
\sum_{m=M'+1}^{\infty} \left(\alpha_{\bar{F}}(m, \mu) - \frac{1 - c}{\mu} \right) < \epsilon \implies \sum_{m=0}^{M'} \left(\alpha_{\bar{F}}(m, \mu) - \frac{1 - c}{\mu} \right) = \mathbb{E}[X] - \epsilon
$$

using ϵ as an absolute error for the summation. Therefore, a method can be devised that ensures that both sequences are truncated below the requested accuracy ϵ using $R = \max(M, M')$ as the truncation point. This requires to know both the 0-moment c and the first moment $\mathbb{E}[X]$ of the random var. X.

6   Alpha-Factors Computation Algorithm

After having introduced all the required elements, it is now possible to show the core alpha-factors computation function. The pseudo-code of the method is shown in Algorithm 1.

Algorithm 1. Pseudocode of the alpha-factors generation function.

Function `compute_alpha_factors_dbl`(f, μ, ϵ):

    $e \leftarrow$ `simplify(parse`$(f))$

    $c \leftarrow$ `moment`$(e, 0)$

    $\mathbb{E}[X] \leftarrow$ `moment`$(e, 1)$

    $err_f \leftarrow c$

    $err_{\bar{F}} \leftarrow \mathbb{E}[X]$ **if** $\mathbb{E}[X] \neq \infty$ **else** 0

    $m \leftarrow 0$

    $value_{\bar{F}} \leftarrow \frac{1}{\mu}$

    while $(err_f > \epsilon \vee err_{\bar{F}} > \epsilon)$:

        $\alpha_f(m, \mu) \leftarrow$ `alpha`(m, μ, e)

        $value_{\bar{F}} \leftarrow value_{\bar{F}} - \frac{\alpha_f(m, \mu)}{\mu}$

        $\alpha_{\bar{F}}(m, \mu) \leftarrow value_{\bar{F}}$

        $err_f \leftarrow err_f - \alpha_f(m, \mu)$

        $err_{\bar{F}} \leftarrow err_{\bar{F}} - \left(\alpha_{\bar{F}}(m, \mu) - \frac{1-c}{\mu}\right)$

        $m \leftarrow m + 1$

    return $\langle \alpha_f, \alpha_{\bar{F}} \rangle$

The algorithm is an implementation of the method defined in [15, p. 394], with the new bound R described in Sect. 5. It starts by parsing the function and building the AST. It then computes the first two moments, initializing the error control variables err_f and $err_{\bar{F}}$ with the moment values. The function than iterates until both error thresholds are below ϵ. At each iteration, the alpha-factor $\alpha_f(m, \mu)$ is computed. The accumulated alpha-factor $\alpha_{\bar{F}}(m, \mu)$ is derived implicitly by subtracting incrementally all the instantaneous alpha-factors, starting from the initial value $\frac{1}{\mu}$. This follows Property 4. The algorithm than subtracts the alpha-factors from the err_f and $err_{\bar{F}}$ control variables, and repeats. Even if it is true that the sequence of $\alpha_{\bar{F}}(m, \mu)$ can be completely derived from the sequence of $\alpha_f(m, \mu)$, it is important to compute both together, in order to establish the single truncation point R that guarantees that both sequence have an absolute error below ϵ.

Numerical Precision. The alphaFactory implementation uses multi-precision floating point. Floating point precision is controlled using the MPFLOAT_PRECISION constant, which is defaulted to 1024 bits. This allows to treat factors with large difference in magnitude, without a dangerous loss of precision. Of course, a different strategy (like the one used by the previously mentioned Fox-Glynn method) could be devised to improve the accuracy without resorting to multi-precision arithmetic. In particular, the error control variables are subject to multiple subtractions, which could result in numerical instability without enough precision. The strategy used by the Fox-Glynn method and many other Uniformization methods is that of starting from the central value of the $\beta(m, \mu)$ series, and the computing the left and right tails independently. Unfortunately, it is hard to derive a single computational strategy that is at the same time general w.r.t $f(x)$ and that computes the values in a non-sequential order. However, if we restrict to some specific classes of general functions (like expolynomials), a better strategy could be devised.

Tool Validation. The tool has been validated against the results computed by SPNica and by a direct evaluation of the alpha-factor formulas in Mathematica. Results confirm the correctness of the tool on a benchmark of distributions. The tool is also equipped with a small unit test, that runs using the `test` command line argument. The unit test verifies that the tool re-computes correctly the alpha-factors from a small set of PDFs, and compares the obtained results with the values computed by evaluating the corresponding formula integrals in Mathematica.

6.1   Example of Running alphaFactory

Figure 1 shows four by three plots obtained running alphaFactory on four PDFs.

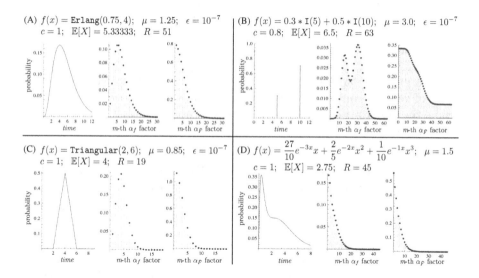

Fig. 1. Alpha-factor distributions generated by alphaFactory from four PDFs.

For each PDF three graphs are shown: the PDF itself and instantaneous and accumulated alpha-factors (left to right). The plot header reports the computed values for c, $\mathbb{E}[X]$, and the truncation point R. PDF (A) is an Erlang distribution of rate 0.75 and 4 phases. Alpha factors are computed for a CTMC of rate $\mu = 1.25$, with accuracy $\epsilon = 10^{-7}$. Values for (A) are obtained running the tool from the command line:

```
./alphaFactory 'Erlang(0.75, 4)' 1.25 0.0000001
```

The second line in the header of plot (A) reports the computed values for c and $\mathbb{E}[X]$, and the truncation point R. PDF (B) is a linear combination of two Dirac impulses, which is intuitively a random choice between two deterministic events. The combination is defective, as can be seen by the computed integral value $c = 0.8$. PDF (C) is a triangular distribution, which is actually decomposed into a polynomial combination of two rectangular signals. Finally, PDF (D) is an expolynomial distribution. When run from the command line as a standalone tool, alphaFactory first writes the two moments of the function, followed by the number of factors and by a list of one factor per line. The generated alpha-factors can be used directly inside a Extended Uniformization method, following Eq. (1).

7   Integration of **alphaFactory** into Other Softwares

As mentioned before, alphaFactory can be included in another software project as a static library or as a C++ compilation unit, to be used non-interactively. The Application Programming Interface of alphaFactory is minimal and it is made by just two exported functions:

- verify_alpha_factors_expr takes in input a character strings and verifies if it is a valid input expression for the tool, if it is defective and if the tool is capable of integrating that function. This function is useful for expression validation, for instance during model loading, or within a graphical editor.
- compute_alpha_factors_dbl(const char* fg, double mu, double eps) computes the alpha factor distributions as in Eq. (2). The function returns a pair of vectors, containing the values of $\alpha_f(m, \mu)$ and $\alpha_{\bar{F}}(m, \mu)$. Argument mu is the uniformized CTMC rate. Argument eps specifies the computation accuracy ϵ.

The minimalist API allows to integrate the tool into other softwares that use the Uniformization method with a minimal effort. We have integrated alphaFactory in our DSPN solver [3], making it capable of solving MRSPN models.

7.1   Integration of **alphaFactory** into GreatSPN

We now show a small example of an application of generalized functions in MRSPN, solved with the help of alphaFactory. The tool has been integrated inside the DSPN solver of GreatSPN [3], which is now capable of solving MRSPN Petri net models with general transitions with the usual *enabling restriction*. We consider the case [5] of a multi-utility company, who works in a specified geographical area of about $2200\,\mathrm{km}^2$, with 531K inhabitants. The problem the company is interested in is the optimal allocation of human resources, in order to comply with the national regulation authority rules, which require that in case of call from a client of a detected leak of gas, a technician must be on-site in less than 1 h. When on site, the technician first secures the problem. Then, he may decide to actually fix the problem, if there are no other open requests. Otherwise, he leaves the site, sending an external plumber to do the fix.

Figure 2 shows the travelling time distribution and securing time distribution, extracted from the company log (about 600 samples). We used the Erlang distribution for data fitting, deriving using the company logs an Erlang$(1.15, 3)$

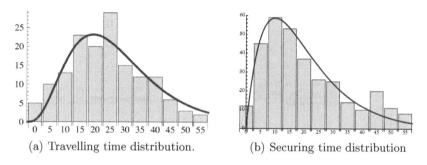

(a) Travelling time distribution.   (b) Securing time distribution

Fig. 2. Data samples from which the general distributions were derived.

for the travelling time, and Erlang(1.6, 4) for the securing time. We do not have precise timing for the repairing phase, but the company told us that the time is usually in the order of 10–30 min. Hence, we modeled the repairing time with a Uniform(10, 30).

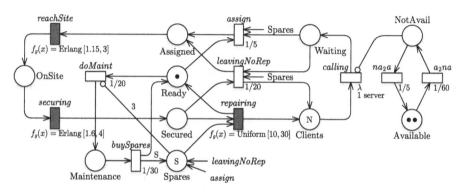

Fig. 3. Simplified MRSPN of the multi-utility company repairmen problem used in [5].

The MRSPN model of the simplified multi-utility company is depicted in Fig. 3. Distribution functions are written as annotations of the general transition (black rectangles) in the graphical representation of the MRSPN. The functions are passed verbatim to alphaFactory during the solution process. We run the steady-state MRSPN solver on the model, for various client *inter-arrival times* (IAT), to study the behaviour of the system and the load balance. The goal of the analysis is to find the client IAT value where the probability of having another client being served is below 10%.

Figure 4 shows the expected number of clients in the system (left) and the probability distribution of finding the technician idle (right), for increasing values of the client IAT. We observe that with a client IAT of ∼28 min the probability of finding another client in queue is about 1. The probability drops rapidly, and is below 0.1 with an inter-arrival time of 100 min. The probability of finding the technician idle is about ∼74% when the client IAT is 100 min. We therefore conclude that the target system load is at a client IAT value of about 100 min,

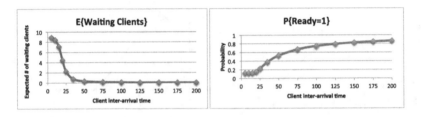

Fig. 4. MRSPN performance indexes computed in steady-state.

considering the provided travelling time distributions and securing time distributions, when a single technician is available.

Overall, the MRSNP has 1116 markings. The steady state solution time for a single run takes ~0.2 s. In general, we may say that in most cases the cost of running alphaFactory is negligible, since it needs to be run just once for each $f(x)$ and μ pair.

8 Conclusions and Future Works

This paper describes the tool alphaFactory, whose purpose is the generation of the alpha-factors of a general distribution $f(x)$. Alpha-factors are used by the Extended Uniformization method to compute instantaneous/accumulated transient probabilities at time t, with t distributed as $f(x)$. The main purpose of alphaFactory is to make a standalone, re-usable component to ease the implementation of Extended Uniformization. This is mostly of interest for MRgP solvers, MRSPN and DSPN tools, and tools that compute transient CTMC probabilities.

We plan to extend also our Phase-Mission systems [4,20] tools with alphaFactory, to allow the definition of phases of general duration. This extension is useful for many processes (like workflow processes) where phases are more naturally described with uniform distributions or distributions fitted from data. We also plan to investigate the use of alphaFactory inside other tools that support more sophisticated data structure representations, like state-spaces in Kronecker form [9].

Availability. The source code of alphaFactory is distributed under a modified BSD license, and can be found at: https://github.com/amparore/alphaFactory.

References

1. Marsan, M.A., Chiola, G.: On Petri nets with deterministic and exponentially distributed firing times. In: Rozenberg, G. (ed.) APN 1986. LNCS, vol. 266, pp. 132–145. Springer, Heidelberg (1987). doi:10.1007/3-540-18086-9_23
2. Amparore, E.G., Balbo, G., Beccuti, M., Donatelli, S., Franceschinis, G.: 30 years of GreatSPN. In: Fiondella, L., Puliafito, A. (eds.) Principles of Performance and Reliability Modeling and Evaluation. SSRE, pp. 227–254. Springer, Cham (2016). doi:10.1007/978-3-319-30599-8_9
3. Amparore, E.G., Donatelli, S.: DSPN-tool: a new DSPN and GSPN solver for GreatSPN. In: International Conference on Quantitative Evaluation of Systems, Los Alamitos, CA, USA, pp. 79–80. IEEE Computer Society (2010)

4. Amparore, E.G., Donatelli, S.: Efficient solution of extended multiple-phased systems. In: 10th ValueTools Conference, pp. 125–132. EAI (2016)
5. Amparore, E.G., Donatelli, S., Landini, E.: Modelling and evaluation of a control room application. In: van der Aalst, W., Best, E. (eds.) PETRI NETS 2017. LNCS, vol. 10258, pp. 243–263. Springer, Cham (2017). doi:10.1007/978-3-319-57861-3_15
6. Bobbio, A., Puliafito, A., Scarpa, M., Telek, M.: WebSPN: non-Markovian stochastic Petri net tool. In: 18th Conferecne on Application and Theory of Petri Nets (1997)
7. Bodenstein, C., Zimmermann, A.: TimeNET optimization environment: batch simulation and heuristic optimization of SCPNs with TimeNET 4.2. In: 8th International Conference on Performance Evaluation Methodologies and Tools, pp. 129–133. ICST (2014)
8. Bucci, G., Carnevali, L., Ridi, L., Vicario, E.: Oris: a tool for modeling, verification and evaluation of real-time systems. Int. J. Soft. Tools Technol. Transf. **12**(5), 391–403 (2010)
9. Buchholz, P.: Markov matrix market. http://ls4-www.cs.tu-dortmund.de/download/buchholz/struct-matrix-market.html
10. Carnevali, L., Ridi, L., Vicario, E.: A framework for simulation and symbolic state space analysis of non-Markovian models. In: Flammini, F., Bologna, S., Vittorini, V. (eds.) SAFECOMP 2011. LNCS, vol. 6894, pp. 409–422. Springer, Heidelberg (2011). doi:10.1007/978-3-642-24270-0_30
11. Choi, H., Kulkarni, V.G., Trivedi, K.S.: Markov regenerative stochastic Petri nets. Perform. Eval. **20**(1–3), 337–357 (1994)
12. Dingle, N.J., Harrison, P.G., Knottenbelt, W.J.: Response time densities in generalised stochastic Petri nets. In: Workshop on Software and Performance, pp. 46–54 (2002)
13. Fox, B.L., Glynn, P.W.: Computing poisson probabilities. Commun. ACM **31**(4), 440–445 (1988)
14. German, R.: Markov regenerative stochastic Petri nets with general execution policies: supplementary variable analysis and a prototype tool. Perform. Eval. **39**(1–4), 165–188 (2000)
15. German, R.: Performance Analysis of Communication Systems with Non-Markovian Stochastic Petri Nets. Wiley, New York (2000)
16. Grassmann, W.: Transient solutions in Markovian queueing systems. Comput. Oper. Res. **4**(1), 47–53 (1977)
17. Jansen, D.N.: Understanding Fox and Glynn's "Computing Poisson Probabilities". Technical report, Nijmegen: ICIS R11001 (2011)
18. Kulkarni, V.G.: Modeling and Analysis of Stochastic Systems. Chapman & Hall Ltd., London (1995)
19. Longo, F., Scarpa, M.: Two-layer symbolic representation for stochastic models with phase-type distributed events. Int. J. Syst. Sci. **46**(9), 1540–1571 (2015)
20. Mura, I., Bondavalli, A., Zang, X., Trivedi, K.S.: Dependability modeling and evaluation of phased mission systems: a DSPN approach. In: International Conference on Dependable Computing for Critical Applications (DCCA), pp. 299–318. IEEE (1999)
21. Stewart, W.J.: Introduction to the numerical solution of Markov chains. Princeton University Press, Princeton (1994)
22. Trivedi, K.S., Reibman, A.L., Smith, R.: Transient analysis of Markov and Markov reward models. In: Computer Performance and Reliability 1987, pp. 535–545 (1987)
23. Zimmermann, A., Freiheit, J., German, R., Hommel, G.: Petri net modelling and performability evaluation with TimeNET 3.0. In: Haverkort, B.R., Bohnenkamp, H.C., Smith, C.U. (eds.) TOOLS 2000. LNCS, vol. 1786, pp. 188–202. Springer, Heidelberg (2000). doi:10.1007/3-540-46429-8_14

Smart Energy Systems over the Cloud
(Special Session)

Quantitative Model Checking
for a Smart Grid Pricing

YoungMin Kwon[1(✉)], Eunhee Kim[2], Seonghwan Jeong[1], and Arthur H. Lee[1]

[1] Department of Computer Science, The State University of New York, Korea,
119 Songdo Moonhwa-Ro, Yeonsu-Gu, Incheon 21985, Korea
{youngmin.kwon,alee}@sunykorea.ac.kr, seonghwan.jeong@stonybrook.edu
[2] 2e Consulting Corporation, 1710 KnK Digital Tower, 220 Yeongsin-Ro,
Yeongdeungpo-gu, Seoul 07288, Korea
keh@2e.co.kr

Abstract. A quantitative model checking technique is applied to compute a day-ahead pricing for smart grids. In a smart grid system, smart meters enable bidirectional communications between electricity providers and customers. The providers can monitor the customers' detailed electricity usage and by posting dynamically changing prices, they can shape the energy demand. We propose a model checking based day-ahead pricing technique and demonstrate the usefulness of the technique. Specifically, we model the power demand changes of various types of loads by first order differential equations while considering expected loads and price changes as external forces. Complex requirements about the customers' energy usage, the current system state estimate, and the expected load for the next day are described in an LTL based quantitative temporal logic, called LTLC. Day-ahead prices that can satisfy the description are computed through the LTLC model checking.

1 Introduction

A smart grid is an energy delivery system with two-way communications between electricity providers and customers. The two-way communication enabled by smart meters allows the energy providers to measure the customers' energy consumption level and to post time dependent prices to the customers. In other words, the bidirectional communication forms a feedback loop that enables the electricity providers to shape the customers' energy demand levels.

Smart meters are one of the key elements in the smart grid infrastructures. They periodically record the electricity consumption levels and send the information back to the energy provider at least at a daily interval. The provider can build the customers' energy usage profile in hour-to-hour details from the report. In addition to the monitoring and reporting capabilities, smart meters can display the providers' time-dependent electricity prices. Smart devices at home or customers themselves can either automatically or manually adjust their scheduled electricity usage based on this price. These detailed demand monitoring and price posting capabilities enable the electricity providers to shape the customers' energy consumption level to meet various requirements [3,8,17].

© Springer International Publishing AG 2017
N. Bertrand and L. Bortolussi (Eds.): QEST 2017, LNCS 10503, pp. 55–71, 2017.
DOI: 10.1007/978-3-319-66335-7_4

In this paper, we propose a method to compute a pricing scheme for smart grid systems. Assuming that the aggregate behaviors of smart grid systems are expressed in dynamics equations, a pricing scheme that can shape the state trajectories of the system to meet the requirements can be computed through a quantitative model checking. Describing a goal in a temporal logic has several advantages over directly implementing the goal in a code. Some of the merits are: the approach is manageable in that one can easily update a goal without worrying about rewriting the implementation; a complex goal can be intuitively expressed and is less likely to have implementation errors; when the goal is not achievable, its impossibility will be reported. One of the disadvantages in using the technique is that the resulting constraint solving mechanism may be less efficient than handwritten programs. The inefficiency can be somewhat relieved by saving the feasible paths of intersection automata [4] offline.

To demonstrate the usefulness of the proposed technique, we developed a non-trivial mathematical model of a smart grid system and could successfully compute a pricing scheme against the model. Specifically, we model a smart grid system by first order differential equations, describe requirements in a quantitative temporal logic, called *Linear Temporal Logic for Control* (LTLC) [15], and show that the model checking technique can find a pricing scheme that can shape the demand to satisfy the requirements. Speaking of the pricing, we are focusing on the day-ahead pricing scheme, where hourly electricity prices for the next day will be posted together in advance [8]. Compared to the real-time pricing scheme, where the price is determined based on the current measurement and is enforced right away [3,17], the day-ahead pricing scheme has certain advantages. For example, from the customer's perspective, the real-time scheme leaves a lot of uncertainty about the current electricity price and this scheme makes it difficult to schedule their device usages in advance. Moreover, some tasks, like laundry, are not flexible enough to be interrupted in the middle of the operation.

In addition to the day-ahead pricing scheme, we take into account the fact that different types of devices may have different sensitivities to the price changes and to the expected daily usage. For instance, refrigerators are turned on all the time and their usage is insensitive to price changes and to the expected daily usage. On the other hand, we may want to watch TV shows or news at specific times of a day, but we are less likely to leave a TV on without watching when the electricity is expensive. Hence, the demand for TV is sensitive to both. We model the dynamics of each device usage by first order differential equations with the expected device usage and the price change acting as two external forces driving the system states. There is a previous work on finding an optimal day-ahead pricing scheme considering that devices have different sensitivity to prices [10]. In this paper, we will show that more complex requirements can be easily described in our temporal logic, and a feasible pricing scheme or its non-existence can be systematically computed through a model checking process.

Model checking is an automatic process of validating properties of concurrent systems with finite states [5,6,9,16]. It checks whether there are any computations of a system that may violate a specification, commonly written in a temporal logic. One of the benefits of model checking techniques over simulations or testings is that they can give the completeness about the result. In other words, if a model checking result is true, it is based on the complete examination of all possible cases. Moreover, if the result is false, it can report a counterexample witnessing the violation. Recently, we have been progressively extending the model checking techniques to target models with continuous variables. One of the difficulties in handling such systems is that the infinite state space cannot be enumerated. Coping with the difficulty, we have developed a technique that can check the trajectories of *probability mass functions* (pmfs) of Markov chains [14]. Later, we broadened the target to include general linear systems that have external inputs as well as an initial state [13]. More recently, this technique is extended once again to target hybrid systems [15].

Figure 1 shows an overview of the proposed a day-ahead pricing technique for a smart grid system. The smart meters at customers' home can show the electricity price such that the customers or their devices can schedule the electricity usage. Detailed energy consumption levels are frequently reported back to the energy provider. The energy profiler collects this information, builds an electricity load profile [18], and produces expected energy consumption levels for different device types. In our system model, this expected power consumption level is considered as an external force that tends the actual power demand towards it. The measured energy information is also fed into the state estimator, such as Kalman filter [19], where the current state of the system can be estimated. In this paper we assume that the energy profile and the state estimate are already available and we will focus on how these information and the requirements about the demand can be incorporated into an LTLC description and how a day-ahead pricing scheme can be computed by the LTLC model checking.

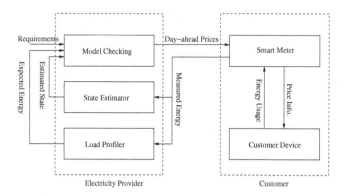

Fig. 1. A diagram for a day-ahead pricing scheme. Based on the state estimate and the expected energy consumption level for a day, day-ahead prices are computed by the LTLC model checking.

2   Smart Grid Model

To demonstrate the usefulness of the proposed method, in this section, we develop a mathematical model for a smart grid system. We model the dynamics of the electricity demand change by first order differential equations. These continuous dynamics are reformulated in the state variable form and discretized. The state variable form is a standard form on which many control techniques, including our LTLC model checking algorithm, are developed. Finally, the resulting discrete state variable model is partitioned into related components.

We assume that the change of the demand is proportional to the difference between the expected load and the current load. For example, suppose that there is a popular TV show then customers are likely to turn on their TVs at that time, which will make the load change proportional to the difference between the current and the expected load. We also assume that the demand is negatively propositional to price changes. That is, if the price is raised the demand will be dropped. Hence, if customers know that the price of the electricity will be cheaper at noon, they would shift their laundry or vacuuming towards noon.

The description above can be expressed in the first order differential equation below

$$\dot{e}(t) = a \cdot (\ell(t) - e(t)) - b \cdot p(t), \tag{1}$$

where $e : \mathbb{R} \to \mathbb{R}$ is the power consumption level, $\dot{e}(t)$ is $\frac{\mathrm{d}}{\mathrm{dt}}e(t)$, $\ell : \mathbb{R} \to \mathbb{R}$ is an input to the system representing the expected load based on the load profile, $p : \mathbb{R} \to \mathbb{R}$ is another input to the system representing the change of the electricity price, $a \in \mathbb{R}$ is a constant representing the sensitivity of the demand to the expected load, and $b \in \mathbb{R}$ is a constant representing the sensitivity of the demand to the price. Observe that the load change $\dot{e}(t)$ is proportional to the difference between the expected load and the current demand $(\ell(t) - e(t))$ and negatively proportional to the price change $p(t)$. Large value of a means that the demand is sensitive to the expected load. TVs and heaters are examples of such devices and refrigerators, laundry machines, and vacuum cleaners are examples at the opposite side. As for the sensitivity to the price changes, b, customers are likely to avoid using the devices with large b when the electricity is expensive. Vacuuming and the laundry are such examples and refrigerators and heaters are the devices at the other side of the spectrum.

Regarding the expected load $\ell(t)$ in Eq. (1), depending on the device, the shape of the expected load throughout a day will vary. This information can be obtained from the load profile. $p(t)$ in Eq. (1) is the change of the price not the price itself. That is, a positive value of $p(t)$ is the amount of price increase from the nominal price and a negative value of $p(t)$ is the amount of discount from the nominal price. $p(t)$ is the control input we want to find to shape the power demand $e(t)$ in the presence of the expected load $\ell(t)$.

In this paper, we are considering the four representative load types $\Lambda = \{KS, Ks, kS, ks\}$, where the capital letter K and the small letter k represent a large and a small sensitivity of $e(t)$ to the expected load respectively. Similarly,

S and s represent $e(t)$'s large and small sensitivity to the price changes. More types of nodes with other sensitivities can be added without difficulty. For each $\sigma \in \Lambda$, the dynamics equations describing the demands are

$$\dot{e}_\sigma(t) = a_\sigma \cdot (\ell_\sigma(t) - e_\sigma(t)) - b_\sigma \cdot p_\sigma(t).$$

One of the objectives of the pricing policy for smart grids is to keep the maximum demand below the maximum capacity of the infrastructure. In other words, for all $t \geq 0$,

$$\sum_{\sigma \in \Lambda} e_\sigma(t) \leq \theta,$$

where θ is the maximum capacity of the infrastructure.

A harsh way to achieve this goal is to simply raise the price and force customers to reduce their electricity usage. However, there is a more graceful solution like keeping the accumulated demands of each device type close to their expected values. In addition, it is desirable to keep the accumulated price change close to zero. In other words, while keeping the price near its nominal value, we want to have the customers shift their electricity demand rather than reduce it.

To specify such requirements, we added the following 9 state variables: for each $\sigma \in \Lambda$, $E_\sigma : \mathbb{R} \to \mathbb{R}$, $L_\sigma : \mathbb{R} \to \mathbb{R}$, and $P : \mathbb{R} \to \mathbb{R}$ representing the accumulated demand, the accumulated expected load, and the accumulated price respectively. These variables satisfy the following relations. For all $\sigma \in \Lambda$,

$$\dot{E}_\sigma(t) = e_\sigma(t), \quad \dot{L}_\sigma(t) = \ell_\sigma(t), \quad \dot{P}(t) = p(t).$$

With the new state variables, the requirements can be written as follows. For all $\sigma \in \Lambda$,

$$E_\sigma(24) = \int_0^{24} e_\sigma(t)\,dt \simeq L_\sigma(24) = \int_0^{24} \ell_\sigma(t)\,dt, \quad P(24) = \int_0^{24} p(t)\,dt \simeq 0.$$

These differential equations can be compactly rewritten in the state variable form [7]. Let us first define a sub-state vector function and sub-system matrices for each $\sigma \in \Lambda$.

$$\mathbf{x}_\sigma(t) = \begin{bmatrix} e_\sigma(t) \\ E_\sigma(t) \\ L_\sigma(t) \end{bmatrix}, \quad A_\sigma = \begin{bmatrix} -a_\sigma & 0 & 0 \\ 1 & 0 & 0 \\ 0 & 0 & 0 \end{bmatrix}, \quad B_\sigma = \begin{bmatrix} a_\sigma \\ 0 \\ 1 \end{bmatrix}.$$

Using the sub-matrices A_σ, and B_σ, the system matrices $A \in \mathbb{R}^{13 \times 13}$ and $B \in \mathbb{R}^{13 \times 5}$ for the whole system can be defined as

$$A = \text{diag}([A_{KS}, \ldots, A_{ks}, 0]), \quad B = \text{diag}([B_{KS}, \ldots, B_{ks}, 1]),$$

where $\text{diag}(\mathbf{v})$ is a diagonal matrix whose diagonal is $\mathbf{v}$.

Let the vector functions $\mathbf{u} : \mathbb{R} \to \mathbb{R}^5$, $\mathbf{y} : \mathbb{R} \to \mathbb{R}^{13}$, and $\mathbf{x} : \mathbb{R} \to \mathbb{R}^{13}$ be input, output, and state trajectories of the whole system defined as

$$\mathbf{u}(t) = [\ell_{KS}(t), \ell_{Ks}(t), \ell_{kS}(t), \ell_{ks}(t), p(t)]^T,$$
$$\mathbf{x}(t) = [\mathbf{x}_{KS}(t)^T, \mathbf{x}_{Ks}(t)^T, \mathbf{x}_{kS}(t)^T, \mathbf{x}_{ks}(t)^T, P(t)]^T,$$
$$\mathbf{y}(t) = \mathbf{x}(t).$$

Then, the input, output, and state trajectories satisfy the following system dynamics equations in the state variable form.

$$\dot{\mathbf{x}}(t) = A \cdot \mathbf{x}(t) + B \cdot \mathbf{u}(t), \quad \mathbf{y}(t) = C \cdot \mathbf{x}(t),$$

where $C \in \mathbb{R}^{13 \times 13}$ is the identity matrix.

We discretized the continuous dynamics equations in the state variable form using the *First Order Hold* (FOH) method.[1] Let the discrete system dynamics equations be

$$\bar{\mathbf{x}}(t+1) = \bar{A} \cdot \bar{\mathbf{x}}(t) + \bar{B} \cdot \bar{\mathbf{u}}(t), \quad \bar{\mathbf{y}}(t) = \bar{C} \cdot \bar{\mathbf{x}}(t) + \bar{D} \cdot \bar{\mathbf{u}}(t),$$

where the vector functions $\bar{\mathbf{u}} : \mathbb{N} \to \mathbb{R}^5$, $\bar{\mathbf{y}} : \mathbb{N} \to \mathbb{R}^{13}$, and $\bar{\mathbf{x}} : \mathbb{N} \to \mathbb{R}^{13}$ are the discrete input, output, and state trajectories respectively.

Because the system matrices $\bar{A} \in \mathbb{R}^{13 \times 13}$, $\bar{B} \in \mathbb{R}^{13 \times 5}$, $\bar{C} \in \mathbb{R}^{13 \times 13}$, and $\bar{D} \in \mathbb{R}^{13 \times 5}$ are large and sparse to compactly describe in LTLC and to show in the paper, we rewrite the non-zero elements into a system of difference equations of related components. The state variable representation can be partitioned into the following set of difference equations.

$$e_\sigma^x(t+1) = a_\sigma^{ee} \cdot e_\sigma^x(t) + b_\sigma^{le} \cdot \ell_\sigma^u(t) - b_\sigma^{pe} \cdot p^u(t), \quad e_\sigma^y(t) = e_\sigma^x(t) + d_\sigma^{le} \cdot \ell_\sigma^u(t) - d_\sigma^{pe} \cdot p^u(t),$$

$$E_\sigma^x(t+1) = E_\sigma^x(t) + a_\sigma^{eE} \cdot e_\sigma^x(t) + b_\sigma^{lE} \cdot \ell_\sigma^u(t) \quad E_\sigma^y(t) = E_\sigma^x(t) + d_\sigma^{lE} \cdot \ell_\sigma^u(t) - d_\sigma^{pE} \cdot p^u(t),$$
$$- b_\sigma^{pE} \cdot p^u(t),$$

$$L_\sigma^x(t+1) = L_\sigma^x(t) + \ell_\sigma^u(t), \qquad\qquad L_\sigma^y(t) = L_\sigma^x(t) + 0.5 \cdot \ell_\sigma^u(t),$$
$$P^x(t+1) = P^x(t) + p^u(t), \qquad\qquad\quad P^y(t) = P^x(t) + 0.5 \cdot p^u(t),$$

where $\sigma \in \Lambda$ and the superscript u, y, and x on the time functions represent the input, output, and state respectively. Hence, $e_\sigma^x : \mathbb{N} \to \mathbb{R}$ and $e_\sigma^y : \mathbb{N} \to \mathbb{R}$ are the state and the output trajectories for the demand, $E_\sigma^x : \mathbb{N} \to \mathbb{R}$ and $E_\sigma^y : \mathbb{N} \to \mathbb{R}$ are the trajectories for the accumulated demand, and so on. Disregarding the superscripts and the subscripts, the constants a, b, c, and d in the equation above are from the matrices $\bar{A}$, $\bar{B}$, $\bar{C}$, and $\bar{D}$ respectively. To identify the constants easily, we annotated them with two letter superscripts such that the first letter represents the source variable and the second letter represents the target variable. Table 1 shows the coefficients for the continuous and the discretized models.

Table 1. Coefficients for the continuous time models and discrete time models.

	$\sigma = KS$	$\sigma = Ks$	$\sigma = kS$	$\sigma = ks$		$\sigma = KS$	$\sigma = Ks$	$\sigma = kS$	$\sigma = ks$
a_σ	7.0000	6.0000	2.0000	0.0500					
b_σ	5.0000	1.0000	7.0000	0.0500					
a_σ^{ee}	0.0009	0.0025	0.1353	0.9512	a_σ^{eE}	0.1427	0.1663	0.4323	0.9754
b_σ^{le}	0.1426	0.1658	0.3738	0.0476	b_σ^{lE}	0.9796	0.9724	0.8131	0.0486
b_σ^{pe}	0.1019	0.0276	1.3084	0.0476	b_σ^{pE}	0.6997	0.1621	2.8458	0.0486
d_σ^{le}	0.8573	0.8337	0.5677	0.0246	d_σ^{lE}	0.3775	0.3610	0.2162	0.0082
d_σ^{pe}	0.6123	0.1390	1.9868	0.0246	d_σ^{pE}	0.2697	0.0602	0.7566	0.0082

---

[1] We used *c2d* function of Matlab® to discretize the continuous dynamics.

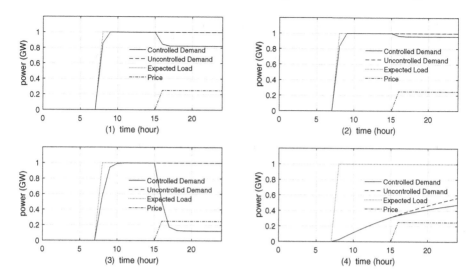

Fig. 2. Response of the system models to the expected power change and to the price change (the unit for the price graph is \$/KWh). The four graphs are $\sigma = KS$, $\sigma = Ks$, $\sigma = kS$, and $\sigma = ks$.

Figure 2 shows how the models respond to the expected load changes and to the price changes. The four graphs are for the load types $\sigma = KS$, $\sigma = Ks$, $\sigma = kS$, and $\sigma = ks$. In the figures, the system is initially in the relaxed state, meaning all state variables are zero, and the expected load is set to 1 GWh at step 7. Later at step 15 the price is increased by 0.25 \$/KWh. As expected, the second graph and the fourth graph show that the controlled demands (sold lines) do not change much from their uncontrolled counterparts (dashed line). Regarding the sensitivity to the expected loads, the first and the second graphs show that the controlled and the uncontrolled demands move quickly in response to the changes in the expected load (dotted lines). However, as the third and the fourth graphs show, the demand lines are slow in responding to the expected load changes.

3 Specification on Smart Grid Models

In the previous section, we built a discrete time LTI model for a smart grid system. In this section, we formally define a discrete time LTI model and define its computation paths which are trajectories of the input, output, and state variables. The syntax and the bounded semantics of the specification logic, called LTLC, will follow the model definitions.

Definition 1. *An LTI system M is a seven tuple $M = \langle U, Y, X, A, B, C, D \rangle$, where $U = \{u_1, \ldots, u_{nu}\}$, $Y = \{y_1, \ldots, y_{ny}\}$, and $X = \{x_1, \ldots, x_{nx}\}$ are the sets of input, output, and state variables, $A \in \mathbb{R}^{nx \times nx}$, $B \in \mathbb{R}^{nx \times nu}$, $C \in \mathbb{R}^{ny \times nx}$, and $D \in \mathbb{R}^{ny \times nu}$ are system matrices.*   □

The input, output, and state *trajectories* can be defined as vector functions $\mathbf{x} : \mathbb{N} \to \mathbb{R}^{nx}$, $\mathbf{u} : \mathbb{N} \to \mathbb{R}^{nu}$, and $\mathbf{y} : \mathbb{N} \to \mathbb{R}^{ny}$ such that $\mathbf{u}(t)_i = u_i$ at t for $i = 1, \ldots, nu$, $\mathbf{y}(t)_i = y_i$ at t for $i = 1, \ldots, ny$, and $\mathbf{x}(t)_i = x_i$ at t for $i = 1, \ldots, nx$. Then, $\mathbf{u}$, $\mathbf{y}$, and $\mathbf{x}$ satisfy the following system dynamics equations:

$$\mathbf{x}(t+1) = A \cdot \mathbf{x}(t) + B \cdot \mathbf{u}(t), \quad \mathbf{y}(t) = C \cdot \mathbf{x}(t) + D \cdot \mathbf{u}(t). \tag{2}$$

Solving these equations, $\mathbf{x}$ and $\mathbf{y}$ can be explicitly expressed in terms of $\mathbf{x}(0)$ and $\mathbf{u}$ as:

$$\mathbf{x}(t) = A^t \cdot \mathbf{x}(0) + \sum_{i=0}^{t-1} A^{t-i-1} \cdot B \cdot \mathbf{u}(i), \quad \mathbf{y}(t) = C \cdot \mathbf{x}(t) + D \cdot \mathbf{u}(t). \tag{3}$$

Definition 2. *A computation path π of an LTI system $M = \langle U, Y, X, A, B, C, D \rangle$ is a function $\pi : \mathbb{N} \to \mathbb{R}^{nu} \times \mathbb{R}^{ny} \times \mathbb{R}^{nx}$ such that $\pi(t) = (\mathbf{u}(t), \mathbf{y}(t), \mathbf{x}(t))$, where $\mathbf{u}$, $\mathbf{y}$, and $\mathbf{x}$ satisfy Eq. (2).* □

Linear Temporal Logic for Control (LTLC) is an LTL with the usual temporal and logical operators, but the atomic propositions are linear (in)equalities about input, output and state variables of a system. The computation paths of LTLC are the trajectories of an LTI system and LTLC model checking is a process of verifying whether all computation paths of a system satisfy a specification. Observe that the state space is in $\mathbb{R}^n$ and there are uncountably many computation paths.[2] The LTLC model checking algorithm converts the quantitative model checking problem into a series of feasibility checking problems and checks if there is a computation path that violates the specification.

Definition 3. *The syntax of LTLC is*

$$\phi ::= \mathbf{T} \,|\, \mathbf{F} \,|\, AP \,|\, (\,\phi\,) \,|\, \neg \phi \,|\, \phi \wedge \varphi \,|\, \phi \vee \varphi \,|\, \phi \to \varphi \,|\, \phi \leftrightarrow \varphi \,|$$
$$\mathbf{X}\phi \,|\, \Diamond\,\phi \,|\, \Box\,\phi \,|\, \phi\mathbf{U}\varphi \,|\, \phi\mathbf{R}\varphi,$$
$$AP ::= c_1 \cdot v_1 + \cdots + c_n \cdot v_n \bowtie d,$$

where $c_i, d \in \mathbb{R}$, $v_i \in U \cup Y \cup X$ for $i = 1, \ldots, n$, $\bowtie$ is one of $<, \leq, =, \geq,$ and $>$. □

Implicitly, the meaning of an LTLC formula is as follows. An *atomic proposition* $c_1 \cdot v_1 + \cdots + c_n \cdot v_n \bowtie d$ at a state $\pi(t)$ is true iff the (in)equality is true when the variables are assigned with their corresponding values in $\pi(t)$. In other words, $c_1 \cdot v_1 + \cdots + c_n \cdot v_n \bowtie d$ is true iff $c_1 \cdot \theta_{\pi(t)}(v_1) + \cdots + c_n \cdot \theta_{\pi(t)}(v_n) \bowtie d$, where $\theta_{\pi(t)}$ is the *assignment function* for a variable v such that $\theta_{\pi(t)}$ is $\mathbf{u}(t)_i$ of $\pi(t)$ if $v = u_i$, $\mathbf{y}(t)_i$ of $\pi(t)$ if $v = y_i$, or $\mathbf{x}(t)_i$ of $\pi(t)$ if $v = x_i$.

Logical operators have their usual meaning: $\neg\phi$ is true at t iff ϕ is false at t; $\phi \wedge \varphi$ is true at t iff ϕ is true at t and φ is true at t; and $\phi \vee \varphi$ is true at t iff ϕ is

---

[2] A state here is in the range of a computation path, the product of the input, output, and state variables ($\mathbb{R}^{nu} \times \mathbb{R}^{ny} \times \mathbb{R}^{nx}$), not just the state variables of the LTI system.

true at t or φ is true at t. $\phi \to \varphi$ is equivalent to $\neg\phi \vee \varphi$ and $\phi \leftrightarrow \varphi$ is equivalent to $(\phi \to \varphi) \wedge (\varphi \to \phi)$.

Temporal operators can be interpreted as follows. $\mathbf{X}\phi$ is true at t iff ϕ is true at the next time step $t+1$; $\Box\phi$ is true at t iff ϕ is always true from t; and $\Diamond\phi$ is true at t iff ϕ eventually becomes true at some time $t' \geq t$. $\phi\mathbf{U}\varphi$ is true at t iff ϕ is true until φ eventually becomes true. To be specific, there is a time $t' \geq t$ when φ is true and ϕ is true at τ for $t \leq \tau < t'$. $\phi\mathbf{R}\varphi$ is equivalent to $\varphi\mathbf{U}(\phi\wedge\varphi)$, except that ϕ is not required to hold eventually.

Formally, the semantics of an LTLC formula ϕ can be defined by the ternary satisfaction relation $\models \subset \Pi \times \mathbb{N} \times \Phi$ and the binary satisfaction relation $\models \subset \mathcal{M} \times \Phi$, where Π is the set of computation paths, Φ is the set of LTLC formulas, and $\mathcal{M}$ is the set of LTI models.

Definition 4. *The* ternary satisfaction relation $\models \subset \Pi \times \mathbb{N} \times \Phi$ *is defined in Fig. 3(a).* □

$$
\begin{array}{ll}
\pi, t \models \mathsf{T}, & \pi, t \models_b \mathsf{T}, \\
\pi, t \not\models \mathsf{F}, & \pi, t \not\models_b \mathsf{F}, \\
\pi, t \models c_1 \cdot v_1 + \cdots + c_n \cdot v_n \bowtie d & \pi, t \models_b c_1 \cdot v_1 + \cdots + c_n \cdot v_n \bowtie d \\
\quad \Leftrightarrow \sum_{i=1}^n c_i \cdot \theta_{\pi(t)}(v_i) \bowtie d, & \quad \Leftrightarrow \sum_{i=1}^n c_i \cdot \theta_{\pi(t)}(v_i) \bowtie d, \\
\pi, t \models \neg\phi \;\Leftrightarrow\; \pi, t \not\models \phi, & \pi, t \models_b \neg\phi \;\Leftrightarrow\; \pi, t \not\models_b \phi, \\
\pi, t \models \phi \wedge \varphi \Leftrightarrow \pi, t \models \phi \text{ and } \pi, t \models \varphi, & \pi, t \models_b \phi \wedge \varphi \Leftrightarrow \pi, t \models_b \phi \text{ and } \pi, t \models_b \varphi, \\
\pi, t \models \phi \vee \varphi \Leftrightarrow \pi, t \models \phi \text{ or } \pi, t \models \varphi, & \pi, t \models_b \phi \vee \varphi \Leftrightarrow \pi, t \models_b \phi \text{ or } \pi, t \models_b \varphi, \\
\pi, t \models \mathbf{X}\phi \;\Leftrightarrow\; \pi, t+1 \models \phi, & \pi, t \models_b \mathbf{X}\phi \;\Leftrightarrow\; t < b \text{ and } \pi, t+1 \models_b \phi, \\
\pi, t \models \phi \mathbf{U} \varphi \Leftrightarrow \pi, i \models \varphi \text{ for some } i \geq t \text{ and} & \pi, t \models_b \phi \mathbf{U} \varphi \Leftrightarrow \pi, i \models_b \varphi \text{ for some } t \leq i \leq b \text{ and} \\
\quad \pi, j \models \phi \text{ for all } t \leq j < i, & \quad \pi, j \models_b \phi \text{ for all } t \leq j < i, \\
\pi, t \models \phi \mathbf{R} \varphi \Leftrightarrow \pi, t \models \varphi \text{ and for } i > t, \pi, i \models \varphi & \pi, t \models_b \phi \mathbf{R} \varphi \Leftrightarrow \pi, i \models_b \phi \text{ for some } t \leq i \leq b \text{ and} \\
\quad \text{if } \pi, j \not\models \phi \text{ for all } t \leq j < i. & \quad \pi, j \models_b \varphi \text{ for all } t \leq j \leq i.
\end{array}
$$

(a)   (b)

Fig. 3. (a) Ternary satisfaction relation $\models$ and (b) bounded ternary satisfaction relation $\models_b$.

The missing operators in Definition 4 can be explained by the equivalence relations. The following well-known relations are for the logical operators. $\phi \to \varphi \equiv \neg\phi \vee \varphi$, $\phi \leftrightarrow \varphi \equiv (\phi \to \varphi) \wedge (\varphi \to \phi)$, and $(\phi) \equiv \phi$. For the temporal operators $\Box$ and $\Diamond$, these relations hold: $\Box\phi \equiv \mathsf{F}\,\mathbf{R}\,\phi$ and $\Diamond\phi \equiv \mathsf{T}\,\mathbf{U}\,\phi$.

Using the ternary satisfaction relation, we can define the binary satisfaction relation.

Definition 5. *The* binary satisfaction relation $\models \subset \mathcal{M} \times \Phi$ *is*

$$M \models \phi \Leftrightarrow \pi, 0 \models \phi \text{ for all computation path } \pi \text{ of } M.$$

□

The satisfaction relation $\models$ is about computation paths with infinite length. However, by allowing an arbitrary input, it is difficult to check whether all computation paths satisfy a given specification to their infinite length. For practical reasons, the LTLC model checking algorithm supports the bounded semantics [2]. In the bounded semantics, given a bound b, a computation path π satisfies a specification ϕ, if (1) a finite prefix of π with length up to b is enough to decide that π satisfies ϕ or (2) π ends with a loop whose second cycle starts before or at the bound b and π satisfies ϕ.

Formally, given a bound b, the bounded semantics of an LTLC formula ϕ can be defined by the ternary satisfaction relation $\models_b \subset \Pi \times \mathbb{N} \times \Phi$ and the binary satisfaction relation $\models_b \subset \mathcal{M} \times \Phi$.

Definition 6. *The* ternary bounded satisfaction relation $\models_b \subset \Pi \times \mathbb{N} \times \Phi$ *is defined in Fig. 3(b).* □

To define the bounded binary satisfaction relation $\models_b$, let us define a helper function $\tau_{s,p} : \mathbb{N} \to [0, s + p - 1]$ that maps a time step to an index for a loop-ending computation path: $\tau_{s,p}(t) = t$ if $t < s + p$; otherwise $\tau_{s,p}(t) = s + (t - s) \bmod p$, where $s, p \in \mathbb{N}$ such that $s \geq 0$ and $p \geq 1$. $\tau_{s,p}$ generates a loop-ending index of period p starting from s. As an illustration, $\tau_{1,2}(t) = 0, 1, 2, 1, 2, \ldots$ when $t = 0, 1, 2, 3, 4, \ldots$. A computation path π is a *loop-ending computation path* if $\pi(t) = \pi(\tau_{s,p}(t))$ for $t \geq 0$, where p is the period of the loop and s is its starting index.

With the loop index function τ, the bounded binary satisfaction relation can be defined as follows.

Definition 7. *Given a bound b, the* binary bounded satisfaction relation $\models_b \subset \mathcal{M} \times \Phi$ *is:*

$$M \models_b \phi \Leftrightarrow \begin{cases} \pi, 0 \models \phi & \text{if } \pi(t) = \pi(\tau_{s,p}(t)) \text{ for } t \geq 0 \text{ and } s + p \leq b \\ \pi, 0 \models_b \phi & \text{otherwise.} \end{cases}$$

□

Given a model M, an LTLC specification ϕ, and a bound b, an LTLC model checking is a process of deciding if $M \models_b \phi$. If the model checker found a counterexample, it reports *false* with the counterexample witnessed. The counterexample contains an initial state and a sequence of input that can lead to the violation of the specification.

4   Day-Ahead Pricing by Model Checking

In smart grid systems, electricity providers can shape the customers' energy demand by posting different electricity prices at different times of a day. In this section we will explain how to describe the discrete time model developed in Sect. 2 and a goal about the electricity usage in LTLC-Checker [1]. Day-ahead prices can be found by model checking the description.

Suppose that the goal we want to achieve is (1) the total power demand of all devices never exceed 1.65 GW, the capacity of the infrastructure, (2) the price changes are always within $\pm 0.1\,\$/\text{KWh}$ range of the nominal value, (3) the accumulated price changes for a day never exceed $\pm 0.1\,\$/\text{KWh}$ range, and (4) the accumulated energy demand of each load type is within $\pm 0.5\,\text{GWh}$ range of the accumulated expected load of the device for the day. We assume that initially, the power demand for each device is close to its expected load and choose the initial state estimates as $e^x_{KS}(0) = 0.01$, $e^x_{Ks}(0) = 0.12$, $e^x_{kS}(0) = 0$, $e^x_{ks}(0) = 0.44$, and all the other states are 0.

We will divide the goal into smaller, manageable pieces and describe them in LTLC. The pieces will be combined together to formulate the original goal later. For the first condition, we introduce a new output variable e^y representing the total power consumption of all devices:

$$e^y(t) = \sum_{\sigma \in \Lambda} e^y_\sigma(t) = \sum_{\sigma \in \Lambda} e^x_\sigma(t) + d^{\ell e}_\sigma \cdot \ell^u_\sigma(t) - d^{pe}_\sigma \cdot p^u(t).$$

The first condition can be expressed in LTLC as

$$\phi_e = \Box\,\varphi_e, \text{ where } \varphi_e = e^y \le 1.65.$$

Observe that the time step t is dropped in the LTLC formula because t will be provided during the model checking process like the ternary satisfaction relations $\models$ and $\models_b$. The second condition can be expressed similarly as

$$\phi_p = \Box\,\varphi_p, \text{ where } \varphi_p = -0.1 \le p^u \wedge p^u \le 0.1.$$

The third and the fourth condition is about the accumulated price and the accumulated energy at the end of the day. Let us define the conditions without the temporal operators for now.

$$\varphi_P = -0.1 \le P^x \wedge P^x \le 0.1, \quad \varphi_E = \bigwedge_{\sigma \in \Lambda} \left(L^x_\sigma - 0.5 \le E^y_\sigma \wedge E^y_\sigma \le L^x_\sigma + 0.5\right).$$

The initial conditions can be written as

$$\phi_i = e^x_{KS} = 0.01 \wedge e^x_{Ks} = 0.12 \wedge e^x_{kS} = 0 \wedge e^x_{ks} = 0.44 \wedge P^x = 0 \wedge c^x = 0 \wedge$$
$$\bigwedge_{\sigma \in \Lambda} \left(E^x_\sigma = 0 \wedge L^x_\sigma = 0\right),$$

where c^x is a clock state that will be explained shortly after.

In addition to the conditions above, there are conditions enforced by the physical constraints of the system: the power consumptions of each device type should never be negative. The mathematical model of the system can produce a negative power demand by increasing the price large. To avoid such anomalies, we added the following LTLC formula

$$\phi_+ = \Box\,\varphi_+, \text{ where } \varphi_+ = \bigwedge_{\sigma \in \Lambda} e^y_\sigma \ge 0.$$

The always formulas in the first two conditions and in the physical constraints prevent the use of the more efficient bounded ternary relation $\models_b$. Because we are only interested in 24 h, we can rewrite the always formulas using the U operator and enjoy the efficiency of checking $\models_b$ as oppose to $\models$. For this purpose we introduced a clock state

$$c^x(t+1) = c^x(t) + dc^u(t),$$

where $dc^u(t) = 1$ for $0 \leq t \leq 24$ and $c^x(0) = 0$. With the clock, the three conditions with the always operator can be collectively rewritten as

$$\phi_s = \varphi_s \, \mathsf{U} \, (c^x \geq 24), \text{ where } \varphi_s = \varphi_e \wedge \varphi_p \wedge \varphi_+ \wedge (dc = 1).$$

One problem with this specification is that the *Right Hand Side* (RHS) of U formula is not enforced when its *Left Hand Side* (LHS) is satisfied. That is, at the moment when c^x becomes 24, φ_s is not enforced to be true. To fix this issue, we separately enforced φ_s at the 24th step.

We need to specify the expected electricity usage of each device type for the day. Because these numbers are given from the customer profile and there are no simple formulas to describe them, we manually specify the expected loads for each time step. The formula is

$$\phi_\ell^t = \mathsf{X}^t \left(\bigwedge_{\sigma \in \Lambda} \ell_\sigma^u = \hat{\ell}_\sigma(t) \right),$$

where X^t is the string of X of length t, $\hat{\ell}_\sigma(t)$ is the expected load from the profile at time t for each $\sigma \in \Lambda$.

Finally, combining the subformulas together, we can write the goal as

$$\phi_{goal} = \phi_i \wedge \phi_s \wedge \bigwedge_{t=0}^{24} \phi_\ell^t \wedge \mathsf{X}^{24}(\varphi_P \wedge \varphi_E \wedge \varphi_s).$$

Figure 4 shows an LTLC description of the discrete model built in Sect. 2. In the description a letter $\mathtt{t}$, $\mathtt{h}$, $\mathtt{v}$ or $\mathtt{f}$ representing TV, heater, vacuuming, and refrigerator respectively, is suffixed for the device types KS, Ks, kS, and ks. For example, $\mathtt{et}$, $\mathtt{eh}$, $\mathtt{ev}$, and $\mathtt{ef}$ are for the variables e^y_{KS}, e^y_{Ks}, e^y_{kS}, and e^y_{ks}. In addition, we put a letter $\mathtt{x}$ at the second letter position of the variable names to indicate that the variables are a state variable. For instance, $\mathtt{ext}$, $\mathtt{exh}$, $\mathtt{exv}$, and $\mathtt{exf}$ are for the state variables e^x_{KS}, e^x_{Ks}, e^x_{kS}, and e^x_{ks}.

On the top level, LTLC-Checker description has a $\mathtt{model}$ section for a model description and a $\mathtt{specification}$ section for the model checking problem. The $\mathtt{model}$ section begins with optional constant definitions followed by system variable definitions. Each system variable is annotated with the type of the variable, one of *input*, *output*, and *state*.

With the system variables, one can define modes that comprise a set of system dynamics equations in the $\mathtt{mode}$ section. The LHS of a dynamics equation can

```
model:                                          specification:
var    # power and accumulated energy demand    condition
  ext:state, et:output, Ext:state, Et:output,   # Initial conditions
  exh:state, eh:output, Exh:state, Eh:output,   Init=(Ext=0 /\ Exh=0 /\ Exv=0 /\ Exf=0 /\
  exv:state, ev:output, Exv:state, Ev:output,         Lxt=0 /\ Lxh=0 /\ Lxv=0 /\ Lxf=0 /\
  exf:state, ef:output, Exf:state, Ef:output,         ext=0.01 /\ exh=0.12 /\ exv=0 /\ exf=0.44 /\
                                                      Px=0 /\ c=0),
  # total demand of all devices:et + eh + ev + ef
  e: output,                                    # Final conditions
                                                Final=(
  # expected load, acc. expected load             # accumulated energy demand of each device should
  lt: input, lh: input, lv: input, lf: input,     # be close to its expected accumulated energy
  Lxt: state, Lxh: state, Lxv: state, Lxf: state, Lt - 0.5 <= Et /\ Et <= Lt + 0.5 /\
  Lt: output, Lh: output, Lv: output, Lf: otuput, Lh - 0.5 <= Eh /\ Eh <= Lh + 0.5 /\
                                                  Lv - 0.5 <= Ev /\ Ev <= Lv + 0.5 /\
  # price change, acc. price change               Lf - 0.5 <= Ef /\ Ef <= Lf + 0.5 /\
  p: input, Px: state, P: output,                 # Bound for acc. price change
                                                  -0.1 <= P /\ P <= 0.1),
  # clock and tick
  c: state, dc: input;                          # Always enforced conditions
                                                Safety=(
mode                                              # total power demand should be less than 1.65 GW
  M = {                                           e <= 1.65 /\
  # individual device's power demand             # price change should be within +- 0.1 $/KWh range
  ext = 0.0009*ext + 0.1426*lt - 0.1019*p,       -0.1 <= p /\ p <= 0.1 /\
  exh = 0.0025*exh + 0.1658*lh - 0.0276*p,        # positive power demand and dc
  exv = 0.1353*exv + 0.3738*lv - 1.3084*p,        et >= 0 /\ eh >= 0 /\ ev >= 0 /\ ef >= 0 /\
  exf = 0.9512*exf + 0.0476*lf - 0.0476*p,        dc = 1.0),
  et =         ext + 0.8573*lt - 0.6123*p,
  eh =         exh + 0.8337*lh - 0.1390*p,      # Expected load for the day
  ev =         exv + 0.5677*lv - 1.9868*p,      Load=Init /\
  ef =         exf + 0.0246*lf - 0.0246*p,        lt = 0.10 /\ lh = 0.70 /\ lv = 0.00 /\ lf = 0.40 /\
                                                  X(lt = 0.00 /\ lh = 0.70 /\ lv = 0.00 /\ lf = 0.40 /\
  # total power demand of all devices             X(lt = 0.00 /\ lh = 0.70 /\ lv = 0.00 /\ lf = 0.40 /\
  e = ext + exh + exv + exf                       X(lt = 0.00 /\ lh = 0.70 /\ lv = 0.00 /\ lf = 0.40 /\
    + 0.8573*lt + 0.8337*lh + 0.5677*lv           X(lt = 0.00 /\ lh = 0.60 /\ lv = 0.00 /\ lf = 0.40 /\
    + 0.0246*lf - 2.7627*p,                       X(lt = 0.25 /\ lh = 0.50 /\ lv = 0.00 /\ lf = 0.45 /\
                                                  X(lt = 0.30 /\ lh = 0.10 /\ lv = 0.00 /\ lf = 0.45 /\
  # accumulated power demand                      X(lt = 0.30 /\ lh = 0.00 /\ lv = 0.00 /\ lf = 0.45 /\
  Ext = Ext + 0.1427*ext + 0.9796*lt - 0.6997*p, X(lt = 0.20 /\ lh = 0.00 /\ lv = 0.10 /\ lf = 0.45 /\
  Exh = Exh + 0.1663*exh + 0.9724*lh - 0.1621*p, X(lt = 0.00 /\ lh = 0.00 /\ lv = 0.30 /\ lf = 0.50 /\
  Exv = Exv + 0.4323*exv + 0.8131*lv - 2.8458*p, X(lt = 0.00 /\ lh = 0.00 /\ lv = 0.30 /\ lf = 0.50 /\
  Exf = Exf + 0.9754*exf + 0.0486*lf - 0.0486*p, X(lt = 0.00 /\ lh = 0.00 /\ lv = 0.10 /\ lf = 0.50 /\
  Et = Ext + 0.3775*lt  - 0.2697*p,              X(lt = 0.00 /\ lh = 0.00 /\ lv = 0.00 /\ lf = 0.50 /\
  Eh = Exh + 0.3610*lh  - 0.0602*p,              X(lt = 0.00 /\ lh = 0.00 /\ lv = 0.20 /\ lf = 0.50 /\
  Ev = Exv + 0.2162*lv  - 0.7566*p,              X(lt = 0.00 /\ lh = 0.00 /\ lv = 0.30 /\ lf = 0.50 /\
  Ef = Exf + 0.0082*lf  - 0.0082*p,              X(lt = 0.00 /\ lh = 0.00 /\ lv = 0.40 /\ lf = 0.50 /\
                                                  X(lt = 0.20 /\ lh = 0.00 /\ lv = 0.30 /\ lf = 0.50 /\
  # accumulated expected energy                   X(lt = 0.40 /\ lh = 0.10 /\ lv = 0.10 /\ lf = 0.45 /\
  Lxt = Lxt + lt,    Lt = Lxt + 0.5*lt,          X(lt = 0.50 /\ lh = 0.20 /\ lv = 0.10 /\ lf = 0.45 /\
  Lxh = Lxh + lh,    Lh = Lxh + 0.5*lh,          X(lt = 0.50 /\ lh = 0.50 /\ lv = 0.30 /\ lf = 0.45 /\
  Lxv = Lxv + lv,    Lv = Lxv + 0.5*lv,          X(lt = 0.40 /\ lh = 0.70 /\ lv = 0.30 /\ lf = 0.45 /\
  Lxf = Lxf + lf,    Lf = Lxf + 0.5*lf,          X(lt = 0.40 /\ lh = 0.70 /\ lv = 0.10 /\ lf = 0.45 /\
                                                  X(lt = 0.30 /\ lh = 0.70 /\ lv = 0.00 /\ lf = 0.45 /\
  # accumulated price change                      X(lt = 0.10 /\ lh = 0.70 /\ lv = 0.00 /\ lf = 0.40 /\
  Px = Px + p,      P = Px + 0.5*p,               Final /\ Safety
                                                )))))) )))))) )))))) )))))) ))))),
  # clock
  c = c + dc                                    # The final goal: initial, final, expected load for
  };                                            #              the day, and safety conditions
                                                Goal = (
edge                                                Load /\
  # loop to the single mode                         (Safety U (c >= 24)) );
  E = M -> M;
                                                # Check if the Goal cannot be achieved
system                                          check
  # the hybrid system model with a single mode    Sys |= ~Goal in 24;
  Sys = ( {M}, {E} );
```

Fig. 4. LTLC description of the discrete dynamics of each device built in Sect. 2 (left) and specification of the goal and expected electricity consumption level of each device (right).

be a state variable or an output variable. If a state variable is on the LHS of a dynamics equation, its implicit time index is $t + 1$. Other types of variables and state variables appearing on the RHS of an equation have time index t. For example, the equation `Lxt = Lxt + lt` in Fig. 4 is for the difference equation $L_{KS}^x(t + 1) = L_{KS}^x(t) + \ell_{KS}^u(t)$, and `et = ext + 0.8573*lt - 0.6123*p` is for the difference equation $e_{KS}^y(t) = e_{KS}^x(t) + 0.8573\ell_{KS}^u(t) - 0.6123p^u(t)$.

Definitions of edges can follow the mode definitions in the `edge` section. For a general hybrid system model, the edges are decorated with a predicate formula describing the mode switch condition. However, in this example with a single mode, the only edge is the unconditional self-loop `M -> M`.

With a set of modes and a set of edges connecting the modes, a hybrid system can be defined as a labeled directed graph. In the `system` block of Fig. 4, `Sys = ({M}, {E})` defines a hybrid system as a pair of the set of modes and the set of edges.

In Fig. 4, the `specification` section begins with an optional `condition` block, where subformulas can be defined to build a more complex formula. The logical operators $\land$, $\lor$, $\neg$, $\rightarrow$, and $\leftrightarrow$ are `/\`, `\/`, `~`, `->`, and `<->` respectively in the LTLC-Checker description. The temporal operators `X`, `U`, `R`, $\Diamond$, and $\Box$ are `X`, `U`, `R`, `<>`, and `[]` respectively in the description of the checker. The first condition `Init` defines the initial condition ϕ_i, `Final` defines the final condition $\varphi_P \land \phi_E$, `Safety` defines the safety condition φ_s that needs to be satisfied all the time, and `Load` defines the expected loads for the 24 h. Observe that in `Load`, the `X` formulas are nested and the expected loads for the day can be sequentially defined in columns. `Load` also piggybacked φ_P, φ_E, and φ_s at the 24th step. Using the subformulas, `Goal` defines the goal ϕ_{goal}.

The `check` block is the last element of the specification. In this block, the bounded model checking problem is finally defined. In the last line of Fig. 4, `Sys |= ~Goal in 24` describes the model checking problem $M \models_{24} \neg\phi_{goal}$.

The model checking result was false and by applying the prices in the counterexample, we could get the power consumption levels of Figs. 5 and 6. The third graph in Fig. 6 shows the price changes from the counterexample. Figure 5 shows how the electricity demand was shifted (solid line) from the uncontrolled consumption level (dashed line) when the pricing scheme was applied. The dotted lines in the graphs are the expected power consumption. The four graphs show the electricity demand of load types KS, Ks, kS, and ks. The third graph has the most dramatic load shift because this type is sensitive to the price changes but insensitive to the expected loads.

The differences between the accumulated demands and the accumulated expected loads $(E_\sigma^y(24) - L_\sigma^y(24))$ were 0.045, -0.0825, 0.2157, and -0.3065 (GWh) respectively from the top. Figure 6 also shows how these differences are changing over time. The largest decrease occurred at the ks type because with $b_{ks} = 0.05$, this device type is most insensitive to the expected load changes. kS type has the largest increase and it is because this device type is sensitive to the price changes and is insensitive to the expected load changes. As a result, it has the largest load shift. Observe that all the changes in demand were within the

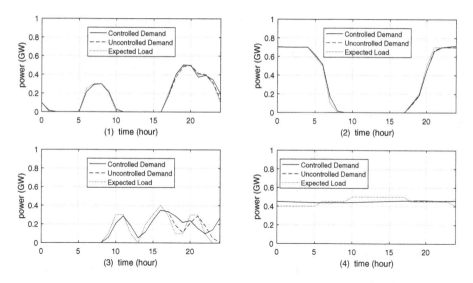

Fig. 5. Power consumption levels of each load types: KS, Ks, kS, and ks. Solid lines are controlled power demand, dashed lines are uncontrolled power consumption, and dotted lines are expected power consumption.

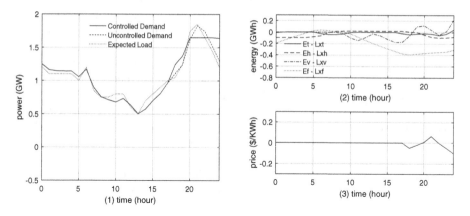

Fig. 6. Total power consumption level (1), the difference between the accumulated demand and the accumulated expected load (2), and the price change (3).

range of φ_E. Moreover, the demand never went below zero. Hence, the trajectories satisfy $\mathbf{X}^{24}\varphi_E$ and ϕ_+.

The first graph of Fig. 6 shows how the total power consumption levels of all device change. Observe that the expected load (dotted line) and uncontrolled demand (dashed line) exceeded the capacity limit of 1.65 GW, but the controlled demand stayed below the limit and satisfied ϕ_e. The third graph shows the price changes and they were always within the ± 0.1 \$/KWh range and satisfied ϕ_p.

The accumulated price change was -0.1 \$/KWh and it was also within the range of φ_P, meaning $X^{24}\varphi_P$ was satisfied.

Combining the results together, the computation path π obtained by applying the price changes in the counterexample satisfied the goal. In other words $\pi, 0 \models_{24} \phi_{goal}$.

5   Conclusion

We have developed a method to compute a pricing scheme for smart grid systems using a quantitative model checking technique. To demonstrate the usefulness of the method, a smart grid system is modeled as a first order differential equation and a day-ahead pricing scheme is computed successfully using a quantitative model checking technique. Apart from the flexibility and the correctness, an attractive merit of using a model checking technique is that it can provide the impossibility of achieving the goal.

The example linear system model is designed such that the current electricity demand level moves towards the expected power consumption level of that time. The tendency is proportional to the difference between the expected power consumption level and the current demand. Furthermore, the price change is negatively affecting the power consumption level. We assumed some reasonable sensitivity coefficients for different device types. We are developing techniques that can estimate the system matrices directly from the price changes and the measured electricity load changes [11, 12].

With the progressive use of smart devices that can schedule their own operation time, the pricing scheme in controlling the customers' energy consumption level can only be more effective. We cannot overemphasize the importance of evaluating the trajectories of system states systematically and finding a way to control them to meet the requirements. Moreover, the current trend of installing energy generating devices at the customers' home can add another dimension to consider in managing the smart grid systems. We believe, the proposed pricing technique based on a quantitative model checking method can provide a flexible and systematic way of controlling the complex smart grid systems.

Acknowledgement. The authors thank the anonymous referees for their helpful comments. This work was supported by MSIP, Korea under the ITCCP program (IITP-2015-R0346-15-1007) and by KEIT under the GATC program (10077300).

References

1. LTLC-Checker. https://sites.google.com/site/youngminkwon
2. Biere, A., Cimatti, A., Clarke, E., Zhu, Y.: Symbolic model checking without BDDs. In: Cleaveland, W.R. (ed.) TACAS 1999. LNCS, vol. 1579, pp. 193–207. Springer, Heidelberg (1999). doi:10.1007/3-540-49059-0_14
3. Borenstein, S., Jaske, M., Rosenfeld, A.: Dynamic pricing, advanced metering, and demand response in electricity markets. In: Center for the Study of Energy Markets (2002)

4. Büchi, J.: On a decision method in restricted second order arithmetic. In: Proceedings of the International Conference on Logic, Methodology and Philosophy of Science, pp. 1–11. Stanford University Press (1960)

5. Clarke, E.M., Emerson, E.A.: Design and synthesis of synchronization skeletons using branching time temporal logic. In: Kozen, D. (ed.) Logic of Programs 1981. LNCS, vol. 131, pp. 52–71. Springer, Heidelberg (1982). doi:10.1007/BFb0025774

6. Clarke, E., Grumberg, O., Peled, D.: Model Checking. MIT Press, Cambridge (2000)

7. Franklin, G.F., Powell, J.D., Emami-Naeini, A.: Feedback Control of Dynamic Systems, 3rd edn. Addison Wesley, Reading (1994)

8. Hirst, E.: The financial and physical insurance benefits of price-responsive demand. Electr. J. **15**, 66–73 (2002)

9. Holzmann, G.J.: The model checker spin. IEEE Trans. Softw. Eng. **23**, 279–295 (1997)

10. Joe-Wong, C., Sen, S., Ha, S., Chiang, M.: Optimized day-ahead pricing for smart grids with device-specific scheduling flexibility. IEEE J. Sel. Areas Commun. **30**, 1075–1085 (2012). IEEE

11. Juang, J.-N., Pappa, R.S.: An eigensystem realization algorithm for modal parameter identification and model reduction. J. Guid. Control Dyn. **8**, 620–627 (1985). AIAA

12. Juang, J.-N., Phan, M., Horta, L.G., Longman, R.W.: Identification of observer/Kalman filter Markov parameters: theory and experiments. J. Guid. Control Dyn. **16**, 320–329 (1993). AIAA

13. Kwon, Y.M., Agha, G.: LTLC: linear temporal logic for control. In: Egerstedt, M., Mishra, B. (eds.) HSCC 2008. LNCS, vol. 4981, pp. 316–329. Springer, Heidelberg (2008). doi:10.1007/978-3-540-78929-1_23

14. Kwon, Y., Agha, G.: Verifying the evolution of probability distributions governed by a DTMC. IEEE Trans. Softw. Eng. **37**, 126–141 (2011). IEEE Computer Society

15. Kwon, Y., Kim, E.: Bounded model checking of hybrid systems for control. IEEE Trans. Autom. Control **60**, 2961–2976 (2015). IEEE

16. Lichtenstein, O., Pnueli, A.: Checking that finite state concurrent programs satisfy their linear specification. In: POPL, pp. 97–107 (1985)

17. Mohsenian-Rad, A.-H., Leon-Garcia, A.: Optimal residential load control with price prediction in real-time electricity pricing environments. IEEE Trans. Smart Grid **1**, 120–133 (2010). IEEE

18. Pipattanasomporn, M., Kuzlu, M., Rahman, S., Teklu, Y.: Load profiles of selected major household appliances and their demand response opportunities. IEEE Trans. Smart Grid **5**, 742–750 (2014). IEEE

19. Start, H., Woods, J.W.: Probability and Random Processes with Applications to Signal Processing, 3rd edn. Prentice-Hall, Upper Saddle River (2002)

Aggregated Markov Models of a Heterogeneous Population of Photovoltaic Panels

Andrea Peruffo[1], Emeline Guiu[2],
Patrick Panciatici[2], and Alessandro Abate[1(✉)]

[1] Department of Computer Science, University of Oxford, Oxford, UK
alessandro.abate@cs.ox.ac.uk
[2] Réseau de Transport d'Electricité, Paris, France

Abstract. We present a new framework for aggregated quantitative modelling of a heterogeneous population of photovoltaic panels. We are interested in the behaviour of photovoltaic panels as electric power sources, and in an aggregated model that can capture how such a population behaves when connected to the power grid. After an initial analysis of the characteristics and behaviour of a single device, we propose two Markov chain models for the aggregation of a heterogeneous population of such devices. We study the dynamical behaviours of the aggregated models, embedded within the dynamics of the grid frequency. A simulation study shows the effectiveness of the aggregated models when compared to the physical system, and leads to conclude that population heterogeneity is a desirable feature for the overall system dynamics.

1 Introduction

In recent years, both academia and industry have increased their interest and attention on renewable energy sources. The growing trend toward environmental preservation, witnessed by the famous Kyoto protocol, the Paris Agreement and the so called 2-degrees challenge, is leading scientists to focus on new technologies and their applications in power generation. Well known technologies are e.g. wind power, solar energy and geothermal energy. Devices using these energy sources are typically distributed over a large area rather than being concentrated in a small production area, which leads to consider the issue of distributed generation. In the field of renewables, many studies are now focused on photovoltaics (PV), which is nowadays the third most important renewable energy source, after hydro and wind power, in terms on total installed capacity [1]. PV panels produce electrical current from solar irradiation by virtue of the photovoltaic effect of semiconductor materials. In this work we will interchangeably denote photovoltaic panels as PV or solar panels.

Although models for a single PV cell or panel are well known in literature [2], up to the authors' knowledge there has not been any model encompassing the connection between panels and grid. Large PV farms have been modelled and studied as a whole, but a model dealing with distributed generation of power in a large city, or a country, from household applications is still missing. This can

© Springer International Publishing AG 2017
N. Bertrand and L. Bortolussi (Eds.): QEST 2017, LNCS 10503, pp. 72–87, 2017.
DOI: 10.1007/978-3-319-66335-7_5

be due to many reasons: first of all because photovoltaic application for a single home has a limited power to inject into the grid and therefore can be neglected; secondly because often the power produced is consumed at the source (the house itself), hence the net contribution to the wider grid is nearly null.

Nevertheless, the growing population of PV panels justifies the study, modelling and control of this energy source. The power network, at a regional or state level, must be prepared to deal with the volatility and unpredictability intrinsically related to the production of renewable energy. As an example, PV generated power – distributed over 1.5 million PV setups – has provided for approximately 7.5% of Germany's electricity demand, with peaks of 50% during weekends and holidays [3]. Often the power production of such devices do not follow usual demand patterns: the power production of a panel, in a clear day, follows the irradiance of the sun, with a maximum at around midday, when the power is usually injected into the grid, due to a lack of consumption. This unbalanced flow, if not handled correctly, could lead to network issues, such as blackouts [4]. Great care has been taken to proactively cope with the eclipse occurred in March 2015: it was anticipated that a reduction of around 30 GW could have been caused by the sun occlusion [5], which is about 10 times the size of a blackout accident that can be resiliently handled by the grid.

In this paper, we present a new framework for modelling and abstraction of a large population of photovoltaic panels. We firstly analyse the behaviour of a single physical device when connected to the grid. The PV panel is equipped with a sensor, to sample the network frequency, and with an internal counter in order to ensure that the network frequency remains inside a certain range of admissible values for a defined amount of time, before an action is taken. Two quantities seem to be key to model the PV behaviour: the working interval of network frequencies, and the internal time delay required for a safe connection to the network. Each device in principle can have a different admissible frequency range and a different time delay. In order to model this heterogeneity we present two discrete-time Markov chains, one closer to the physical description with $(n+2)$ states, and the other more abstract with 3 states. These models are then connected to the dynamics of the network frequency, giving an extended model for the whole system, for which stability properties are studied.

The remainder of the paper is organised as follows. In Sect. 2 we describe the behaviour of a single photovoltaic panel in frequency. The Markov chains models and the frequency evolution are explained in Sect. 3. The performance comparisons between the two models and a realistic model taken as benchmark in Sect. 4. Conclusions are drawn in Sect. 5.

2   Description of the Physical System

In order to provide an aggregated model for a population of PV panels, we start with the description of the behaviour of a single panel, as a function of the local dynamics of the grid.

A panel has two working states, ON and OFF. It switches amongst these two states according to two conditions: the local network frequency $f(\cdot)$ (whose nominal value is taken as $f_0 = 50$ Hz) and an internal time delay τ_r, usually given in seconds (cf. Table 1). The panel is connected to the grid and senses it by sampling its frequency discretely in time. It can be in the ON mode if the network frequency lies within a given local frequency interval $\mathcal{I}_f$, otherwise it must disconnect and transition onto the OFF mode. It is sensible to assume that the ON-to-OFF switch happens within a negligible time interval [6,7], whereas the OFF-to-ON switch cannot happen before a time delay τ_r, during which the frequency f must dwell within $\mathcal{I}_f$: this is in order to ensure that the network frequency is stable enough to render the panel connection to the grid safe, and to avoid chattering behaviours that can lead to overall network instability. During this interval of time the panel keeps sampling the frequency and if it measures it to be outside the working interval, the internal counter is reset. In order to describe this functioning, one can imagine a PV panel as a device equipped with an internal counter $\tau(k)$ at time k and a time threshold τ_r. This counter is set off when the device is in the OFF state and the frequency enters the working interval $\mathcal{I}_f$. If the counter value reaches τ_r as $f \in \mathcal{I}_f$, then the device can turn to the ON state. If instead the frequency goes outside $\mathcal{I}_f$ while $\tau(k) < \tau_r$, then the device resets its counter. As the PV panel senses the network frequency via a digital sensor with a defined sampling rate, τ_r is a value given in number of samples. In practice, the sampling time of a PV panel is in the order of 200 ms, and τ_r is around 20 s. The internal clock τ can then be thought of as a counter, and as such it will be modelled in this work.

Table 1. Behaviour of single photovoltaic panel within the power network at time k. Key quantities are: panel state $q \in \{ON, OFF\}$; network frequency f; operating frequency band $\mathcal{I}$; clock/counter τ and re-connection delay τ_r.

Current state $q(k)$	Frequency	Delay	Next state $q(k+1)$
OFF	$f(k) \in \mathcal{I}_f$	$\tau(k) \geq \tau_r$	ON
ON	$f(k) \in \mathcal{I}_f$	n.a.	ON
ON	$f(k) \notin \mathcal{I}_f$	n.a.	OFF
OFF	$f(k) \in \mathcal{I}_f$	$\tau(k) < \tau_r$	OFF

The power injected into the grid by the population of panels has non negligible dynamical effects to the network frequency, which directly influences the behaviour of single panels in a feedback loop.

Our aim is to develop a model for a large population of photovoltaic panels. This cannot be modelled considering N identical PV panels, simply because in reality it is not the case: different regulations, manufacturers, makes and age, all render the population highly heterogeneous [6–8]. As such, this work focuses on parameters heterogeneity as a key aspect: we deal with network frequency thresholds and time delays, to model the more realistic situation where each panel has a different working interval as well as a different counter. Another interesting part of the real system is its power output: the production of electrical power is subject to many external factors, e.g. weather conditions, light occlusions. However, here we will focus on a power production that is constant in time. Time varying features can be added at a second stage, for example modelling a power production as a stochastic process [9].

3   Markov Chain Model of a Population of PV Panels

Since the behaviour of each panel depends on a discrete sampling of the frequency, we refer to a discrete-time framework and model the aggregation of photovoltaic panels with a discrete-time Markov chain. Since the focus is on the frequency thresholds and the delays of the panels, for simplicity we consider the power production to be at its maximum nominal value whenever a panel is ON.

As an early simplifying assumption (to be shortly repealed in the next subsection), assume exact population homogeneity, meaning that every device shows the same behaviour in time and is characterised by the same parameters. This allows us to define a quantity P expressing the weighted power production as

$$P = \frac{1}{N} \sum_{i=1}^{N} P_i,$$

where N is the total number of panels in the population, and P_i is the power output of the single i-th panel. We can consider the normalised power production $R(k)$, at time k, as

$$R(k) = \frac{1}{NP} \sum_{i=1}^{N} q_i(k) P_i,$$

where $q_i(k) \in \{0, 1\}$ denotes whether the i-th device at time k is in the OFF or ON state, respectively. Since we have assumed population homogeneity, all the devices behave in accord, and $R(k) \in \{0, 1\}$ at each time step k: we can then consider $R(k)$ as a Bernoulli random variable, and introduce

$$x(k) = E[R(k)] = \mathbb{P}[R(k) = 1],$$

a variable defined as the expected value of $R(k)$ at time k, which denotes the probability of being in the ON state at that time. Furthermore, by the law of total probability,

$$x(k+1) = \mathbb{P}[R(k+1) = 1] = \mathbb{P}[R(k+1) = 1|R(k) = 1] \cdot \mathbb{P}[R(k) = 1] +$$
$$+ \mathbb{P}[R(k+1) = 1|R(k) = 0] \cdot \mathbb{P}[R(k) = 0].$$

Let us define $a(k) = \mathbb{P}[R(k+1) = 0|R(k) = 1]$, so that $\mathbb{P}[R(k+1) = 1|R(k) = 1] = (1 - a(k))$, and let us introduce $b(k) = \mathbb{P}[R(k+1) = 1|R(k) = 0]$: the previous equation can be rewritten as

$$x(k+1) = \mathbb{P}[R(k+1) = 1] = (1 - a(k))x(k) + b(k)(1 - x(k)). \tag{1}$$

This relation describes how the probability of being ON gets updated at time k. In the framework that we have adopted, the transition probability ON-to-OFF $(a(k))$ and OFF-to-ON $(b(k))$ are governed by the value of the network frequency $f(k)$, namely whether or not $f(k) \in \mathcal{I}_f$. Since we assumed population homogeneity, these values binary: for instance, when $f(k) \in \mathcal{I}_f$ then $a(k) = 0$ whilst $b(k) = 1$. Since the power output is the sum of the PV devices turned ON, it can be expressed as $P_{out}(k) = NPx(k)$.

A Markov Model Without Delays

Let us now introduce heterogeneity over the frequency behaviours of different panels, as expected in reality. Let us suppose that each panel has different frequency thresholds (which we take to be constant in time): each panel reacts to the network frequency distinctively, namely it disconnects/reconnects at a different frequencies than other panels. We assume that these thresholds are distributed continuously according to a known probability distribution - such continuous statistics can be interpolated from discrete population data. Similarly to the homogeneous case, we introduce $a(k)$ and $b(k)$ as the probability of turning to the ON state or to the OFF state at the $k+1$-th time step, starting from the opposite condition at the k-th time step, respectively. Unlike the homogeneous case where a and b were binary, in order to encompass population heterogeneity we integrate the probability distribution function comprising the frequency thresholds using the current value of network frequency $f(k)$ as one of the extrema: we thus obtain the portion of panels that are enabled to change their state. Formally,

$$a(k) = \int_{-\infty}^{f(k)} pdf^{0|1}(u)du, \qquad b(k) = \int_{f(k)}^{\infty} pdf^{1|0}(u)du,$$

where $pdf^{0|1}$ is the probability density function of a random variable modelling the transition from $R(k) = 1$ to $R(k + 1) = 0$, and analogously for $pdf^{1|0}$. Alternatively, $a(k)$ and $b(k)$ can be expressed via the cumulative distribution function of the known probability distribution for the frequency thresholds.

The expression in (1) can be interpreted as a Markov chain with two states and time varying transition probabilities, as depicted in Fig. 1. Here the edges

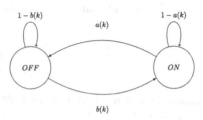

Fig. 1. A time in homogeneous Markov chain model for the aggregated dynamics, without delays.

representing the transitions between state ON and OFF are labelled with $a(k)$ and $b(k)$, while the self loops are simply $1 - a(k)$ and $1 - b(k)$.

A Markov Model with Delays

We now introduce a framework to encompass delays in the aggregate model (cf. Fig. 2): as observed in practice, panels cannot switch to the ON state instantaneously. As stated above, each panel has an internal counter for the OFF-to-ON transition. We assume to sample the delays with a coarse sampling time of 1 s: in this way we can lump together the panels which show delays in the same sampling time. We then obtain probability transition values τ_i, which represent the probability of switching on after i seconds. In order to pin down this idea, we utilise n states, defined as w_i $i = 1, \dots, n$, representing the i-th time step when the network frequency sampled by the panel is within the given threshold, but when the panel has still to turn to the ON state. In other words, the w_i state describes a device that has been waiting to turn on for i time steps. In view of the discussions in Sect. 2, we focus on the case $n \gg 1$. In the i-th delay state, there are three outgoing transitions: one towards the ON state, a second towards state $i + 1$, and one back to the OFF state. The probability associated with the third transition is $1 - b(k)$, which is the probability of sensing the network frequency outside the working interval. The first outgoing probability is $\tau_i b(k)$: τ_i is the probability to have a time delay that permits the panel to go from state w_i to state ON, which can happen only if the frequency is within the working interval (hence the multiplication by $b(k)$). We have tacitly assumed that the probability distributions of the frequency thresholds and time delay are independent. There can be, in reality, correlation between these two quantities, in which case we need to compute integrals of joint probability distributions.

We assume that $\forall i, \tau_i \geq \tau_{i+1}$ and that $\sum_i \tau_i = 1$, so the terms resemble a geometric distribution that can describe an arrival process or a waiting-time random variable.

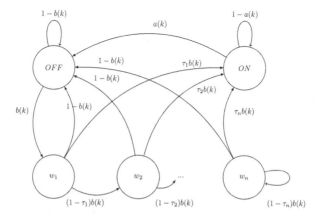

Fig. 2. A Markovian model for the aggregated dynamics, with delays.

The dynamics of the Markov chain in Fig. 2 can be summarised as

$$
\begin{cases}
x(k+1) = (1 - a(k))x(k) + b(k)\sum_{i=1}^{n} \tau_i w_i(k) \\
w_1(k+1) = b(k)[1 - x(k) - \sum_{i=1}^{n} w_i(k)] \\
w_i(k+1) = b(k)(1 - \tau_{i-1})w_{i-1}(k) \\
w_n(k+1) = b(k)[(1 - \tau_{n-1})w_{n-1}(k) + (1 - \tau_n)w_n(k)],
\end{cases}
\tag{2}
$$

where $x(k)$ represents the probability of being in the ON state at time k. We use this value as the *portion of panels* ON at time k; similarly, $w_i(k)$ is the portion of panels waiting to turn ON for i time steps at time k; and $a(k), b(k)$ are the *cdf*'s of the distributions of frequency thresholds in the population of panels. Let us further explain the details of the model in (2). To model real-life applications, we deal with an interval of time in which no panel switches on and a subsequent time interval in which panels switch on according to a geometric probability distribution. The latter time interval is described above with the w_i states. In order to encompass the time interval in which panels do not switch on (which is around 20 s in real-life applications) we must include new states that are "pure waiting states", denoted $pw_j(\cdot)$. Their number depends on the desired delay and on the given sampling time: e.g. if the sampling time is 1 s and the minimum desired delay is 20 s we add 20 new states. These states are such that $\tau_j = 0$, so to prevent the possibility to switch on, which boils down to a time delay, as desired. These new equations do not invalidate the previous analysis and for simplicity we will continue our analysis without them.

Abstraction of the Markov Model with Delays

Towards a simplified and more insightful analysis of the dynamics of the model with delays, we aggregate the n waiting states into a single location, which thus represents the sum of the portion of devices that are waiting to turn ON and is associated with a new variable $y(k) = \sum_i w_i(k)$. To express the overall dynamics, we rewrite the term $\sum_i \tau_i w_i(k)$ as a function of $y(k)$, and introduce a term $\varepsilon_k \in [0, 1]$ $\forall k$, such that

$$
\sum_{i=1}^{n} \tau_i w_i(k) = \varepsilon_k \sum_{i=1}^{n} w_i(k), \quad \text{thus} \quad \varepsilon_k = \frac{\sum_{i=1}^{n} \tau_i w_i(k)}{\sum_{i=1}^{n} w_i(k)}.
$$

The model now presents only three states, as depicted in Fig. 3, whose transition equations are

$$
\begin{cases}
x(k+1) = (1 - a(k))x(k) + b(k)\varepsilon_k y(k) \\
y(k+1) = b(k)(1 - x(k)) - b(k)\varepsilon_k y(k)
\end{cases}
\tag{3}
$$

The new model is smaller and easier to analyse. However, in general we do not know the exact value of ε_k, so we seek a value for it that ensures that the error between the two models decreases to zero with time. Define the abstraction error $e(k)$ as the difference between the element $x(k)$ of each model: $e(k) = x^{n-s}(k) - x^{3-s}(k)$, where x^{n-s} and x^{3-s} denote the x component of the model with n delay states and that with 3 states, respectively. We obtain

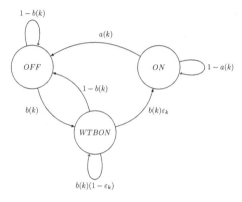

Fig. 3. Abstraction of the Markov model with delays for aggregated dynamics.

$$|e(k+1)| = |x^{n-s}(k+1) - x^{3-s}(k+1)|$$
$$= |(1 - a(k))x(k)^{n-s} - b(k)\sum_{i=1}^{n} \tau_i w_i(k) - (1 - a(k))x(k)^{3-s} + b(k)\varepsilon_k y(k)|$$
$$= |(1 - a(k))e(k) + b(k)d(k)| \leq (1 - a(k))|e(k)| + b(k)|d(k)|,$$

where $d(k) = \varepsilon(k)y(k) - \sum_{i=1}^{n} \tau_i w_i(k)$ Notice that $(1 - a(k)) \in [0,1]$ $\forall k$, by definition, which results in a term that does not increase with time. We study the evolution of the second term, $d(k)$: whilst $d(k) \in [-1,1]$ by definition, we do not know its sign, which could change at each time step k. It can be shown that $|d(k)|$ is a contracting map if we select the value $\varepsilon_k = \tau_1$. This result can be interpreted recalling the meaning of ε_k: it represents a weighted convex combination of the w_i; considering $\varepsilon = \tau_1$ results in the worst-case scenario, where we utilise the maximum value among the τ_i.

In other words, choosing ε_k to be the constant value τ_1 asymptotically decreases the error $e(k)$ between the two models to zero. Note that, knowing the values τ_i, we could estimate ε_k by estimating the states $w_i(k)$. The estimation of the states $w_i(k)$ can be attained, as will be discussed in Appendix.

We now illustrate more clearly how the frequency value affects the Markov chains: $a(k)$ can be written formally as

$$1 - a(k) = \int_{f(k)}^{\infty} pdf^{0|1}(u)du = 1 - cdf^{0|1}(f(k)),$$

and similarly for the term $b(k)$. In order to simplify this nonlinear term, let us linearise it as

$$1 - a(k) = 1 - cdf^{0|1}(f(k)) \simeq 1 - (k_1 f(k) - k_2') = -k_1 f(k) + k_2,$$
$$b(k) = 1 - cdf^{1|0}(f(k)) \simeq -k_3 f(k) + k_4,$$

where $k_2 = k'_2 + 1$. We then obtain terms that are easier to study, which allows us to derive conditions to guarantee the stability of the whole system. Note that this approximation will require saturation within the interval $[0, 1]$, and that this is the *exact* form of the cdf in case of a uniform distribution. Model (3) then becomes

$$\begin{cases} x(k+1) = (-k_1 f(k) + k_2)x(k) + (-k_3 f(k) + k_4)\varepsilon_k y(k) \\ y(k+1) = (-k_3 f(k) + k_4)(1 - x(k) - \varepsilon_k y(k)) \end{cases}. \tag{4}$$

Overall Closed-Loop Model

As previously mentioned, the overall dynamical system under study comprises both the population of panels and the network frequency. So far, we have built aggregated models only for the population of solar panels: we are interested in studying how the network frequency is influenced by the power production of PV panels. We focus on an approximate version of frequency dynamics, proposing a more realistic extension in a remark below. Consider

$$\Delta f(k+1) = \alpha_1 \Delta f(k) + \beta'_1 \Delta P(k),$$

where $\Delta f(k) = f(k) - f_0$, and f_0 represents the nominal value of the network frequency, and $f(k)$ the value of the frequency at time k, and where

$$\Delta P(k) = PN x(k) - P_0 = PN(x(k) - x_0),$$

with P_0 representing the power output of the population at time $k = 0$. We obtain

$$f(k+1) - f_0 = \alpha_1(f(k) - f_0) + \beta'_1 PN(x(k) - x_0).$$

Introducing terms $\beta_1 = \beta'_1 PN$, $\gamma = (1 - \alpha)f_0 - \beta_1 x_0$, we get

$$f(k+1) = \alpha_1 f(k) + \beta_1 x(k) + \gamma. \tag{5}$$

Embedding the frequency description Eq. (5) within the dynamics of the Markov chain with n waiting states (2), results in

$$\begin{cases} f(k+1) = \alpha_1 f(k) + \beta_1 x(k) + \gamma \\ x(k+1) = (1 - a(k))x(k) + b(k)\sum_{i=1}^n \tau_i w_i(k) \\ w_1(k+1) = b(k)[1 - x(k) - \sum_{i=1}^n w_i(k)] \\ w_i(k+1) = b(k)(1 - \tau_{i-1})w_{i-1}(k) \\ w_n(k+1) = b(k)[(1 - \tau_{n-1})w_{n-1}(k) + (1 - \tau_n)w_n(k)], \end{cases} \tag{6}$$

and into the Markov chain with 3 states (4) to obtain

$$\begin{cases} f(k+1) = \alpha_1 f(k) + \beta_1 x(k) + \gamma \\ x(k+1) = (-k_1 f(k) + k_2)x(k) + (-k_3 f(k) + k_4)\varepsilon_k y(k) \\ y(k+1) = (-k_3 f(k) + k_4)(1 - x(k) - \varepsilon_k y(k)) \end{cases}. \tag{7}$$

Dynamical Analysis of Closed-Loop Model

We study the stability of the model in (7) with techniques that come from control theory [10,11]. We are interested in the stability of the model, in the sense that we want to understand the asymptotics of the model, and under which conditions the network frequency will remain within certain (safe) operational bounds. In particular, from Lyapunov stability theory we know that if the Jacobian of a nonlinear system has stable eigenvalues (i.e. absolute value less than 1), then we can decide the asymptotic stability of the system.

Besides stability, we investigate what characteristics the panels population distributions of the frequency working thresholds must have to lead to disturbances rejection. More precisely, in an actual setup, if the grid frequency goes below a certain safety threshold, a black out is forced by the network operator in order to avoid severe damage to the network itself. The intuition leads to believe that robustness is associated with high variance of population distributions: in particular, dispersed values of disconnection thresholds means that less panels will disconnect from a given external disturbance.

Let us compute the Jacobian of the vector field in (7), which is a matrix formed by its partial derivatives, as

$$J(f, x, y) = \begin{bmatrix} \alpha_1 & \beta_1 & 0 \\ -k_1 x - k_3 \varepsilon_k y & -k_1 f + k_2 \, \varepsilon_k(-k_3 f + k_4) \\ -k_3(1 - x - \varepsilon_k y) & k_3 f - k_4 & \varepsilon_k(k_3 f - k_4) \end{bmatrix}. \tag{8}$$

Its determinant and characteristic polynomial may be computed analytically, however with a nontrivial algebraic expression. As such, in order to obtain insight, we consider an identical distribution for $a(k)$ and $b(k)$, which leads to parameters $k_1 = k_3$ and $k_2 = k_4$. This is a reasonable assumption if we consider the semantics of these distributions: they describe the probability to switch ON or OFF, which happens whenever the network frequency is greater or less than a threshold. In practical terms, this means that the threshold related to the ON switch on is the same as that of the OFF switch. The overall model becomes

$$\begin{cases} f(k+1) = \alpha_1 f(k) + \beta_1 x(k) + \gamma \\ x(k+1) = (-k_1 f(k) + k_2)(x(k) + \varepsilon_k y(k)) \\ y(k+1) = (-k_1 f(k) + k_2)(1 - x(k) - \varepsilon_k y(k)) \end{cases},$$

which admits two equilibrium points $(f_{1,2}^E, x_{1,2}^E, y_{1,2}^E)$. The Jacobian is

$$J(f, x, y) = \begin{bmatrix} \alpha_1 & \beta_1 & 0 \\ -k_1(x - \varepsilon y) & -k_1 f + k_2 & (-k_1 f + k_2)\varepsilon \\ -k_1(1 - x - \varepsilon y) & k_1 f - k_2 & (k_1 f - k_2)\varepsilon \end{bmatrix}, \tag{9}$$

and its associated characteristic 3-rd order polynomial is

$$z^3 + ((k_1 f^* + k_2)(1 - \varepsilon) - \alpha_1)z^2 +$$
$$+ [\beta_1 k_1(x^* + \varepsilon y^*) + \alpha_1(k_1 f^* + k_2)(\varepsilon - 1)]z - \beta_1 \varepsilon k_1(k_1 f^* - k_2).$$

We now set conditions on parameters k_1 and k_2 to study the attractivity of the equilibria. Leveraging Rouché arguments [12], we can synthesise the following sufficient condition on the values of k_1 and k_2 where asymptotic stability is guaranteed:

$$|(k_1 f^* + k_2)(1 - \varepsilon) - \alpha_1| + |\beta_1 k_1 (x^* + \varepsilon y^*) + \alpha_1 (k_1 f^* + k_2)(\varepsilon - 1)| +$$
$$+ |-\beta_1 \varepsilon k_1 (k_1 f^* - k_2)| < 1.$$

In order to practically reason on this condition, we need to define at least some of the numerical values of its unknowns: in the following section we provide approximate values to the unknowns, and accordingly manage to draw conclusions on the stability of the characteristic polynomial.

Remark 1. A more realistic transfer function for the grid frequency would be a second order model, namely

$$f(k) = a_1 f(k - 1) + a_2 f(k - 2) + b_1 x(k - 1) + b_2 x(k - 2).$$

In order to obtain this second-order model, we have referred to [13], and developed a simple model reproduce the frequency response, taking into account the inertia of the system, the self-regulation of the load, and the primary regulation. This was then discretised in time in order to be compatible with the current framework. The stability structure and the following extended dynamical analysis can be carried out in a similar way. For the sake of brevity and clarity we will stick to the first order model in the following analysis, while in the simulation we will use the second-order dynamics.   □

Extension of Dynamical Analysis to the Entire State Space

The analysis above holds as long as the frequency remains within certain bounds, namely where the linear approximation of the cdf holds; we need also to take into account other configurations of the system, when the frequency is outside the interval and the evolution of the system changes. Given the switching nature of the system, a hybrid system setup is a rather natural framework. The linearisation of the *cdf* is in fact defined within an interval of frequencies, outside of which $a(k)$ and $b(k)$ have a steady value of 0 or 1. We study those cases in the following.

We focus our attention on the system with five configurations. Under the assumption of the distributions of $a(k)$ and $b(k)$ to be identical, we argue that we model a hybrid system with 5 different configurations; in case of different distributions, the number of configurations may increase, so we do not study them in full in this paper for brevity. Figure 4 shows the divisions into the five different configurations.

Our interest is to analyse the connection between the various configurations of the system and to understand the conditions to keep the frequency close to the nominal value and to avoid its drift to zero.

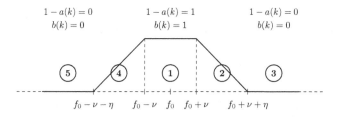

Fig. 4. Schematic diagram of the 5 frequency configurations, indicated as circled numbers.

1. **Configuration 1:** this configuration describes the case where the network frequency belongs to the set of values closest to the nominal value. Our system becomes an LTI system, given the conditions $1 - a(k) \equiv 1 \equiv b(k)$, $f(k) \in [f_0 - \nu, f_0 + \nu]$:

$$\begin{cases} f(k+1) = \alpha_1 f_k + \beta_1 x(k) + \gamma \\ x(k+1) = 1 \cdot (x(k) + \varepsilon_k y(k)) \\ y(k+1) = 1 \cdot (1 - x(k) - \varepsilon_k y(k)) \end{cases} \tag{10}$$

There is only one equilibrium point, $p^E = \left(\frac{\beta_1}{1-\alpha_1}(1 - x_0) + f_0, 1, 0\right)$. From the analysis of the evolution matrix, the equilibrium is asymptotically stable. If the contribution coming from $\frac{\beta_1}{1-\alpha_1}(1-x_0)$ is small enough, the equilibrium remains inside the interval $[f_0 - \nu, f_0 + \nu]$, whereas if the contribution is too large the system will switch to Configuration 2.

2. **Configuration 2:** This is the case where the evolution matrix of the system becomes non linear and time varying, with $f(k) \in [f_0 + \nu, f_0 + \nu + \eta]$. The system may be written

$$\begin{cases} f(k+1) = \alpha_1 f(k) + \beta_1 x(k) + \gamma \\ x(k+1) = (1 - a(k))x(k) + b(k)\varepsilon_k y(k) \\ y(k+1) = b(k)(1 - x(k)) - b(k)\varepsilon_k y(k) \end{cases}$$

and as we stated, we assume the same distribution for $a(k)$ and $b(k)$, which we explicit $1 - a(k) = -k_1 f(k) + k_2 = b(k)$ where $k_1, k_2 > 0$. To find the equilibria we set that

$$f = \frac{\beta_1}{1 - \alpha_1} x + \frac{\gamma}{1 - \alpha_1}, \quad x + y = -k_1 f + k_2,$$

and from these two equations we obtain

$$y = \left(\frac{-k_1\beta_1}{1 - \alpha_1} - 1\right) x - \frac{k_1}{1 - \alpha_1}\gamma + k_2.$$

Substituting this value into the previous equation, we have

$$\frac{-k_1\beta_1}{1 - \alpha_1}\left(1 + \varepsilon\left(\frac{-k_1\beta_1}{1 - \alpha_1} - 1\right)\right) x^2 +$$

$$+ \left[\left(-\frac{k_1}{1 - \alpha_1}\gamma + k_2\right)\left(1 - \varepsilon + 2\varepsilon\left(\frac{-k_1\beta_1}{1 - \alpha_1}\right)\right) - 1\right] x + \varepsilon\left(-\frac{k_1}{1 - \alpha_1}\gamma + k_2\right)^2.$$

The solution of this 2-nd order equation can be found symbolically, however in order to garner insight on its solutions, we set some indicative (and practically meaningful) values for the variables ranges:

$$k_1 \sim 10,\, k_2 \sim 10^2,\, \alpha_1 \sim 1 - \alpha_1 \sim 10^{-1}, \varepsilon \sim 10^{-1},\, \beta_1 \sim N,$$

where N represents the number of solar panels, assumed to be $N \gg 1$. This results in equation $10^3 N^2 x^2 - 2 \cdot 10^3 N x + 10^3$, which is endowed with two coincident solutions, $x_{1,2}^E = \frac{1}{N}$. This results in the following approximated equilibrium point:

$$p_{1,2}^E \simeq \left(\frac{1+\gamma}{1-\alpha_1}, x_{1,2}^E, 0 \right),$$

where the value of x is less than 1, giving a frequency to be *less* than the value we found in configuration 1. The equilibrium could, in this case, be inside the thresholds of Configuration 1 or remain inside Configuration 2 depending on the numerical values of the several parameters; we will discuss this in Appendix.

We state that the equilibrium is attractive because we argue that the system in configuration 1 presents variables f, x, y which are always greater or equal than the ones in this configuration: $a(k)$ and $b(k) \in [0,1]\ \forall k$. Since the stability is asymptotic in the first case, we conclude the asymptotic stability also in this configuration.

3. **Configuration 3:** this configuration describes the behaviour of the system when it is completely outside the range of the frequency thresholds, i.e. $1 - a(k) = 0 = b(k)$. The system becomes null, the frequency only evolves with a decreasing exponential, i.e. the frequency will move towards the cases discussed above.

4. **Configuration 4:** the analysis is similar to the one of configuration 2. In this case it is important to notice that, since the frequency at the equilibrium point is less than the nominal value, the system could move towards the configuration 1, remain in the current configuration or slide towards a decreasing frequency. This only depends on the numerical value of the parameters.

5. **Configuration 5:** the analysis is analogous to the configuration 3. The frequency will drop to zero.

4   Experimental Evaluation of the Aggregated Models

In order to show the precision of the aggregated models, we set up rounds of simulations comparing the two alternative models (that with n waiting states in Fig. 2 and the one with three locations in Fig. 3) with the ground truth obtained from an explicit simulation of the entire population of PV panels within the power network (which we denote as the *explicit* model).

For the *explicit* model, each of the N panels has been given four different frequency thresholds (disconnection and reconnection in over- and under-frequency) and a time delay (number of time steps the devices need to wait before turning to ON). These parameters have been generated according to set probability distributions (see below) for the population, which are then used in the computations for the abstract models. We have set up the distributions of frequency thresholds as uniform, and that for the time delay as geometric, and we set $N = 10^6$. The assumption of a uniform pdf allows us to exactly define the variables ν, η mentioned in the previous section, where more generally we utilised a linear approximation of the cdf, e.g. $b(k) = cdf^{1|0}(f(k)) \simeq -k_1 f(k) + k_2$. For simplicity, P_i was set to be constant and equal to $\bar{p} = 3$ kW for each device. With reference to the discussion in Remark 1, the frequency dynamics have been described by a second-order difference equation, as

$$f(k+1) = \alpha_1 f(k) + \alpha_2 f(k-2) + \beta_1 x(k) + \beta_2 x(k-1),$$

where the constants $\alpha_1, \alpha_2, \beta_1, \beta_2$ are set to make the transfer function stable.

Our two aggregated models are derived from the following simplifications:

1. Frequency thresholds: we have an exact description of the frequency thresholds and a linear dependency from $f(k)$ in the extended system;
2. Time delays: we have introduced 1 s time delays and defined waiting states accordingly. We have defined a maximum possible delay as n, and in order to cope with delays longer than n, we have set a self loop on the n-th state;
3. Lumped delays: we have lumped together the n waiting states, introducing an approximation encoded in ε_k.

As motivated in the Introduction to this work, we have set up two specific simulation scenarios:

1. Panels disconnecting in view of an external disturbance;
2. Effect of the distribution of frequency disconnection/reconnection thresholds, in response to an external disturbance.

First Scenario. We set up a stable network with N devices initially in the ON mode, when the grid relies on their power production to be stable and to guarantee a reliable service to the load. The initial condition is set to $x_0 = 1$, and the grid frequency is set to the stable value, $f_0 = 50$ Hz. At time $t = 10$ s we inject an external disturbance, in order to create a frequency peak of 50.16 Hz, and to observe the ensuing dynamics in the network.

The comparison among the three models is in Figs. 5 and 6, in terms of network frequency response and portion of ON panels. On the one hand, we observe from Fig. 5 that both the two abstract models seem to be a low-pass filtered signal of the explicit one. The model with n delay states follows the dynamics of the explicit one, and the model with one delay state follows that with n states, thanks to the estimation of the quantity ε_k (as discussed in Appendix). On the other hand, looking at Fig. 5 we note that the difference in terms of percentage of active panels is always less than 2%. This difference is due to the low-pass action of the abstract models, which leads to slower convergence to the same equilibrium point as the explicit model. These figures show that our abstract models can reflect the evolution of the explicit model in a reliable way.

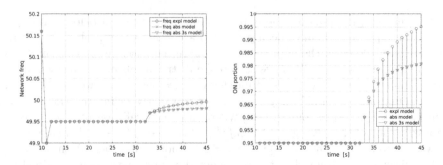

Fig. 5. Comparison of network frequency behaviour (left) and of ON population (right) for the explicit model (blue, circles), the abstract model with n delay states (red, crosses) and the abstract model with 3 states (green, triangles). (Color figure online)

Second Scenario. We run simulations to test if more variance in the reconnection/disconnection thresholds brings a more stable system, as intuition suggests. Figure 6 shows that, as expected, a higher variance is desirable for the model under consideration. In the explicit system, at time $t = 10\,$s an external disturbance was injected in order to make the frequency jump to $50.16\,$Hz, value for which a portion of devices will disconnect. Simulations run with 10 different values of variance (v) for the uniform distributions, resulting in bigger oscillations and longer time to get frequency back to the nominal value for small v; values of v close to 5 makes the system almost insensible to the disturbance.

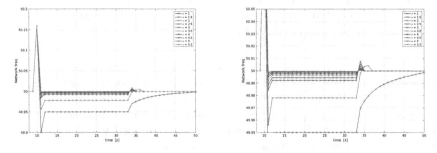

Fig. 6. Comparison of network frequency behaviour with different degrees of variance in the frequency thresholds: smaller variance produces bigger oscillations. Simulations run with the explicit model.

We have finally run simulations to display the effect of several frequency evolution parameters in the Appendix, where we show how the different position of the equilibria, as discussed in Sect. 3, affect the response of the system.

5   Conclusions

We have presented a modelling framework for the aggregation of a population of photovoltaic panels, and studied its dynamics when interacting within the electric grid and its frequency behaviour. We have provided experiments comparing two models against a ground truth, where N photovoltaic panels were singularly simulated: a scenario has shown how reliably the models behave in presence of a massive network failure, another how population heterogeneity influences the performances of the system.

Current research emphasis concerns variable power outputs (modelling uncertainties such as weather, occlusions), and novel global control schemes.

References

1. Solar Power Europe: Global Market Outlook for Solar Power 2016–2020. Technical report (2016)
2. Ellis, K., Tiam, H., Mancilla-David, F.: A detailed performance model for photovoltaic systems. Technical report, NREL (2012)
3. Wirth, H.: Recent facts about photovoltaics in Germany. Technical report, Fraunhofer ISE (2016)
4. Dispersed Generation Impact on CE Region, Dynamic study. Technical report, ENTSOE (2014)
5. Regional Group of Continental Europe and Synchronous Area Great Britain. Solar Eclipse 2015 - Impact Analysis. Technical report, ENTSOE (2015)
6. Lazpita, B., Jung, M., Wiss, O.: Analyses et conclusions "Tests en sous-frequence". Technical report, RTE (2016)
7. Lazpita, B., Jung, M., Wiss, O.: Analyses et conclusions "Tests en sur-frequence". Technical report, RTE (2016)
8. Legal Comparison of PV Regulations. http://www.res-legal.eu/comparison-tool/. Accessed 04 Apr 2017
9. Moller, J.K., Iversen, E.B., Morales, J.M.: Probabilistic forecasts of solar irradiance by stochastic differential equations. Envirometrics 25, 152–164 (2014)
10. Krener, A.J., Hermann, R.: Nonlinear controllability and observability. IEEE Trans. Autom. Control 22, 728–740 (1977)
11. Khalil, H.K.: Nonlinear Systems. Prentice Hall, Upper Saddle River (2002)
12. Beardon, A.F.: Complex Analysis: The Argument Principle in Analysis and Topology. Wiley, New York (1979)
13. Policy 1: Load-Frequency Control and Performance. Technical report, ENTSOE (2009)

Battery Aging, Battery Charging and the Kinetic Battery Model: A First Exploration

Marijn R. Jongerden[✉] and Boudewijn R. Haverkort

Design and Analysis of Communication Systems,
University of Twente, Enschede, The Netherlands
{m.r.jongerden,b.h.r.m.haverkort}@utwente.nl

Abstract. Rechargeable batteries are omnipresent and will be used more and more, for instance for wearables devices, electric vehicles or domestic energy storage. However, batteries can deliver power only for a limited time span. They slowly degrade with every charge-discharge cycle. This degradation needs to be taken into account when considering the battery in long lasting applications. Some detailed models that describe battery degradation processes do exist, however, these are complex models and require detailed knowledge of many (physical) parameters. Furthermore, these models are in general computationally intensive, thus rendering them less suitable for use in larger system-wide models. A model better suited for this purpose is the so-called Kinetic Battery Model. In this paper, we explore how this model could be enhanced to also cope with battery degradation, and with charging. Up till now, battery degradation nor battery charging has been addressed in this context. Using an experimental set-up, we explore how the KiBaM can be used and extended for these purposes as well, thus allowing for better integrated modeling studies.

Keywords: Kinetic battery model · Battery aging · Battery charging · Battery discharging · Measurements

1 Introduction

Batteries-powered devices are everywhere; smart-phones, laptops, wireless sensors, wearables, electric cars and for local energy storage. According to McKinsey, the Internet-of-Things (IoT) is expected to connect 1 trillion (10^{12}) devices by 2025, many of which will be battery powered. According to the International Energy Agency (IAE), in 2016, some 6.4% of Dutch cars was fully or hybrid electric; in Norway this was even 29%! Throughout 2016, the world's electric car population grew to 2 million cars, almost a doubling compared to the end of 2015. These developments clearly underline the importance of understanding battery charging and discharging, as well as battery degradation processes.

Batteries are needed to provide portable power to all these devices. However, batteries have a limited life span. Obviously, non-rechargeable batteries can be

© Springer International Publishing AG 2017
N. Bertrand and L. Bortolussi (Eds.): QEST 2017, LNCS 10503, pp. 88–103, 2017.
DOI: 10.1007/978-3-319-66335-7_6

discharged only once before they need to be replaced. But, even rechargeable batteries will not be usable after some time. How long a battery can be used depends on many factors, such as battery type, discharge and charge current, depth of discharge and temperature. It is hard to predict the lifetime of a battery for any given workload pattern. Electro-chemical and electrical circuit models, that require detailed knowledge of the used batteries, are available in the literature, see for example [1,2]. In recent work, Wognsen et al. [3] propose an approach to compare the impact workload patterns have on the battery life through the Fourier Transform of the workload.

Although some theoretical work exists, little practical work is available in the scientific literature on measuring battery degradation over time, and how such degradation effects models or model parameters. In this paper we present the results of an extensive measurement study on battery cells of the type are used in nano-satellites of GomSpace (lithium ion 18650 cells) [4], which are also used in Tesla electric vehicles [5]. These measurements are analyzed in the context of a widely used battery model, the Kinetic Battery Model. The analysis gives insight on how the degradation of the battery impacts the model parameters, and on how to possibly extend this model to cope with the effects of degradation. Furthermore, we also explicitly address the charging of such batteries; up till now, in the literature, it has been assumed that the charging process proceeds the same as discharging, with "just the flow of current flipped". We show that this is not exactly the case.

The rest of this paper is structured as follows. Section 2 gives a brief overview of related work on battery degradation modeling. Section 3 introduces the Kinetic Battery Model. In Sect. 4 the experimental set-up and the performed experiments are described. The results of the experiments are presented in Sect. 5 and discussed in Sect. 6. Section 7 concludes the paper.

2   Battery (Degradation) Models

There are several types of battery models available in the scientific literature. We provided an overview of the most widely used models, such as electro-chemical models, electrical circuit models and analytical models in [6], with a focus on predicting the duration of a single discharge cycle. These types of models are also used to describe the long-term effects of battery degradation.

In [1], so-called capacity fading is modeled with an electro-chemical battery model for a lithium-ion battery. This type of model requires a very detailed knowledge of the physical characteristics of the battery, and is computationally very intensive to use.

In [2] an electrical circuit model is made that models capacity fading due to cycling (repeated charge–discharge), as well as the increase of the internal resistance due to cycling. The model should be configured with data from the battery data sheets. However, as also the authors mention, in general, it is very hard to obtain all required parameters.

High-level analytical models, such as the Kinetic Battery Model (KiBaM) [7], require much less knowledge of the battery, and can be easily combined with other models. For example, in [8], the KiBaM is extended to a random KiBaM and combined with a Markovian task process that models the battery load. With the combined model, one can compute the probability the battery is depleted due to the defined load pattern. The KiBaM, nor the proposed extensions, do take into account how the battery degrades; it is not known how the essential parameters are effected.

In [3], a generic method for comparing the impact of different load profiles on the wear of the battery is proposed. The load profiles are rated by analyzing the Fourier transform of the load. With this analysis different load profiles can be ranked from little impact on battery wear to large impact.

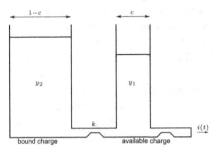

Fig. 1. The two-well Kinetic Battery Model.

However, it is not possible to quantify the wear with this method. In order to do this, many more measurements need to be performed. In this paper, we investigate how the KiBaM-parameters change when the battery is repeatedly discharged. We take an experimental approach. We wear the battery by applying a relatively heavy load to the battery. This gives us the practical insight in how the battery degrades over time.

3   The Kinetic Battery Model

The kinetic battery model (KiBaM) is a compact battery model that includes the most important features of batteries, i.e., the rate-capacity effect and the recovery effect. The model has been originally developed by Manwell and McGowan in 1993 [7] for lead-acid batteries, but analysis has shown that it can also be used in battery discharge modeling for other battery types [9].

3.1   Basic Dynamics

In the model, the battery charge is distributed over two wells: the *available-charge* well and the *bound-charge* well (cf. Fig. 1). A fraction c of the total capacity is considered to be in the available-charge well (denoted $y_1(t)$), and a fraction $1 - c$ in the bound-charge well (denoted $y_2(t)$). The available-charge well supplies electrons directly to the load ($i(t)$), whereas the bound-charge well supplies electrons only to the available-charge well. The charge flows from the bound-charge well to the available-charge well through a "valve" with fixed conductance, k. The parameter k has the dimension $1/$time and limits the rate at which the charge can flow between the two charge wells. Next to this parameter, the rate at which charge flows between the wells depends on the height difference between the two wells. The heights of the two wells are given by: $h_1(t) = y_1(t)/c$

and $h_2(t) = y_2(t)/1 - c$. The change of the charge in both wells is given by the following system of differential equations:

$$\begin{cases} \dfrac{dy_1(t)}{dt} = -i(t) + k(h_2(t) - h_1(t)), \\ \dfrac{dy_2(t)}{dt} = -k(h_2(t) - h_1(t)), \end{cases} \tag{1}$$

with initial conditions $y_1(0) = c \cdot C$ and $y_2(0) = (1 - c) \cdot C$, where C is the total battery capacity. The battery is considered empty when it is observed that there is no charge left in the available-charge well. As shown in [9], we can transform the above equations to

$$\begin{cases} \dfrac{d\gamma(t)}{dt} = -i(t), \\ \dfrac{d\delta(t)}{dt} = \frac{1}{c}i(t) - k'\delta(t), \end{cases} \tag{2}$$

where $k' = k/(c(1 - c))$, $\gamma(t) = y_1(t) + y_2(t)$ and $\delta(t) = y_2(t)/(1 - c) - y_1(t)/c$. We can interpret $\gamma(t)$ as the total charge remaining in the battery, and $\delta(t)$ as the height difference between the the charge levels of the two wells. The initial conditions transform into $\gamma(0) = C$ and $\delta(0) = 0$. The battery is empty when $\gamma(t) = (1 - c)\delta(t)$.

3.2   KiBaM Constant Current Discharge

When we consider a constant current discharge, i.e., $i(t) = I_d$, the differential equations can easily be solved:

$$\begin{cases} \gamma(t) = C - I_d t, \\ \delta(t) = \dfrac{I_d}{ck'}\left(1 - e^{-k't}\right). \end{cases} \tag{3}$$

The battery lifetime L, i.e., the time to empty the available charge well, for a constant current discharge is given by:

$$L = \frac{C}{I_d} - \frac{1}{k'}\left(\frac{1 - c}{c} + \mathcal{W}\left(\frac{1 - c}{c}e^{\frac{1-c}{c} - \frac{Ck'}{I_d}}\right)\right), \tag{4}$$

where $\mathcal{W}(.)$ is the so-called Lambert $\mathcal{W}$ function [10]. By measuring the battery lifetime, and the delivered energy, as a function of the discharge current, we can determine the KiBaM parameters k, c and C by fitting (4) to the data.

3.3   KiBaM Charging

Battery charging normally is performed in two phases. First, the battery is charged at a constant current. In this phase the voltage will slowly rise. When the voltage reaches the maximum level, V_{max}, the second phase starts, during which the voltage is kept constant at V_{max} and the charging current will drop. We discuss the two charging phases in the context of the KiBaM model in the following sections.

KiBaM Constant Current Charging. In the KiBaM, the charging with a constant current is very similar to discharging with a constant current. For a constant charging current I_{ch} the KiBaM equations are:

$$\begin{cases} \dfrac{dy_1(t)}{dt} = I_{ch} - k\left(\dfrac{y_1(t)}{c} - \dfrac{y_2(t)}{1-c}\right), \\ \dfrac{dy_2(t)}{dt} = k\left(\dfrac{y_1(t)}{c} - \dfrac{y_2(t)}{1-c}\right). \end{cases} \tag{5}$$

When we consider the battery fully empty at the start of the charging, the initial conditions are $y_1(0) = 0$ and $y_2(0) = 0$. The constant current charging phase ends when the available charge well is filled, thus $y_1 = cC$. In terms of $\delta_{ch}(t) = \frac{y_1(t)}{c} - \frac{y_2(t)}{1-c}$ ($\delta_{ch}(t) = -\delta(t)$) and $\gamma(t) = y_1(t) + y_2(t)$, the equations are:

$$\begin{cases} \dfrac{d\gamma(t)}{dt} = I_{ch}(t), \\ \dfrac{d\delta_{ch}(t)}{dt} = \dfrac{I_{ch}(t)}{c} - k'\delta_{ch}(t), \end{cases} \tag{6}$$

The initial conditions transform into $\delta_{ch}(0) = 0$ and $\gamma(0) = 0$. The condition for the end of the constant current charging phase is $\gamma(t_{lin}) + (1 - c)\delta_{ch}(t_{lin}) = C$. This condition can be interpreted as follows, at time $t = t_{lin}$, the amount of energy put into the battery is $\gamma(t_{lin})$ and still $(1-c)\delta_{ch}(t_{lin})$ needs to be charged. The solutions for $\gamma(t)$ and $\delta_{ch}(t)$ are again easily obtained:

$$\begin{cases} \gamma(t) = I_{ch}t, \\ \delta_{ch}(t) = \dfrac{I_{ch}}{ck'}(1 - e^{-k't}), \end{cases} \tag{7}$$

where we see that the equation for δ is the same as for discharging, cf. (3).

Under the above described conditions, the time it takes to fill the available charge well, t_{lin}, is similar to the discharging lifetime, cf. (4):

$$t_{lin} = \frac{C}{I_{ch}} - \frac{1}{k'}\left(\frac{1-c}{c} + W\left(\frac{1-c}{c} \cdot e^{\frac{1-c}{c} - \frac{Ck'}{I_{ch}}}\right)\right). \tag{8}$$

We can estimate the charging parameters by measuring the duration of the linear charging phase for different charge currents, and fitting the equation to the results.

KiBaM Non-linear Charging. After the linear charging phase, the battery is charged with a constant voltage and a decreasing current. In the KiBaM we can interpret this as follows. The constant voltage keeps the level of the available charge at its maximum. The rate at which the battery can accept additional charge is limited by the flow between the two charge wells. This rate depends on the height difference between the two wells, and thus will decrease when the battery is further charged.

Since the available charge does not change, we have $\frac{dy_1(t)}{dt} = 0$. From the KiBaM equations we therefore obtain:

$$i(t) = k\left(\frac{y_1(t)}{c} - \frac{y_2(t)}{1-c}\right). \tag{9}$$

In terms of $\delta_{ch}(t) = \frac{y_1(t)}{c} - \frac{y_2(t)}{1-c}$ this yields:

$$i(t) = k\delta_{ch}(t) = k'c\,(1-c)\,\delta_{ch}(t). \tag{10}$$

The KiBaM equations in terms of $\delta_{ch}(t)$ and $\gamma(t)$ now are,

$$\begin{cases} \frac{d\gamma(t)}{dt} = i(t) = k'c(1-c)\delta_{ch}(t), \\ \frac{d\delta_{ch}(t)}{dt} = \frac{i(t)}{c} - k'\delta ch(t) = -k'c\delta_{ch}(t), \end{cases} \tag{11}$$

From these equations it follows that

$$\delta(t) = \delta_0 e^{-ck't}, \tag{12}$$

where δ_0 is the height difference between the two wells at the start of the non-linear charging phase (I_{lin}); δ_0 depends on the charging current in the linear phase. From Eqs. (7) and (8) it follows that

$$\delta_0 = \frac{I_{lin}}{ck'}\left(1 - e^{-ck't_{lin}}\right). \tag{13}$$

If $k't_{lin}$ is large, that is, if the height difference has approached its maximum value during the linear charging phase, we obtain

$$\delta_0 = \frac{I_{lin}}{ck'}. \tag{14}$$

The height difference decreases exponentially, and thus the charging current should decrease exponentially. By fitting an exponential function to the measured current we can estimate the factor ck'. This gives additional information on how the KiBaM performs for charging the battery.

4   Experimental Set-Up

In the experiments we analyze 4 lithium-ion battery cells with a capacity of 2600 mAh, obtained from GomSpace (www.gomspace.com). The nano-satelite battery packs consist of 4 to 8 of these battery cells. Table 1 gives an overview of the key parameters, as provided in the datasheets.

Table 1. Parameters of the GomSpace lithium-ion batteries [4]

Parameter	Value
Nominal capacity	2600 mAh
Maximum charge voltage	4.2 V
End of discharge voltage	3.0 V
Maximum discharge current	3.75 A
Maximum charge current	2.5 A
End of charge current	1.3 A
Charge temperature range	-5–$45\,^{\circ}$C
Discharge temperature range	-20–$60\,^{\circ}$C

The measurements are performed with the Cadex C8000 battery testing system, cf. Fig. 2, which can test four batteries simultaneously. The tester is programmed to discharge and charge the cells in a controlled fashion according to a user-defined load profile, while measuring the voltage, current and temperature. This data is logged each second, and is used for the analysis of the battery properties. The experiments are conducted in a number of steps (phases):

1. In the first phase, *KiBaM estimation measurements*, the cells are discharged and charged at various constant rates. The charge rates vary from 0.1 C to 0.9 C, while the discharge rates vary from 0.1 C to 1.4 C. Table 2 gives an overview of the discharge and charge currents of the individual measurement cycles. The data from these measurements will be used to estimate the parameters for the Kinetic Battery Model.
2. In the second phase, *the degradation measurements*, the cells are repeatedly fully discharged at 1 C and charged at 0.5 C. This high load will result in

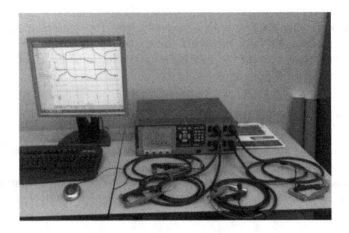

Fig. 2. Experimental set-up with the Cadex C8000 battery tester.

Table 2. Discharge and charge currents for the parameter estimation measurements.

Test	Discharge current	Charge current	Test	Discharge current	Charge current
1	$0.1\,C = 0.26\,A$	$0.1\,C = 0.26\,A$	7	$0.7\,C = 1.82\,A$	$0.7\,C = 1.82\,A$
2	$0.2\,C = 0.52\,A$	$0.2\,C = 0.52\,A$	8	$0.8\,C = 2.08\,A$	$0.8\,C = 2.08\,A$
3	$0.3\,C = 0.78\,A$	$0.3\,C = 0.78\,A$	9	$0.9\,C = 2.34\,A$	$0.9\,C = 2.34\,A$
4	$0.4\,C = 1.04\,A$	$0.4\,C = 1.04\,A$	10	$1.0\,C = 2.60\,A$	$0.6\,C = 1.56\,A$
5	$0.5\,C = 1.3\,A$	$0.5\,C = 1.3\,A$	11	$1.2\,C = 2.86\,A$	$70.7\,C = 1.82\,A$
6	$0.6\,C = 1.56\,A$	$0.6\,C = 1.56\,A$	12	$1.4\,C = 3.64\,A$	$70.9\,C = 2.34\,A$

a relative fast degradation of the cells. After 50 discharge-charge cycles, the cycles of the first phase are repeated, in order to see whether and how the battery parameters have changed.

3. The battery parameters will be determined after every such 50 repetitions, until the cell capacity has dropped below 80% of its initial value. The results of these experiments give an indication on how the cells degrade over time.

5   Measurement Results

In this section we discuss the results of the performed measurements. We start with the degradation measurements in Sect. 5.1, since these results provide a clear view on how the battery slowly degrades during the experiments. Then, we analyze the change of the KiBaM parameters for discharging and charging in Sects. 5.2 and 5.3, respectively.

5.1   Degradation Measurements

Figure 3 shows how the discharge capacity decreases as a function of the discharge cycle number. In the first discharge cycle, on average, the batteries deliver 92.8% of the nominal capacity (2600 Ah). In the subsequent cycles the discharge capacity slowly drops. The decrease in capacity is more or less linear. We fit a linear function, $Cap^{(1-100)} = \alpha \cdot cycle + \beta$, to the first 100 measurements with using the nonlinear least squares method built in the Matlab fit function. The fit yields the following estimates and 95% confidence intervals: $\alpha = -0.057 \pm 0.0025$ and $\beta = 92.8 \pm 0.14$. This means that the capacity, on average, drops 0.057% point with every discharge-charge cycle.

After approximately 140 cycles the capacity decreases more rapidly. Battery 3 (yellow) now degrades clearly faster than the other 3 batteries. We fit another line, $Cap^{(151-200)} = \alpha \cdot cycle + \beta$, to the last 50 measurements, cycle 151 to 200. This yields, $\alpha = -0.41 \pm 0.027$ and $\beta = 144.2 \pm 4.8$. This means that the degradation is more than a factor 7 faster than in the first phase, with an average of 0.41% point per cycle.

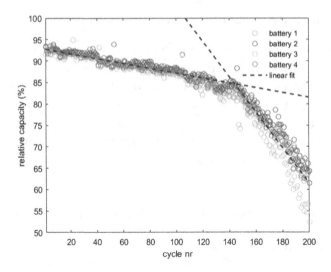

Fig. 3. Capacity relative to the nominal capacity as a function of the cycle number. (Color figure online)

Next to the capacity we investigate how the efficiency evolves when the battery is used. The efficiency, in percent, is determined as $100 \cdot E_{dis,n}/E_{ch,n-1}$, where $E_{dis,n}$ is the delivered energy in cycle n, and $E_{ch,n-1}$ is the charging energy of cycle $n - 1$. The results are shown in Fig. 4. As for the capacity, we see that the efficiency also degrades in two phases. Again we fit two lines to the data. The first line is fit to the first 100 cycles. The efficiency starts at $89.3\% \pm 0.17$. The efficiency degrades linearly with a rate of $0.020 \pm 0.0028\%$ point per cycle. The second line is fit to the last 50 cycles. Here we see that the efficiency degrades at a rate of $0.061 \pm 0.022\%$ point per cycle. This means that the efficiency degrades 3 times faster at the end of the battery life than at the beginning. Furthermore, we see that the variation of the measured efficiency is much larger at the end of the battery lifetime.

Finally, we investigate the non-linear charge phase of the degradation measurements. According to the KiBaM theory, the charge current should drop exponentially during the non-linear charge phase, cf. Eq. (12). We fit a negative exponential curve to the measured current. In Fig. 5, the exponent, which corresponds to $k'c$, is plotted as a function of the cycle number. We see that the exponent decreases as the number of discharge-charge cycles increases. We have fitted a linear curve, $y = \alpha \cdot x + \beta$ to the data. This fit yields $\alpha = -1.71 \cdot 10^{-6} \pm 0.05 \cdot 10^{-6}$ and $\beta = 1.03 \cdot 10^{-3} \pm 0.005 \cdot 10^{-3}$. In the KiBaM, the decrease of the exponent $k'c$ is either caused by a decrease in k, i.e., the conductance between the available and bound charge well, or by a decrease in c, i.e., the size of the available charge well. The slower exponential drop of the charging current may also be a result of the drop in the charging efficiency, which we discussed above. The charging efficiency, however, is currently not included in the KiBaM.

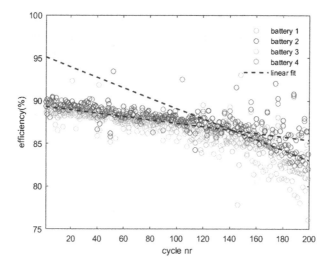

Fig. 4. Efficiency of charge discharge cycle as a function of the cycle number.

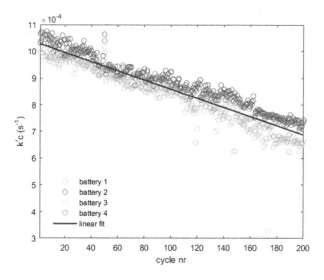

Fig. 5. The exponent for the non-linear charge phase as a function of the cycle number.

5.2 KiBaM Discharging Parameter Estimation

We started the battery degradation analysis with a series of measurements for determining the KiBaM parameters. In these measurements the batteries are discharged and charged at various constant currents, cf. Table 2. These measurements have been repeated after every 50 cycles in the degradation measurements. Figure 6(a) shows the measured discharge capacity of the four batteries for the different discharge currents of the first series.

The measurements at $0.9\,C = 2.34$ A discharging current have been performed twice. The first run, which was the first experiment that was performed, resulted for all batteries in a discharge capacity that was higher than expected. The second run resulted in a capacity that was in line with the other experiments. The reason for these results remains unclear.

For battery 3, we see a relative low capacity at the low discharge currents. We expect that this is due to some internal damage or lower quality of the battery. Battery 3 has a slightly lower performance throughout the experiments, as we will see in the later results.

The measured delivered capacity (C_{del}) in As as a function of the discharge current (I_d) is fitted to the function (cf. 4):

$$C_{del} = C_{nom} - \frac{I_d}{k'} \left(\frac{1 - c}{c} + \mathcal{W} \left(\frac{1 - c}{c} e^{\frac{1-c}{c} - \frac{C_{nom}k'}{I_d}} \right) \right) \tag{15}$$

In the fitting procedure we use the parameter $\kappa = 1/k'$ instead of k', since the fitting algorithm was not stable when k' was used directly. In the fit we ignored the outliers of the first measurement and battery 3. The result is included in Fig. 6(a). From the fit we obtained $C = 9.67 \cdot 10^3 As \pm 220\ As$, which is higher than the nominal capacity of 2600 mAh $= 9360$ As. The other parameters are: $c = 0.90 \pm 0.015$ and $\kappa = 9.36 \cdot 10^3 s \pm 9.12 \cdot 10^3 s$. The parameter κ has a very large confidence interval, thus we cannot draw any strong conclusions on the actual value of this parameter, nor for the parameter $k = 1/\kappa$.

After every 50 discharge-charge cycles another series of measurements is done to determine the KiBaM parameters. The results are given in Figs. 6(b)–(e). In these figures we see that, like in the degradation measurements, the capacity first drops slowly in Figs. 6(b) to 6(d), and then drops dramatically in Fig. 6(e). In all these measurement series, as in the results of the first series, battery 3 shows a lower capacity for the low discharge currents. At high discharge currents, i.e., larger than 2.5 A, all batteries perform less good than expected. When we include these measurements in the fitting procedure the results for the parameters c and κ are nearly meaningless, with extremely large confidence intervals. The degradation of the battery clearly has a larger impact when high discharge currents are applied.

When we discard the high current measurements in the fitting procedure, the results are more in line with the analysis of the first measurement series (cf. Sect. 5.2). The values of the fitted parameters and their confidence intervals are given in Table 3(a). We see a decrease in the capacity of the battery, as expected. Also, the parameter c slowly decreases, as the battery ages. This means that the decrease in capacity affects the available charge more than the bound charge. For the parameter κ it is, statistically speaking, impossible to tell whether the battery degradation has any real impact, due to the large confidence intervals.

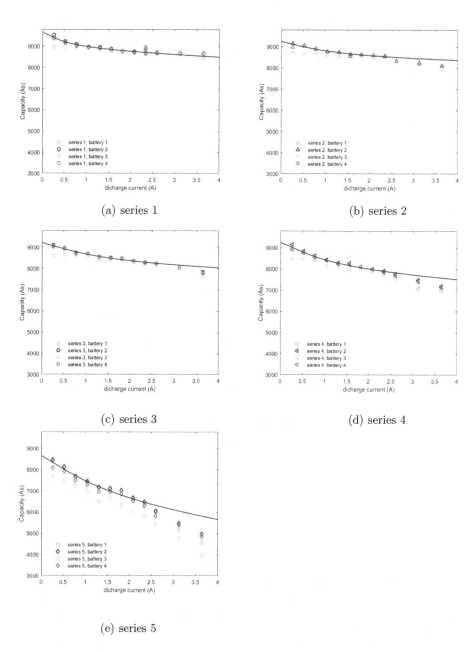

(a) series 1

(b) series 2

(c) series 3

(d) series 4

(e) series 5

Fig. 6. Measured discharge capacity as function of the discharge currents and a nonlinear least squares fit of the KiBaM for the 5 measurements series.

5.3   KiBaM Charging Parameter Estimation

Next to the parameters for discharging, we also fit the KiBaM parameters to the charging measurements. Figure 7 shows the energy put into the battery during the linear charge phase of the five series. In all five figures we notice some deviating measurements. These measurements coincide with the deviations in the discharge results. Battery 3 again deviates at low currents, however, the linear charge capacity is larger than for the other batteries at low currents, whereas the discharge capacity was lower.

Table 3. KiBaM parameters and the 95% confidence intervals based on a non-linear least squares fit of the (a) discharge measurements (upper half), and (b) charge measurements (lower half).

The outliers are again discarded in the fitting procedure. The curves we fitted are given in Fig. 7, and the parameters are given in Table 3(b). Again, we see that the capacity decreases. The estimated capacity is, however, smaller than for discharging. The parameter c is much smaller during charging than during discharging. This implies that the available charge well is much smaller when the battery is charged. For the parameter κ it is again hard to draw firm conclusions. The estimated values for κ are lower for charging than for discharging. This suggests that the flow between bound and available charge is faster

Experiment	$C(10^3 \text{ As})$	c	$\kappa \ (10^3 \text{ s})$
(a) Discharge measurements			
Series 1	9.67 ± 0.22	0.90 ± 0.015	9.36 ± 9.12
Series 2	9.25 ± 0.10	0.90 ± 0.019	4.37 ± 2.66
Series 3	9.23 ± 0.08	0.86 ± 0.019	3.76 ± 1.56
Series 4	9.26 ± 0.15	0.83 ± 0.027	4.43 ± 2.24
Series 5	8.67 ± 0.26	0.70 ± 0.080	2.85 ± 2.05
(b) Charge measurements			
Series 1	9.38 ± 0.12	0.579 ± 0.076	1.74 ± 0.73
Series 2	9.22 ± 0.09	0.646 ± 0.031	2.57 ± 0.59
Series 3	9.18 ± 0.12	0.599 ± 0.045	2.37 ± 0.70
Series 4	9.09 ± 0.15	0.548 ± 0.057	2.22 ± 0.78
Series 5	8.57 ± 0.27	0.504 ± 0.071	2.62 ± 1.21

during charging than during discharging. It difficult to interpret the differences between the KiBaM parameters for discharging and charging within the context of the chemical battery processes. However, our experiments do show that when the KiBaM model is used, it appears **not justified** to just reverse the flow of the current and keep the parameters the same when we switch from discharging to charging.

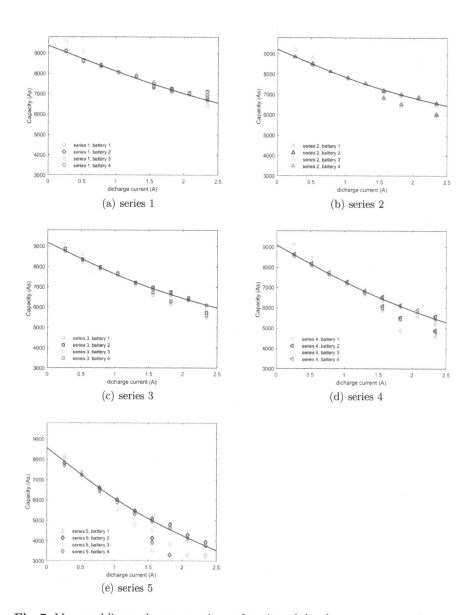

Fig. 7. Measured linear charge capacity as function of the charge currents and a non-linear least squares fit of the KiBaM for the 5 measurements series.

6 Discussion

The measurements do bring forward three main points for improvements of the KiBaM. First of all, the KiBaM does not take into account the efficiency, ϵ, of the battery. This may be corrected by multiplying the charge current with a factor ϵ in the KiBaM equations for charging, cf. Eq. (5). This correction will account all losses to the charging process. In an equivalent manner, all losses can be accounted to the discharging process by multiplying the discharge current with $1/\epsilon$. Since it is not possible to determine the efficiency of charging and discharging process separately either solution, or a mix, is valid.

The second improvement of the KiBaM is to use different parameters c and k for charging and discharging. In the analysis we see that the parameter c is clearly larger for discharging. One challenge in changing the parameter c, when switching from charging to discharging, and vice versa, is how to redistribute the charge in the battery over the available and bound charge wells. Since the KiBaM model actually is a first order approximation of the continuous diffusion model by Rakhmatov and Vrudhula [9,11], the most natural option seems to be to keep the height of the available charge well constant, and redistribute the charge accordingly. Due to the large confidence intervals for the parameter k, we cannot draw any strong conclusions on how this parameter should change between charging and discharging. However, changing this parameter can be done without any additional challenges arising.

The third improvement to the KiBaM model deals with the battery degradation. The experiments show a degradation of the capacity and efficiency of the battery, as well as a change in the parameter c, both for charging and discharging. It is not straightforward how to incorporate the degradation into the KiBaM. As stated earlier, the rate of the battery degradation highly depends on the discharge and charge rate and the depth of discharge. Since the degradation is a slow process, one can use the KiBaM with constant parameters when considering a time scale of a couple of charge-discharge cycles. However, when the battery is modeled over a longer period, the degradation must be taken into account. This might be done with a multi-modal KiBaM, in which one switches between different constant parameter sets as the battery degrades. In order to know when to change parameter sets, more experiments are needed, in which the batteries are discharged at various rates and to different depths of discharge.

Finally, our experiments clearly show the degradation of the battery over time. Note that in the experiments we applied a relatively heavy load to the battery, by discharging it fully in just one hour. Both this high discharge rate and discharging to a very low state of charge have a negative impact on the overall battery lifetime [3]. In most practical scenarios, a battery will not be discharged at such high rates nor to such a low state of charge. Commercial devices often discharge lithium-ion batteries only to 20% state of charge, in order to preserve the battery. In order to translate the measured degradation to a practical scenario further measurements and analysis are needed.

7   Conclusion

In this paper we presented the results of a first experimental analysis of the aging process for a set of lithium-ion 18650 cells. In the analysis we see that these batteries degrade in two phases. In the first phase, of approximately 140 cycles, the capacity drops slowly at a rate of 0.057% point with every cycle. In the second phase the degradation increases with a factor 7. Next to measuring the degradation of the battery, we also estimated the KiBaM parameters at several points during the degradation process, in order to learn how the parameters change as the battery ages. Furthermore, our experiments show that the KiBaM parameters are different for charging and discharging. The analysis resulted in a number of proposals on how to extend the KiBaM to take into account the results of the experiments. We do note, however, that more experimental work is needed, with different workload scenarios and battery types, to make more concrete proposals for such model extensions.

Acknowledgements. The work in this paper has been supported through the FP7 projects Sensation (318490) and e-balance (609132).

References

1. Ning, G., Popov, B.N.: Cycle life modeling of lithium-ion batteries. J. Electrochem. Soc. **151**(10), A1584–A1591 (2004)
2. Petricca, M., Shin, D., Bocca, A., Macii, A., Macii, E., Poncino, M.: Automated generation of battery aging models from datasheets. In: Proceedings of the 32nd IEEE International Conference on Computer Design (ICCD), pp. 483–488. IEEE (2014)
3. Wognsen, E.R., Haverkort, B.R., Jongerden, M., Hansen, R.R., Larsen, K.G.: A score function for optimizing the cycle-life of battery-powered embedded systems. In: Sankaranarayanan, S., Vicario, E. (eds.) FORMATS 2015. LNCS, vol. 9268, pp. 305–320. Springer, Cham (2015). doi:10.1007/978-3-319-22975-1_20
4. GomSpace: NanoPower Battery Datasheet, Lithium Ion 18650 cells for space flight products (2012) supplied by GomSpace (2015)
5. Berdichevsky, G., Kelty, K., Straubel, J., Toomre, E.: The Tesla roadster battery system. Tesla Motors **1**(5), 1–5 (2006). http://large.stanford.edu/publications/coal/references/docs/tesla.pdf
6. Jongerden, M.R., Haverkort, B.R.: Which battery model to use? IET Softw. **3**(6), 445–457 (2009)
7. Manwell, J.F., McGowan, J.G.: Lead acid battery storage model for hybrid energy systems. Sol. Energy **50**(5), 399–405 (1993)
8. Hermanns, H., Krčál, J., Nies, G.: Recharging probably keeps batteries alive. In: Berger, C., Mousavi, M.R. (eds.) CyPhy 2015. LNCS, vol. 9361, pp. 83–98. Springer, Cham (2015). doi:10.1007/978-3-319-25141-7_7
9. Jongerden, M.R.: Model-based energy analysis of battery powered systems. Ph.D. thesis, University of Twente, Enschede, December 2010
10. Wolfram Mathworld (2015). http://mathworld.wolfram.com/LambertW-Function.html. Accessed November 2016
11. Rakhmatov, D.N., Vrudhula, S.B.K.: An analytical high-level battery model for use in energy management of portable electronic systems. In: Proceedings of the 2001 IEEE/ACM International Conference on Computer-Aided Design, ICCAD 2001, 488–493. IEEE Press, Piscataway (2001)

Petri Nets and Performance Modelling

A Hybrid Multi-trajectory Simulation Algorithm for the Performance Evaluation of Stochastic Petri Nets

Armin Zimmermann[1(✉)], Thomas Hotz[2], and Andrés Canabal Lavista[1]

[1] Department of Computer Science and Automation,
Systems and Software Engineering Group, Technische Universität Ilmenau,
Ilmenau, Germany
armin.zimmermann@tu-ilmenau.de
[2] Institute for Mathematics, Stochastics Group,
Technische Universität Ilmenau, Ilmenau, Germany
thomas.hotz@tu-ilmenau.de

Abstract. Standard performance evaluation methods for discrete-state stochastic models such as Petri nets either generate the reachability graph followed by a numerical solution of equations, or use some variant of simulation. Both methods have characteristic advantages and disadvantages depending on the size of the reachability graph and type of performance measure. The paper proposes a hybrid performance evaluation algorithm for Stochastic Petri Nets that integrates elements of both methods. It automatically adapts its behavior depending on the available size of main memory and number of model states. As such, the algorithm unifies simulation and numerical analysis in a joint framework. It is proved to result in an unbiased estimator whose variance tends to zero with increasing simulation time; furthermore, its applicability is demonstrated through case studies.

Keywords: Stochastic Petri nets · Multi-trajectory simulation · Hybrid numerical analysis/simulation method

1 Introduction

Model-based systems engineering is an important tool for complex system design, especially to predict non-functional properties in early design stages. This requires a model such as a stochastic Petri net as well as an efficient evaluation algorithm implemented in a user-friendly tool. This paper considers Stochastic Petri Nets (SPN, [2]). A variety of evaluation methods are known from the literature, all with certain individual advantages, but for a non-expert user it is not obvious which method should be used for his or her problem.

Perhaps the most significant classification of algorithms is the one between: (1) numerical analysis that explores and stores the full state space and thus suffers from large memory requirements that may exceed the given hardware;

© Springer International Publishing AG 2017
N. Bertrand and L. Bortolussi (Eds.): QEST 2017, LNCS 10503, pp. 107–122, 2017.
DOI: 10.1007/978-3-319-66335-7_7

as well as (2) simulation that stores and follows just one state and trajectory of the system and thus may generate samples of interest only rarely for certain performance measures. For the latter, there is no restriction on large state spaces thus, but some simulation problems will take unacceptable computation time.

The idea behind the algorithm presented in this paper is to find a hybrid mix of both methods: working similar to a numerical analysis as long as the memory is sufficient, but not storing all states if a certain maximum is reached and thus being able to handle any size like a simulation. We introduce a new algorithm that follows many but not all simulation trajectories, and stores them internally as particles with a certain weight. Different settings of maximum particle numbers lead to either an (adapted) standard numerical analysis or a standard simulation. This allows to explore new trade-offs between considering all or a single state in contrast to the two existing standard methods, which may be seen then as the extreme cases of the proposed algorithm. Moreover, the algorithm automatically adapts its behavior depending on whether the size of the underlying state space fits into the main memory. It thus combines the advantages of simulation and numerical analysis without a-priori in-depth knowledge of the modeler. However, the aim is not an algorithm that is faster than the existing approaches, but a method that integrates their behavior and thus avoids the decision which algorithm to use. The algorithm has been implemented as an extension of TimeNET [19]. Experiments show that it is competitive both in run time and accuracy in comparison to simulation and numerical analysis.

Considering a mathematical framework which allows treating simulation, numerical analysis as well as the proposed multi-trajectory algorithm, unbiasedness and convergence of the resulting estimator are proved. As a side effect, the possibility of configuring our algorithm with one simple numerical parameter to behave either like a simulation or a numerical analysis can be seen as a step towards a unified understanding of the two main methods in performance evaluation.

We use the name *multi-trajectory simulation* here as there are multiple trajectories of the same system model evaluated concurrently (and not multiple particles of one system as done in multi-particle simulation, for instance). This name has been coined in previous work which keeps several possible trajectories of a combat simulation [7]. However, the cited approach does not cover performance evaluation in a rigorous mathematical way. Approximation and discretization are used to decrease the number of trajectories (or particles). Heuristics merge particles that are similar, and delete trajectories viewed as being less significant.

In the literature, the term "hybrid simulation" has been used both for simulation of discrete/continuous state models as well as for methods that combine analytical and simulative methods. In the latter related work, parts of the model are evaluated by numerical analysis, and the local results are then fed into a simulation algorithm [3,14] using a decomposition / aggregation approach quite different from our algorithm. A related method working with several internal states and trajectories (coined proxels) has been proposed in [10]. It aims at transient evaluation of non-Markovian Petri net models and uses a time discretization.

Weighted ensemble simulation [9] considers simulation of trajectories that rarely pass from one attractor to another in a state space, and is applied to system models in natural sciences. The approach is used to manage uncertainty in input data as well as probabilistic model evolutions common in weather forecasting or robotics motion planning. Importance splitting techniques in rare-event simulation [11] are another area in which multiple trajectories are being considered, which lead to the idea of the algorithm presented here. The RESTART algorithm [17], for instance, uses a depth-first-like algorithm to search for promising paths towards the state(s) of interest, and by discarding others that are assumed to be ineffective.

The paper is structured as follows: some terminology for the later explanation is introduced in Sect. 2, covering stochastic Petri nets and their quantitative evaluation. Section 3 introduces the multi-trajectory method for the performance evaluation of stochastic Petri nets. Subsequently, convergence results for the estimator are proved in Sect. 4, after which Sect. 5 reports numerical results obtained for a series of application examples, that were analyzed with an implementation in the TimeNET tool. Finally, conclusions and future work are pointed out.

2   Performance Evaluation of Stochastic Petri Nets

Stochastic Petri nets can be defined as $\mathsf{SPN} = (P, \mathcal{T}, \mathbf{Pre}, \mathbf{Post}, \lambda, \boldsymbol{m}_0, RV)$. We denote by P the (finite) set of places (i.e., state variables, denoted by circles), which may contain tokens. Each marking $\boldsymbol{m}$ of the Petri net is a vector of non-negative, integer numbers of tokens for each place $\boldsymbol{m} \in \mathbb{N}^{|P|}$. The initial state of the system is given by $\boldsymbol{m}_0$.

$\mathcal{T}$ specifies the (finite) set of transitions (depicted as rectangles). $\mathbf{Pre}$ describes the multiplicities of the input arcs connecting places to transitions $\mathbf{Pre} : P \times \mathcal{T} \to \mathbb{N}$. Similarly, output arcs $\mathbf{Post}$ from transitions to places are defined with their cardinality. Each transition tr has a firing rate $\lambda(tr)$ of an underlying exponential distribution. Finally, the measure(s) of interest to be computed are given by reward variables RV.

The behavior of a Petri net is defined as follows: A transition tr is enabled in a marking $\boldsymbol{m}$ if there are enough tokens available in each of its input places, i.e., $\forall p \in P : \boldsymbol{m}(p) \geq \mathbf{Pre}(p, tr)$. Whenever a transition becomes newly enabled, a remaining firing time (RFT) is randomly drawn from its associated exponential firing time distribution. The RFTs of all enabled transitions decrease with identical speed until one of them reaches zero.

The fastest transition tr is fired, changing the current marking $\boldsymbol{m}$ to a new one $\boldsymbol{m}'$ denoted as $\boldsymbol{m} \xrightarrow{tr} \boldsymbol{m}'$. The new marking is derived by removing the necessary number of tokens from the input places and adding tokens to output places with $\forall p \in P : \boldsymbol{m}'(p) = \boldsymbol{m}(p) - \mathbf{Pre}(p, tr) + \mathbf{Post}(p, tr)$.

If there is a transition tr enabled in marking $\boldsymbol{m}$ and its firing leads to marking $\boldsymbol{m}'$, we say that $\boldsymbol{m}'$ is directly reachable from $\boldsymbol{m}$. The set of all directly or indirectly reachable states from $\boldsymbol{m}_0$ is the reachability set RS or state space of the model. We assume models with a finite state space where the initial state

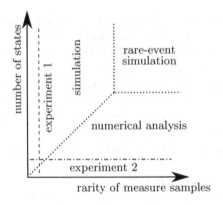

Fig. 1. Suggested application regions of evaluation methods

m_0 is directly or indirectly reachable from every other state such that the state space is irreducible.

Performance measures $rvar(t) \in RV$ are defined here as functions of the stochastic process at time t and given by a rate part $RateReward^{rvar}(m_t)$ returning the amount of reward gained in marking m per time unit, plus the impulse part $ImpulseReward^{rvar}(m_t, tr)$ [13] specifying the amount of reward associated to firing transition tr in marking m_t if the Petri net is in state m_t at time t.

The stochastic process defined by such a model is a continuous-time, irreducible Markov chain with finite state space and transitions isomorphic to the reachability graph of the Petri net model [1,5]. Its infinitesimal generator matrix Q with entries $q_{i,j}$ is given by the sum of all rates of exponential transitions tr for which $m_i \xrightarrow{tr} m_j$ (or zero if there is none). Diagonal entries of Q denote outflow rates set to $q_{j,j} = -\sum_{i \neq j} q_{j,i}$.

Stationary (steady-state) evaluation of performance measures is considered in this paper, for which there is a set of standard methods known (cf. [18]). Direct numerical analysis considers the full state space, solves for the invariant measure π of the underlying Markov chain via $\pi Q = 0, 1 = \sum_i \pi_i$, and derives the performance measure values simply from π. The alternative is simulation, estimating results by $\lim_{T \to \infty} \frac{1}{T} \int_0^T rvar(t) \, dt$. Numerical analysis is hard or impossible when the state space size becomes huge, while simulation runs into unacceptable execution times for models in which significant samples can be generated only rarely. Figure 1 depicts regions of performance evaluation problems and sketches suggested evaluation methods as well as the problem types covered by application examples in Sect. 5.

3   A Hybrid Multi-trajectory Simulation Algorithm

The algorithm proposed in this paper is intended to cover the areas of simulation and numerical analysis without a-priori knowledge about the problem region. The idea is the following: instead of the two extremes of either following just one trajectory in the state space as done by simulation versus considering all states at once as done in a numerical analysis, there should be a hybrid approach in between that stores and follows *some* trajectories. If possible, this should allow new trade-offs between available memory space, numerical accuracy and speed.

Our proposed multi-trajectory algorithm follows this scheme: we start with one simulation particle[1] for state m_0 at simulation time $t = 0$ with weight 1. The weight of each particle equals the probability that a simulation would have arrived at the corresponding state until the current simulation step, given the previous probabilistic decisions. It is thus similar to trajectory weights considered in some rare-event simulation methods (e.g., [16]) to keep track of the amount of splitting.

The algorithm maintains two sets (current *Particles* and next *Particles'*) with elements of the form $p = $ (marking m, particle weight w). In each step of the main simulation loop, the algorithm iterates over the current set of particles *Particles*. For each particle $p \in Particles$ stored, two treatments are possible:

- **Propagate**: the particle is simply followed like in a standard simulation by probabilistically choosing one of the subsequent states. The weight is kept constant.
- **Split**: all possible subsequent states are computed similarly to an iterative step in a numerical transient solution of a discrete-time Markov chain. The weight of the particle is distributed over all descendant particles resulting from a split by multiplying it with the enabled transition's firing probabilities.

All created particles are stored in the next particle set *Particles'*; if a particle with an identical state (marking) already exists, their weights are simply added (the particles are *merged*). For both cases, the simulation time is updated according to the average sojourn time in each particle, which can be computed from the delays of the enabled transitions.

For practical implementation reasons, the size of the particle sets needs to be bounded by a number of N particles to be stored. If the state space size of the model is larger than N, not every possible split will be executable. Obviously, the question of whether to use propagation or splitting influences the algorithm's performance. Some simple heuristics have been considered in our experiments so far [4]. The numerical results in Sect. 5 have been achieved with a heuristic that splits if one of the enabled transition has been fired less than average so far, if the weight of the particle is bigger than the average weight, or if the number of existing particles is less than $\frac{3}{4}N$ [4]. We see considerable room for improvements towards better heuristics in future work.

Algorithm 1 sketches the proposed program structure. It takes as input SPN, the stochastic Petri net model including performance measure definitions *rvar* as well as an initial state m_0, and the maximum number of particles N. Its output is the estimated value of performance measure *rvar*.

---

[1] We use the term particle here to denote one simulation state as part of a larger set, not in the sense of a multi-particle simulation in physics.

MULTITRAJECTORY (SPN, N)

(* initializations *)
State($rvar$) := 0
$SimTime$:= $Reward$:= 0
$Particles$:= $\{(\boldsymbol{m}_0, 1)\}$; $Particles'$:= {}

repeat (* main simulation loop *)
    while $|Particles| > 0$ **do**
        Select any $p \in Particles$; $Particles$:= $Particles \setminus \{p\}$
        w := $p.weight$; m := $p.marking$
        $\mathcal{T}_{ena}$:= set of all transitions enabled in marking $\boldsymbol{m}$

        (* rate reward and sojourn time *)
        $WeightSum$:= $\sum_{tr \in \mathcal{T}_{ena}} \lambda(tr)$
        $Reward$ += $\frac{w}{WeightSum} RateReward^{rvar}(m)$
        $SimTime$ += $\frac{w}{WeightSum}$

        (* decision heuristic, here: only split if enough space *)
        if $|Particles| + |Particles'| + |\mathcal{T}_{ena}| > N$ **then** (* don't split *)
            Select any $tr \in \mathcal{T}_{ena}$ randomly
            $\mathcal{T}_{ena}$:= $\{tr\}$; $WeightSum$:= $\lambda(tr)$

        (* fire transition(s) *)
        for $\forall tr \in \mathcal{T}_{ena}$ **do**
            m' := $FireTransition(m, tr)$
            if $m' \notin Particles'$ **then** (* add new particle *)
                $Particles'$:= $Particles' \cup \{(\boldsymbol{m'}, 0)\}$
            (* merge particle weight *)
            $Particles'(\boldsymbol{m'}).weight$ += $\frac{w}{WeightSum} \lambda(tr)$
            (* impulse reward for fired transition *)
            $Reward$ += $\frac{w}{WeightSum} ImpulseReward^{rvar}(m, tr)$

    (* full particle set finished *)
    $Particles$:= $Particles'$; $Particles'$:= {}
    Collect measure sample with value $\frac{Reward}{SimTime}$

until simulation stop criterion is reached (confidence interval estimation)
return average of samples

Algorithm 1: Multi-trajectory simulation algorithm

The behavior of the algorithm depends significantly on the number of particles and size of state space: For $N = 1$, only one particle will be considered, for which the weight will stay at 1. There will never be a split and the algorithm behaves like a normal simulation with one single trajectory. Numerically exact firing probabilities of all enabled transitions are computed instead and used to randomly select the next state.

If the algorithm is started with $|RS| < N$, particles for all markings $m \in RS$ of the Petri net can be stored, and the algorithm works similar to a numerical algorithm variant as pointed out in Sect. 4. Settings of practical interest are thus $1 \ll |RS| < N$ and $|RS| > N$, which will be considered in the examples and numerical results of Sect. 5.

Algorithm 1 assumes only one performance measure *rvar* for simplicity, but any number can be computed concurrently. Intermediate variable *Reward* stores accumulated reward [13], i.e., the integral over the reward function.

During the evaluation, *SimTime* keeps the current simulation time (starting at 0) passed by all particles together on average; as such, marking holding (sojourn) times are not simply added to it, but weighted by the particle weights and computed for each particle using the enabled transitions' firing rates. The set *Particles'* is organized to be rapidly searchable for a marking m to be part of it; for simplicity in the algorithm we denoted finding a marking by $m \in Particles \longleftrightarrow \exists (m,.) \in Particles$ and the found particle $Particles(m) = (m, w)$ if such a particle exists.

The decision from when on samples should be collected to avoid an initial transient bias, and how long samples have to be collected until a predefined accuracy setting given by some confidence interval and relative error is reached, is based on standard methods from the literature [12].

4   Unbiasedness and Convergence

The underlying stochastic process [5] of a well-specified stochastic Petri net model (c.f. Sect. 2) is a time-homogeneous and ergodic continuous-time Markov chain $MC = \{X_t\}_{t \in T}$ [15] with discrete and finite state space $\mathcal{S}$, $d = |\mathcal{S}| < \infty$. The mathematical analysis will be carried out in a discrete-time setting; this is justified since a well-known alternative to the standard way of deriving the steady-state numerical solution for the Petri net performance measures via $\pi Q = 0, 1 = \sum_{i=1...d} \pi_i$ is to embed a discrete-time Markov chain *EMC* upon each state transition of the original *MC*. While both state spaces $\mathcal{S}$ will be identical, the one-step state transition probability matrix $P = \{p_{i,j}\}$ of the *EMC* can be derived by

$$p_{i,j} = \begin{cases} \frac{q_{i,j}}{\sum_{k \neq i} q_{i,k}} & \text{if } i \neq j \\ 0 & \text{else} \end{cases}$$

and the steady-state solution of the *EMC* μ will be represented as a vector in $[0, 1]^d$, given by $\mu^\top P = \mu^\top, 1 = \sum_{i=1...d} \mu_i$. The solution of the original process can be computed by taking into account the state sojourn (holding) times $h_i = \frac{1}{\sum_{k \neq i} q_{i,k}} = -\frac{1}{q_{i,i}}$ and normalization, leading to $\pi_i = \frac{h_i \mu_i}{\sum_k h_k \mu_k}$.

In the following we consider this time-discrete, irreducible, aperiodic Markov chain $EMC = (X_t)_{t \in \mathbb{N}_0}$ on a finite state space $\mathcal{S} = \{1, \dots, d\}$ with time-invariant transition matrix $P \in [0, 1]^d$ that results from such an embedding.

The essential aim is then to calculate $\int f \, d\mu$ for a given $f : \mathcal{S} \to \mathbf{R}$ where f corresponds to the performance measure *rvar* of the model and is expressed as a vector in $\mathbf{R}^d$ such that the integral is given by $\boldsymbol{\mu}^\top f$. If $\boldsymbol{\mu}$ can not be computed explicitly, one may resort to simulation, using that, by the ergodic theorem if $X_0 \sim \boldsymbol{\mu}$,[2]

$$F_T = \frac{1}{T} \sum_{t=1}^{T} f(X_t) \to \int f \, d\mu \quad \text{a.s.} \tag{1}$$

when the simulation time $T \to \infty$; note that this implies convergence in quadratic mean as well. Since $\sum_{t=1}^{T} f(X_t)$ can be updated while the Markov chain is simulated, the computation of F_T requires only constant memory.

The analysis of the (multi-particle) simulation needs a mathematical framework which allows to bridge numerical analysis and simulation. For this, recall that an alternative to solving $\boldsymbol{\mu}^\top P = \boldsymbol{\mu}^\top$ in the numerical analysis is to start with some $\boldsymbol{\mu}_0 \in [0,1]^d$ whose entries some to 1, and iteratively compute $\boldsymbol{\mu}_{i+1}^\top = \boldsymbol{\mu}_i^\top P$ until convergence (which is guaranteed by the Perron-Frobenius theorem). We will now describe standard simulation and multi-trajectory algorithm in a similar vector-matrix calculus for which we introduce the following notation: $e_i \in \mathbf{R}^d$, $i = 1, \ldots, d$ will denote the canonical basis vectors of $\mathbf{R}^d$ such that $\delta : \mathcal{S} \to \{e_i : i = 1, \ldots, d\}$, $x \mapsto \delta(x) = e_x$ is a bijective mapping which is interpreted as mapping a state to the Dirac measure on $\mathcal{S}$ at that state. Furthermore, we are going to denote the rows of the matrix P by $\boldsymbol{p}_i \in \mathbf{R}^d$, $i = 1, \ldots, d$ – and analogously for other matrices – such that $P = \sum_{i=1}^{d} e_i \boldsymbol{p}_i^\top$.

Now consider independent random variables $Z_{t,i} \sim \boldsymbol{p}_i$, $t \in \mathbf{N}$, $i = 1, \ldots, d$ and form the matrices $R_t = \sum_{i=1}^{d} e_i \delta(Z_{t,i})^\top$. Then, if $X_0 \sim \boldsymbol{\mu}$ is independent of the $Z_{t,i}$, setting $X_t = \delta^{-1}\left(\delta(X_{t-1})^\top R_t\right)$ for $t \in \mathbf{N}$ realizes the stationary Markov chain, i.e. $\delta(X_t)^\top = \delta(X_0)^\top \prod_{s=1}^{t} R_s$ encodes the evolution of the chain through Dirac measures (where here and in the following matrix products expand from left to right). This is identical to a standard simulation that never splits, and serves as the basis for the mathematical formulation of our algorithm.

The splitting can then be modelled as follows: let $D_{t,i} \in \{0,1\}$, $t \in \mathbf{N}$, $i = 1, \ldots, d$ be random decision variables that specify if particle i will be split at time step t or not. We assume them to be random and depending only on the past, i.e., the subsequent proofs will apply to all splitting heuristics which are based only on the history of the process. $D_{t,i}$ may thus depend only on X_0 and R_s for $s < t$; formally, we can form the filtration[3] $\mathcal{F}_0 = \sigma(X_0)$, $\mathcal{F}_t = \sigma(X_0, R_s : s \le t)$ and require that $D_{t,i}$ is $\mathcal{F}_{t-1}$-measurable (i.e. predictable given the past) for

---

[2] Assuming that the initial transient phase of the simulation has passed at $t = 0$.

[3] A filtration is a growing sequence of σ-algebras which may be interpreted as containing the information up to the corresponding time point.

$t \in \mathbf{N}$. Then, based on the rows $r_{t,i} = \delta(Z_{t,i}) \in \mathbf{R}^d$ of $R_t = \sum_{i=1}^d e_i r_{t,i}^\top$ we define random vectors $s_{t,i} \in \mathbf{R}^d$ by[4]

$$
s_{t,i} = \begin{cases} \mathbf{E}\, r_{t,i} = \boldsymbol{p}_i\,, & \text{if } D_{t,i} = 1 \\ r_{t,i}\,, & \text{if } D_{t,i} = 0 \end{cases} \tag{2}
$$

for $t \in \mathbf{N}$, $i = 1, \ldots, d$. Setting $S_t = \sum_{i=1}^d e_i s_{t,i}^\top$ as well as $\rho_0 = \delta(X_0)$ and $\rho_t^\top = \rho_{t-1}^\top S_t = \delta(X_0)^\top \prod_{s=1}^t S_s$ for $t \in \mathbf{N}$, we obtain a sequence of random measures on $\mathcal{S}$, like the deterministic measures $\boldsymbol{\mu}_t$ in the numerical analysis, or the random measures $\delta(X_t)$ encoding the Markov chain in the simulation approach. In fact, S_t contains the one-step probabilities for all particles stored in ρ with their weights, including any splits. Only states with $\rho_i > 0$ are actually stored. If all states should be split, $S = \boldsymbol{P}$ as in the numerical analysis above; if no states should be split, $S = R$ as in the simulation approach. We are now interested in the convergence of $\frac{1}{T} \sum_{t=1}^T \int f \, \mathrm{d}\rho_t$ which should be compared with (1) where we have replaced $f(X_t) = \int f \, \mathrm{d}\delta(X_t)$ by $\int f \, \mathrm{d}\rho_t = \rho_t^\top f$.

Note that in case $S_t = \boldsymbol{P}$ for all $t \in \mathbf{N}$, we have convergence to the equilibrium: $\rho_t \to \boldsymbol{\mu}$ for $t \to \infty$, such that $\int f \, \mathrm{d}\rho_t \to \int f \, \mathrm{d}\boldsymbol{\mu}$ and thus also for its Cesàro means one has $\frac{1}{T} \sum_{t=1}^T \int f \, \mathrm{d}\rho_t \to \int f \, \mathrm{d}\boldsymbol{\mu}$.

The intuition regarding the performance of the multi-trajectory simulation is: for the rows $s_{t,i}$ of S_t we have either $s_{t,i} = r_{t,i}$ if we do not split or $s_{t,i} = \boldsymbol{p}_i$ if we do. The latter is deterministic and thus generates no variance, so randomly deciding whether to split or not will lead to a smaller variance than never splitting at all; since $\boldsymbol{p}_i = \mathbf{E}\, r_{t,i}$ and the decision to split depends only on the past, $\mathbf{E}\, s_{t,i} = \boldsymbol{p}_i$ as well.

Since the sequence of measures ρ_t will not be stationary in general, we will not consider almost sure convergence but we will prove $\frac{1}{T} \sum_{t=1}^T \int f \, \mathrm{d}\rho_t \to \int f \, \mathrm{d}\boldsymbol{\mu}$ in quadratic mean by showing the variance to be smaller when splitting. For this, the following lemma which is straightforward to prove provides the essential estimate.

Lemma 1. *Let $u \in \mathbf{R}^d$ be a deterministic vector, let $v, a_i \in \mathbf{R}^d$, $i = 1, \ldots, d$ be a set of independent, square-integrable vectors, for $i = 1, \ldots, d$ set either $b_i = 0 \in \mathbf{R}^d$ or $b_i = a_i$, and form the matrices $A = \sum_{i=1}^d e_i a_i^\top \in \mathbf{R}^{d \times d}$ and $B = \sum_{i=1}^d e_i b_i^\top$, respectively. Furthermore assume $\mathbf{E}\, A = 0$.*
Then $\mathbf{E}\, u^\top A v = \mathbf{E}\, u^\top B v = 0$ and $\mathbf{Var}(u^\top A v) = \mathbf{E}(u^\top A v)^2 \geq \mathbf{E}(u^\top B v)^2 = \mathbf{Var}(u^\top B v)$.

We are now in the position to state and prove our main result.

Proposition 2. *Let $\boldsymbol{P}$ be the (time-invariant) transition matrix of a time-discrete, irreducible, aperiodic Markov chain on a finite state space $\mathcal{S} = \{1, \ldots, d\}$ with steady-state solution vector $\boldsymbol{\mu}$ and fix $f : \mathcal{S} \to \mathbf{R}$. Using the notation*

---

[4] $\mathbf{E}$ and $\mathbf{Var}$ denote expected value and variance of the subsequent terms.

introduced above, let $X_0 \sim \mu$, $Z_{t,i} \sim p_i$, $t \in \mathbf{N}$, $i = 1, \ldots, d$ be independent random variables, $R_t = \sum_{i=1}^{d} e_i \delta(Z_{t,i})^\top$, $X_t = \delta^{-1}\big(\delta(X_{t-1})^\top R_t\big)$ such that X_t, $t \in \mathbf{N}_0$ realizes the Markov chain, $\mathcal{F}_0 = \sigma(X_0)$, $\mathcal{F}_t = \sigma(X_0, R_s : s \le t)$, $D_{t,i} \in \{0,1\}$ $\mathcal{F}_{t-1}$-measurable random variables for $t \in \mathbf{N}$, $i = 1, \ldots, d$, and let $S_t = \sum_{i=1}^{d} e_i s_{t,i}^\top$, $t \in \mathbf{N}$ be defined via (2) as well as $\rho_t^\top = \delta(X_0)^\top \prod_{s=1}^{t} S_s$ for $t \in \mathbf{N}$.

Then, considering the result $F_T = \frac{1}{T} \sum_{t=1}^{T} f(X_t)$ after $T \in \mathbf{N}$ steps when simulating the Markov chain, and the result $G_T = \frac{1}{T} \sum_{t=1}^{T} \rho_t^\top f$ of the multi-trajectory simulation, we have $\mathbf{E}\, F_T = \mathbf{E}\, G_T = \boldsymbol{\mu}^\top f = \int f \, d\boldsymbol{\mu} = \mathbf{E}\, f(X_0)$ and $\mathbf{Var}\, G_T \le \mathbf{Var}\, F_T$ for all $T \in \mathbf{N}$. Thus, from $\mathbf{Var}\, F_T \to 0$ for $T \to \infty$ one concludes $G_T \to \int f \, d\boldsymbol{\mu}$ in quadratic mean, i.e. $\mathbf{E}(G_T - \boldsymbol{\mu}^\top f)^2 \to 0$ for $T \to \infty$.

Proof. By stationarity of the Markov chain, we have $\mathbf{E}\, F_T = \frac{1}{T} \sum_{t=1}^{T} \mathbf{E}\, f(X_t) = \mathbf{E}\, f(X_0)$. Also, by independence, $\mathbf{E}(r_{t,i} \,|\, \mathcal{F}_{t-1}) = \mathbf{E}(r_{t,i}) = p_i$, so, due to the $\mathcal{F}_{t-1}$-measurability of $D_{t,i}$,

$$\mathbf{E}(s_{t,i} \,|\, \mathcal{F}_{t-1}) = \mathbb{1}\{D_{t,i} = 1\}\, \mathbf{E}(r_{t,i}) + \mathbb{1}\{D_{t,i} = 0\}\, \mathbf{E}(r_{t,i} \,|\, \mathcal{F}_{t-1}) = p_i,$$

thus $\mathbf{E}(S_t \,|\, \mathcal{F}_{t-1}) = \sum_{i=1}^{d} e_i \, \mathbf{E}(s_{t,i} \,|\, \mathcal{F}_{t-1}) = P$ which implies

$$
\begin{aligned}
\mathbf{E}\, \rho_t^\top &= \mathbf{E}\Big(\delta(X_0)^\top \prod_{s=1}^{t} S_s\Big) = \mathbf{E}\,\mathbf{E}\Big(\delta(X_0)^\top \Big(\prod_{s=1}^{t-1} S_s\Big) S_t \,\big|\, \mathcal{F}_{t-1}\Big) \\
&= \mathbf{E}\Big(\delta(X_0)^\top \Big(\prod_{s=1}^{t-1} S_s\Big) \mathbf{E}(S_t \,|\, \mathcal{F}_{t-1})\Big) = \mathbf{E}\,\delta(X_0)^\top \Big(\prod_{s=1}^{t-1} S_s\Big) P = (\mathbf{E}\, \rho_{t-1})^\top P
\end{aligned}
$$

for all $t \in \mathbf{N}$ so that $\mathbf{E}\, \delta(X_0) = \boldsymbol{\mu}$. Induction gives $\mathbf{E}\, \rho_t^\top = \boldsymbol{\mu}^\top P^t = \boldsymbol{\mu}^\top$ and thus $\mathbf{E}\, G_T = \frac{1}{T} \sum_{t=1}^{T} \mathbf{E}\, \rho_t^\top f = \boldsymbol{\mu}^\top f$ which proves that the expected value of our proposed algorithm results will be unbiased.

To derive the variance estimate, we will substitute S_k for R_k in F_T one after the other for $k = 1, \ldots, T$, i.e. we set $H_0 = T\, F_T$ as well as

$$H_k = \sum_{t=1}^{k-1} \Big(\delta(X_0)^\top \prod_{s=1}^{t} S_s\Big) f + \Big(\delta(X_0)^\top \prod_{s=1}^{k-1} S_s\Big) S_k \Big(\mathbf{I} + \sum_{t=k+1}^{T} \prod_{s=k+1}^{t} R_s\Big) f$$

for $k = 1, \ldots, T$ with the last sum being empty for $k = T$ such that $H_T = TG_T$, and prove $\mathbf{Var}\, H_k \le \mathbf{Var}\, H_{k-1}$ for $k = 1, \ldots, T$. For the latter we are of course going to use Lemma 1 applied for the conditional distribution given $\mathcal{F}_{k-1}$.

Observing that (similarly as above) for $k = 1, \ldots, T$, $\mathbf{E}(H_k \,|\, \mathcal{F}_{k-1})$ equals

$$\sum_{t=1}^{k-1} \left(\delta(X_0)^\top \prod_{s=1}^{t} S_s \right) f + \left(\delta(X_0)^\top \prod_{s=1}^{k-1} S_s \right) \mathbf{E}(S_k \,|\, \mathcal{F}_{k-1}) \, \mathbf{E} \left(\mathbf{I} + \sum_{t=k+1}^{T} \prod_{s=k+1}^{t} R_s \right) f$$

$$= \sum_{t=1}^{k-1} \left(\delta(X_0)^\top \prod_{s=1}^{t} S_s \right) f + \left(\delta(X_0)^\top \prod_{s=1}^{k-1} S_s \right) P \left(\mathbf{I} + \sum_{t=k+1}^{T} P^{t-k} \right) f$$

$$= \sum_{t=1}^{k-1} \left(\delta(X_0)^\top \prod_{s=1}^{t} S_s \right) f + \left(\delta(X_0)^\top \prod_{s=1}^{k-1} S_s \right) \mathbf{E}(R_k \,|\, \mathcal{F}_{k-1}) \, \mathbf{E} \left(\mathbf{I} + \sum_{t=k+1}^{T} \prod_{s=k+1}^{t} R_s \right) f$$

$$= \mathbf{E}(H_{k-1} \,|\, \mathcal{F}_{k-1})$$

we may set $u = \delta(X_0)^\top \prod_{s=1}^{t} S_s$, $A = R_k - P$, $B = S_k - P$, and $v = \left(\mathbf{I} + \sum_{t=k+1}^{T} \prod_{s=k+1}^{t} R_s \right) f$ such that A and B have zero (conditional) mean. Lemma 1 (applied conditionally given $\mathcal{F}_{k-1}$) thus gives the desired result $\mathbf{Var}(H_k \,|\, F_{k-1}) \leq \mathbf{Var}(H_{k-1} \,|\, \mathcal{F}_{k-1})$ and therefore also $\mathbf{Var}(H_k) \leq \mathbf{Var}(H_{k-1})$.

Note that Proposition 2 indeed says that both F_T and G_T are unbiased estimators of $\int f \, d\boldsymbol{\mu}$ with the variance of G_T never being larger than that of F_T; however, from the proofs one may expect the variance of G_T to be considerably smaller than that of F_T if one often splits particles.

In particular, symmetric (asymptotic) confidence intervals around F_T derived from the central limit theorem are also valid if translated to be symmetric around G_T but will be quite conservative. As the sequence ρ_t will not be stationary, deriving sharp (asymptotic) confidence intervals around G_T is non-trivial and left for further research, as is a proof of almost sure convergence.

5   Application Examples and Numerical Results

The algorithm introduced in this paper has been implemented in the software tool TimeNET [6, 19] in a Master's thesis [4]. It was based on the existing simulation module for eDSPNs [8], which follows a master-slave architecture with several (6 as a standard) simulation slave processes running concurrently on the host machine. This improves the statistical independence of samples in addition to the used batch means method, and allows to exploit multi-core CPUs. All Petri net analysis parts and computation of statistical measures from the raw samples are reused, including initial transient detection as well as variance analysis for confidence interval evaluation for the stopping criterion. The latter was not changed for the current experiments; a new variance estimation method that better matches our algorithm will have to be developed in the future. Significant changes and additions to the program had to be done only to store the particles.

The goal of this section is to show exemplarily that the proposed algorithm works well both for models that can be analyzed by standard simulation or by numerical analysis methods as depicted in Fig. 1, i.e., the left and lower regions of the picture. Two example models are evaluated for this reason: the first one (dotted red, left side of the figure) showing that for models without rare events and differing state space size, where simulation is usually faster than a numerical

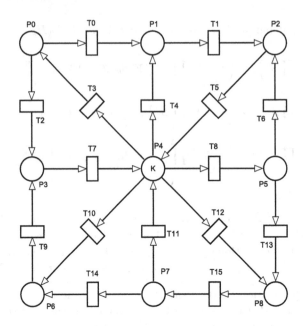

Fig. 2. Example 1: SPN with varying state space size for K

analysis (at least when the state space size increases), the proposed algorithm has the same advantages as a regular simulation: the run time does not increase substantially with the state space size. The second example (chain-dotted blue, bottom of Fig. 1) explores the horizontal dimension: a model with fixed state space size is used, for which one parameter influences the "rareness" of events of interest. For the chosen case it will be shown that our proposed algorithm has the same advantage as numerical analysis: as long as the state space size is manageable (fits into the main memory), the execution time does not increase for harder problems which lead to very long simulation run times. We explicitly do not cover the area marked "rare-event simulation" in Fig. 1, i.e., models which are problematic both for numerical analysis as well as standard simulation. Our future goal is to extend the algorithm with proper heuristics to let it also solve such large rare-event models efficiently, but this is outside the scope of this paper.

All numerical results have been computed on a standard laptop computer under Windows 7 Enterprise 64 Bit on a 2013 Intel Core i7-4600U CPU running at 2.1 GHz, with 12 GByte RAM, 4MB Cache, and an SSD hard disk. This processor has two cores and four logical processors, which are all used by the simulation slave processes, while the numerical analysis (mainly spending its time during the successive over-relaxation (SOR) iterative solution of global balance equations) can only run on one core. All measured times are run times as experienced by the tool user, not pure CPU times.

Figure 2 shows an abridged version of a model introduced in [4] to check the applicability of the algorithm to an arbitrary, not too simplistic stochastic Petri

Table 1. Numerical results and run times in seconds for the first example

K	States	Numeric analysis		Simulation			Multi-trajectory		
		Result	Time	Result	Error	Time	Result	Error	Time
1	9	0.22951	1	0.230	0.4%	1	0.229	0.0%	1
3	165	0.34726	1	0.354	2.1%	1	0.347	0.0%	1
5	1 287	0.36208	1	0.361	0.2%	1	0.363	0.1%	2
7	6435	0.36403	1	0.365	0.4%	1	0.364	0.1%	2
10	43 758	0.36431	7	0.367	0.8%	1	0.360	1.3%	9
12	125970	0.36433	26	0.366	0.5%	1	0.361	1.0%	9
15	490 314	0.36433	161	0.367	0.7%	1	0.371	1.9%	9
17	1081575	0.36433	483	0.367	0.8%	1	0.372	2.2%	14
20	3 108 105	0.36433	2028	0.368	0.9%	1	0.365	0.3%	9
25	13 884 156			0.369		1	0.359		9
30	48 903 492			0.367		1	0.351		11

net. Transitions are set to either single or infinite server semantics with equal probability. The firing delays of the model have been randomly selected in the ranges $[1 \ldots 9]$ and $[100 \ldots 900]$ to avoid simplifying symmetries; only one parametrized model instance is considered[5]. The number of states is controlled by parameter K, the initial number of tokens in place P4. We assume a performance measure of interest as the probability of having at least one token in place P1, being expressed by P{#P1>0} in TimeNET 4.3 syntax.

Precision has been increased for numerical analysis in all experiments to 10^{-15} because of some small result values. For normal and multi-trajectory simulation the following settings are used: confidence level 95% and relative error 5%.

The results for the first example are shown in Table 1, all run times are given in seconds. The model has been chosen such that the state space size grows moderately with increasing number of tokens K. Run time of numerical analysis starts to grow considerably when the size exceeds 20.000 states. Several millions of states can be handled until the memory is exceeded for $K = 25$, but the run time rapidly becomes unacceptable compared to the simulation times. Standard simulation is very fast and independent of the state space size. Its results are in the desired range of accuracy. The proposed multi-trajectory algorithm is very fast and almost exact until $K = 7$, which are the cases where the state space fully fits into the chosen number of trajectories set to 10^4. After that, the speed

---

[5] This and the later second experiment model file as well as more numerical results are available at www.tu-ilmenau.de/sse/timenet/data-for-the-multi-trajectory-algorithm to support reproducibility. TimeNET can be obtained from timenet.tu-ilmenau.de.

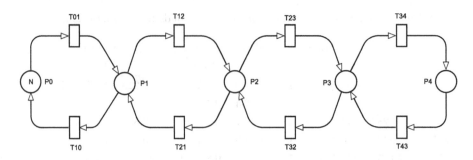

Fig. 3. Example 2: SPN with increasingly hard simulation by adapting transition delays

Table 2. Numerical results and run times in seconds for Example 2

ϵ	Numeric analysis		Simulation			Multi-trajectory		
	Result	Time	Result	Error	Time	Result	Error	Time
1	8.621E−01	1.3	9.18E−01	6.5%	1	9.37E−01	8.7%	2
10	1.000E−04	1.3	9.95E−05	0.5%	2	1.00E−04	0.0%	3
20	6.250E−06	2.2	6.43E−06	2.9%	26	6.25E−06	0.0%	3
50	1.600E−07	1.4	1.54E−07	3.8%	296	1.60E−07	0.0%	2
70	4.165E−08	1.3	3.96E−08	4.9%	838	4.16E−08	0.0%	3
100	1.000E−08	1.2	1.01E−08	0.8%	2418	1.00E−08	0.0%	3
120	4.823E−09	1.4				4.82E−09	0.0%	2
150	2.603E−09	1.2				2.60E−09	0.0%	3
200	6.250E−10	1.2				6.25E−10	0.0%	2
500	1.600E−11	1.2				1.60E−11	0.0%	2
1000	1.000E−12	1.2				1.00E−12	0.0%	2

is slower, but does not depend on the size. Accuracy is also in the desired range of 5% relative error (Table 2).

The measurements support our claim that the run time of the proposed algorithm does not increase substantially for a growing state space.

Figure 3 shows a second example to check how the algorithm behaves in cases where a standard simulation becomes infeasible. It is based on [4,20], and the measure of interest is the probability of having at least one token in place P4. Computing this value by simulation is made increasingly hard by setting the mean delays of the lower exponential transitions (T10, T21, T32, T43) to 1 and the upper transitions' delays to $\epsilon = 1 \dots 1000$. For greater values of ϵ this is a typical rare-event setting. The number of tokens has been chosen as $N = 25$, resulting in a state space size of 23 751.

Experimental results are shown in Fig. 2 with 10^5 maximum particles for the proposed algorithm. Both numerical analysis and multi-trajectory algorithm have no problems in evaluating even hard problems, they are both very fast and the accuracy of the proposed algorithm is perfect except for the case $\epsilon = 1$ which may be due to the stopping criterion. Standard simulation has the expected accuracy, but its run time grows significantly with ϵ. The experiment thus supports the claim that the new algorithm does not suffer from the rare-event simulation run length as long as the state space of the model is not too large, thus inheriting the advantage of numerical analysis for this case.

Other experiments showed that the proposed algorithm can be slower than standard simulation in some cases with medium-size state space. This happens when a simulation is very fast while the multi-trajectory algorithm spends unnecessary time in the particle computation. In the future we will explore algorithm variants that adapt the maximum number of particles N during run time to overcome this effect, as starting with a small N and gradually increasing it over time ought to solve this issue.

6 Conclusion

A new algorithm for the performance evaluation of Markovian stochastic Petri nets has been proposed in the paper. It uses elements of simulation as well as numerical analysis and its convergence has been proved in a unified mathematical framework. The algorithm can be applied to models for which one up to now had to choose a priori which of the standard methods to apply. The two examples show that the algorithm incorporates the advantages of simulation and numerical analysis for models that are either small enough to be handled analytically, or for which the performance measures and event generation is simple enough to result in a fast simulation.

In the future, we plan to use methods from automatic rare-event Petri net simulation to develop splitting heuristics of the algorithm. The goal is to make the algorithm useful for automated rare-event simulation as well, as a step towards a single method that should be applicable and automatically adapting to the two main cases of model evaluation complexity. We will also investigate if the advantages of the algorithm can be transferred to non-Markovian models, and explore possible parallelization of the algorithm.

Acknowledgements. The authors would like to thank Florian Kelma and Thomas Böhme, both from the Institute for Mathematics, Technische Universität Ilmenau, for fruitful discussions on the mathematical treatment.

References

1. Ajmone Marsan, M., Balbo, G., Conte, G., Donatelli, S., Franceschinis, G.: Modelling with Generalized Stochastic Petri Nets. Series in Parallel Computing. Wiley, Hoboken (1995)
2. Marsan, M.A.: Stochastic Petri nets: an elementary introduction. In: Rozenberg, G. (ed.) APN 1988. LNCS, vol. 424, pp. 1–29. Springer, Heidelberg (1990). doi:10. 1007/3-540-52494-0_23

3. Buchholz, P.: A new approach combining simulation and randomization for the analysis of large continuous time Markov chains. ACM Trans. Model. Comput. Simul. **8**, 194–222 (1998)
4. Canabal Lavista, A.: Multi-trajectory rare-event simulation of stochastic Petri nets. Master's thesis, Technische Universität Ilmenau (2015)
5. Ciardo, G., German, R., Lindemann, C.: A characterization of the stochastic process underlying a stochastic Petri net. IEEE Trans. Softw. Eng. **20**, 506–515 (1994)
6. German, R., Kelling, C., Zimmermann, A., Hommel, G.: TimeNET - a toolkit for evaluating non-Markovian stochastic Petri nets. Perform. Eval. **24**, 69–87 (1995)
7. Gilmer Jr., J.B., Sullivan, F.J.: Combat simulation trajectory management. In: Chinni, M. (ed.) Proceedings of Military, Government, and Aerospace Simulation Conference, pp. 236–241. Society for Computer Simulation, San Diego (1996)
8. Kelling, C.: TimeNET$_{sim}$ - a parallel simulator for stochastic Petri nets. In: Proceedings of 28th Annual Simulation Symposium, Phoenix, AZ, USA, pp. 250–258 (1995)
9. Kovalchuk, S.V., Boukhanovsky, A.V.: Towards ensemble simulation of complex systems. Procedia Comput. Sci. **51**, 532–541 (2015)
10. Lazarova-Molnar, S., Horton, G.: Proxel-based simulation of stochastic Petri nets containing immediate transistion. In: Kemper, P. (Hrsg.) Workshop on Stachastic Petri Nets and Related Formalisms: Satellite Workshop of ICALP 2003, Eindhoven, Dortmund, The Netherland 28–29 June 2003 - on-site proceedings, pp. S.203–S.217 (2003)
11. McGeoch, C.: Analysing algorithms by simulation: variance reduction techniques and simulation speedups. ACM Comput. Surv. **24**(2), 195–212 (1992)
12. Pawlikowski, K.: Steady-state simulation of queueing processes: survey of problems and solutions. ACM Comput. Surv. **22**(2), 123–170 (1990)
13. Sanders, W.H., Meyer, J.F.: A unified approach for specifying measures of performance, dependability, and performability. In: Avizienis, A., Laprie, J. (eds.) Dependable Computing for Critical Applications, Dependable Computing and Fault-Tolerant Systems, vol. 4, pp. 215–237. Springer, Heidelberg (1991). doi:10.1007/978-3-7091-9123-1_10
14. Schwetman, H.D.: Hybrid simulation models of computer systems. Commun. ACM **21**(9), 718–723 (1978)
15. Trivedi, K.S.: Probability and Statistics with Reliability, Queuing and Computer Science Applications, 2nd edn. Wiley, Hoboken (2002)
16. Tuffin, B., Trivedi, K.S.: Implementation of importance splitting techniques in stochastic Petri net package. In: Haverkort, B.R., Bohnenkamp, H.C., Smith, C.U. (eds.) TOOLS 2000. LNCS, vol. 1786, pp. 216–229. Springer, Heidelberg (2000). doi:10.1007/3-540-46429-8_16
17. Villén-Altamirano, M., Villén-Altamirano, J.: RESTART: a method for accelerating rare event simulations. In: Queueing, Performance and Control in ATM, pp. 71–76. Elsevier Science Publishers (1991)
18. Zimmermann, A.: Stochastic Discrete Event Systems. Springer, Berlin (2007)
19. Zimmermann, A.: Modeling and evaluation of stochastic Petri nets with TimeNET 4.1. In: Proceedings of 6th International Conference on Performance Evaluation Methodologies and Tools (VALUETOOLS), Corse, France, pp. 54–63 (2012)
20. Zimmermann, A., Reijsbergen, D., Wichmann, A.: Canabal Lavista, A.: Numerical results for the automated rare event simulation of stochastic Petri nets. In: 11th International Workshop on Rare Event Simulation (RESIM 2016), Eindhoven, Netherlands, pp. 1–10 (2016)

A Probabilistic Small Model Theorem to Assess Confidentiality of Dispersed Cloud Storage

Marco Baldi[1], Ezio Bartocci[2], Franco Chiaraluce[1], Alessandro Cucchiarelli[1],
Linda Senigagliesi[1], Luca Spalazzi[1], and Francesco Spegni[1($\boxtimes$)]

[1] Università Politecnica delle Marche, Ancona, Italy
f.spegni@univpm.it
[2] TU Wien, Vienna, Austria

Abstract. Recent developments in cloud architectures have originated new models of online storage clouds based on data dispersal algorithms. According to these algorithms the data is divided into several slices that are distributed among remote and independent storage nodes. Ensuring confidentiality in this context is crucial: only legitimate users should access any part of information they distribute among storage nodes.

To the best of our knowledge, the security analysis and assessment of existing solutions always assume homogeneous networks and honest-but-curious nodes as attacker model. We analyze more complex scenarios with heterogeneous network topologies and a passive attacker eavesdropping the channel between user and storage nodes.

We use *parameterized* Markov Decision Processes to model such a class of systems and Probabilistic Model Checking to assess the likelihood of breaking the confidentiality. Even if, generally speaking, the parameterized model checking is undecidable, in this paper, however, we proved a Small Model Theorem that makes such a problem decidable for the class of models adopted in this work. We discovered that confidentiality is highly affected by parameters such as the number of slices and the number of write and read requests. At design-time, the presented methodology helps to determine the optimal values of parameters affecting the likelihood of a successful attack to confidentiality.

1 Introduction

Recent developments in cloud architectures have originated new models of storage based on data dispersal algorithms, where data is divided into several slices that are dispersed among remote and independent storage node [22,29,32]. The main advantage of these techniques consists in their reliability since the dispersion is usually accompanied by redundancy. However, in this context, ensuring confidentiality is equally important: only legitimate users should have access to any part of the information that was dispersed among the independent storage nodes. For this reason, many proposals for dispersed storage clouds are accompanied by a security analysis [9,10,35,36]. In these cases, the analysis usually determines the probability that an intruder can rebuild a message based on the

© Springer International Publishing AG 2017
N. Bertrand and L. Bortolussi (Eds.): QEST 2017, LNCS 10503, pp. 123–139, 2017.
DOI: 10.1007/978-3-319-66335-7_8

number of captured slices and are usually based on a number of assumptions: (a) the intruder is an honest but curious provider or a dishonest entity which succeeded in compromising one or more independent nodes; (b) all nodes are homogeneous, in other words they all have the same probability to be compromised; (c) no assumption is made about the nature of the communication channel. According to these assumptions several real world scenarios are not considered.

The *main original contribution* of this paper consists in presenting an *assessment methodology for dispersed storage clouds* that takes into account a scenario where some of the previous assumptions have been relaxed while others have been completely changed. Such assumptions are the following:

(1) the intruder is a passive eavesdropper that can spill individual slices from the communication channel; indeed, being passive it does not compromise the storage nodes but can act on the channels,
(2) the communication network is formed by a series of channels with different characteristics (heterogeneous channels).

The proposed methodology relies on *Parameterized Markov Decision Processes* to model such a class of systems and *Probabilistic Model Checking* as a verification tool. First of all, the parameterized nature of such a scenario should be noticed, where a single user decides to store a certain number of slices into an arbitrary number of independent storage nodes through an arbitrary number of communication channels. The model checking problem for parameterized systems is in general undecidable but this paper presents, as *further original contribution*, a *small model theorem for probabilistic systems* that allows, for the models at hand, a verification independent of the number of channels and nodes. As a consequence, the outcome of the analysis is twofold: on one hand, it *assesses confidentiality*, i.e. it produces a set of curves reporting the probability of breaking the confidentiality varying with respect to the number of slices and the number of write and read requests. On the other hand, it makes it possible to determine at design/configuration-time the value of critical parameters, e.g. how often the message must be re-dispersed in order to maintain a given confidentiality level.

Finally, this paper presents as a case study a real world example and this is *another contribution* of this paper. Indeed, the case study consists in analyzing dispersed storage cloud systems based on the AONT-RS schema [29]. This schema has been adopted by real systems as Cleversafe [29] and CDStore [22].

2   Related Work

Formal verification of security requirements in communication protocols is a well-established practice. In particular, model-checking [6,15,26,27] enables to verify automatically all the possible interleaved runs of the protocols in the presence of an adversary that can intercept, remove, modify the original messages as well as inject new messages. In this respect, the Dolev-Yao intruder model [13] is considered the most general model (the worst case) as it assumes a non-deterministic attacker in full control of the communication channels.

Traditionally, model checking can verify whether a system can be attacked or not and are not suitable for verifying security protocols in systems characterized by uncertainty or using randomized algorithms. We *assume*, instead, that every component of a cloud system may be attacked with some probability, and we wish to measure the likelihood of such attacks. This motivated us to define custom probabilistic intruder models, in place of the Dolev-Yao intruder.

Our scenario requires the use of probabilistic model-checking that provides a quantitative measure of security in terms of the probability of reaching a bad state. Examples of probabilistic model-checking applications in security can be found in the recent literature [3,5,21,25,37]. Probabilistic (as also the traditional one) model-checking suffers in general the state-explosion problem, making the verification of real-world security protocols and systems sometimes unfeasible.

One way to address this problem is to trade memory for computation by statistically [19] measuring the probability to satisfy or to violate a property over a set of traces generated by randomly sampling the model. This method in general requires a large number of samples to measure the probability of a rare event such as a security breach.

Another direction, that we have also pursued in this paper, is to find a suitable *abstraction technique* [11] that reduces the description of the system to a feasible state-space, still preserving the properties of interest. For example, distributed systems consisting of several instances of identical communicating components can benefit by proving a *small model theorem* that guarantees the existence of a bound in the number of the identical components for which it is sufficient to solve the verification problem to prove the correctness also for any larger number of components. Although this approach, also referred to as *parameterized verification*, has gained a lot of interest to verify (non-)deterministic systems [1,8, 17,34] to the best of our knowledge it is still scarcely explored in the probabilistic setting [7,20].

This work extends a previous, incomplete, attempt to verify data dispersal algorithms [3] that only considered the writing operation (i.e., the intruder could not take advantage of further read operations of the same file). Furthermore, from a theoretical point of view, the small probabilistic model theorem that we prove here applies also to other systems with a parameterized number of nodes or channels releasing a secret with a given probability.

Finally, let us remark that also the structure of the attacker may determine the feasibility of the verification of security properties. In our work we have employed a passive intruder model, and indeed several authors agree that this is enough when analyzing confidentiality requirements. For example, Li and Pang [23], and Shmatikov [33] used passive intruders to verify anonymity of protocols, a special case of confidentiality. The latter work also considers probabilistic attacks. As far as we know, the use of a probabilistic passive attacker model for the analysis of data confidentiality is original.

3   The AONT-Based Dispersed Storage Clouds

Generally speaking, *dispersed storage clouds* usually deal with reliability and security. Any user has some amount of cloud storage space assigned on independent storage nodes. In order to assure reliability on the one hand and security on the other, several authors have proposed schemata based on fragmentation, erasure coding, and encryption [4,22,24,29].

From a purely abstract point of view, all these algorithms can be characterized by a set of parameters. Let x be the original file size (measured in bits) to be dispersed and let $l \cdot q$ be the size of each fragment called *slice* (when $q = 8$, l is the size of a slice in bytes). The parameters of interest are (n, k), where:

- $n \geq \frac{x}{lq}$ is the the number of slices after the transformation of the original file, i.e. the number of slices to be dispersed;
- $k \leq n$ is the minimum number of slices to recover the original file; i.e., $n - k$ is the maximum number of lost (erased) slices that still enables file recovery;
- $k - 1$ is the maximum number of slices, that an intruder can eavesdrop, still allowing file confidentiality.

One of the most popular schemata is represented by the All-Or-Nothing-Transform Reed-Solomon (AONT-RS) [29]. Basically, AONT-RS consists in applying the Reed-Solomon (RS) erasure code to AONT, as depicted in Fig. 1.

AONT. First of all, the data of x bits are encrypted through an AONT that works as follows. A symmetric random key of α bits is chosen and used to encrypt the message. A digest of the encrypted data is computed and XOR-ed with the random key. The result is appended to the encrypted data, thus forming the transformed message of length $x + \alpha$ bits. Any user who is able to collect all the bits of the transformed message can also be able to retrieve the random key and, thus, to decrypt the transformed message and obtain the plaintext. The miss of any part of the transformed data does not allow the recovery of the random key and, hence, any part of the plaintext. Moreover, AONT does not require any key exchange, since the key is embedded with the transformed data, and it is easily obtainable (only) after retrieving the entire amount of transformed data.

Slicer. It divides the encrypted data produced by AONT into k small fragments, called *slices*, where k is also the dimension of the linear block code used and will be explained later on. Data bits are collected in groups of q bits each; therefore, each slice contains $l = \lceil \frac{x+\alpha}{kq} \rceil$ blocks of q bits.

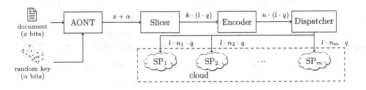

Fig. 1. Block diagram of AONT-RS (with data length expressed in bits)

Encoder. Here, data are subject to the full-length RS code with dimension k, defined over the Galois field of order 2^q with $q > 1$. After encoding, the number of slices increases from k to $n = 2^q - 1$, where $n > k$ due to redundancy added by the code. Thanks to this redundancy, RS codes are able to recover a number of erased symbols smaller than or equal to $n - k$ independently of their positions, while they cannot recover any number of erasures greater than $n - k$.

Dispatcher. It sends each of the n slices to one of the m independent storage nodes (the service providers *SP* of Fig. 1) through the network. Each SP receives a number of slices equal to n_i, with $\sum_{i=1}^{m} n_i = n$.

It is interesting to evidence the differences of this protocol with respect to more standard secret-sharing schemes like that pioneering proposed in [31]. The main difference is in the required redundancy: the protocol in [31] allows to disperse a small secret (typically an encryption key) at the cost of generating a great amount of data. On the contrary, in the AONT-RS techniques the redundancy is fixed by the code rate, that is usually not smaller than 0.5. In this sense, the AONT-RS approach may be more efficient.

4   Attack Scenarios

According to Fig. 1, the m SPs are distributed over the Internet, therefore users must exploit network connectivity in order to dispatch the n slices over them. In this work we assume the user is connected to the Internet from its local area network (LAN) through a gateway or an edge router. Furthermore, the protocol used by the Dispatcher may rely on end-to-end security techniques, like Secure Sockets Layer/Transport Layer Security (SSL/TLS). Nevertheless, let us investigate the worst case scenario where (1) the intruder can sniff the LAN, and (2) end-to-end encryption does not apply or does not work. These assumptions are reasonable because: (i) a LAN, especially if wireless, seems the most exposed one to eavesdropping; (ii) SSL/TLS requires a public key infrastructure, the related certificate management, and this may not be affordable in some cases (e.g., when SPs are small and cheap storage nodes); (iii) SSL/TLS can be affected by implementation bugs or other vulnerabilities [14, 16].

In this work we focus on wireless attacks. The wireless LAN (WLAN) may be open, or the attacker may be an insider of the network and possess all the credentials. This occurs any time the user is in an open wireless network or in a network protected via techniques based on pre-shared keys (e.g., WEP or WPA-PSK), so that all the other users can successfully acquire all her/his packets. Two scenarios will be analyzed: a user connected only to a WLAN, or a user connected to both a WLAN and a wired LAN.

In wireless connections, packets are subjected to channel conditions such as noise and signal degradation. Of course, a wireless attack depends on how far the eavesdropper is from the source, since the channel quality decreases inversely to the distance. Moreover, the legitimate receiver is authorized to ask for retransmission of lost packets (e.g. in presence of bad channel conditions), while the eavesdropper cannot, otherwise it would be revealed.

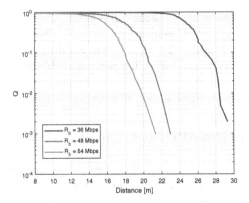

Fig. 2. Probability of correct reception in a WLAN ($L = 200$ bytes, $S_T = 12$ dBm)

Let us consider a setting where errors affecting one slice are independent of errors affecting another slice. This assumption is realistic in many real-world WLAN deployments, e.g., in the presence of relatively fast fading (coherence time not longer than the duration of a slice). Therefore, the probability that a single slice attack is successful corresponds to the probability that a slice is received by the attacker without errors. This probability depends on different parameters which act on the wireless channel, and we are interested in evaluating it as a function of the distance of the eavesdropper from the user. Let us denote by P the channel packet error rate, and by $Q = (1 - P)$ the probability that a slice is received without errors. The value of Q depends on the transmitted power S_T and on the path loss model. Examples are reported in [30], for the case of indoor communications. Numerical values for P are given in [28], where data is collected through a network simulator and packets with $L = 200$ bytes of application data and $S_T = 12$ dBm are considered. By assuming these parameters, Fig. 2 reports the average values of the probability of correct reception for a distance between 8 m and 30 m, considering different data rates R_b, for a typical IEEE 802.11g wireless connection.

5   Probabilistic Model Checking

To model the considered systems and attack scenarios, we use Markov Decision Processes (MDPs, see e.g. [2, Chap. 10.5]).

Definition 1 (Markov Decision Processes). *Assume a finite set of atomic propositions AP. A MDP is a tuple* $\mathcal{M} = (S, Act, Pr, \iota, L)$ *where:*

- $S = \{s_1, s_2, \ldots\}$ *is a finite set of states,*
- $Act = \{\alpha_1, \alpha_2, \ldots\}$ *is a finite set of actions,*
- $Pr : S \times Act \times S \to [0, 1]$ *is a probabilistic transition function such that, for all states* $s \in S$ *and actions* $\alpha \in Act$, $\sum_{s' \in S} Pr(s, \alpha, s') \in \{0, 1\}$;

- $\iota : S \rightarrow [0,1]$ *is the initial distribution probability of states, such that* $\sum_{s \in S} \iota(s) = 1$;
- $L : S \rightarrow 2^{AP}$ *is a labeling function.*

The probabilistic transition function can be extended to sets of states as follows: $\Pr(s, \alpha, T) = \sum_{t \in T} \Pr(s, \alpha, t)$, for all $s \in S$, $\alpha \in Act$, and $T \subseteq S$. Let us call *parametric MDP* any MDP $\mathcal{M}(p_1, \ldots, p_w)$ whose transitions refer to probabilities as parameters $p_1, \ldots, p_w$. We will say that a MDP $\mathcal{M}'$ *instantiates* $\mathcal{M}(p_1, \ldots, p_w)$ if there exist real values $0 \le q_1, \ldots, q_w \le 1$ such that $\mathcal{M}'$ is obtained from $\mathcal{M}$ by replacing p_i with q_i, for all i.

Definition 2 (Parallel composition). *Given MDPs $\mathcal{M}_1 = (S_1, Act_1, Pr_1, \iota_1, L_1)$ and $\mathcal{M}_2 = (S_2, Act_2, Pr_2, \iota_2, L_2)$, we write $\mathcal{M}_1 \parallel \mathcal{M}_2$ to denote the MDP $(S_1 \times S_2, Act_1 \cup Act_2, Pr, \iota, L)$ obtained as follows:*

$$- Pr((s,t), \alpha, (s',t')) = \begin{cases} Pr_1(s, \alpha, s') & \text{if } \alpha \in Act_1 \setminus Act_2, t = t' \\ Pr_2(t, \alpha, t') & \text{if } \alpha \in Act_2 \setminus Act_1, s = s' \\ Pr_1(s, \alpha, s') \cdot Pr_2(t, \alpha, t') & \text{if } \alpha \in Act_1 \cap Act_2 \end{cases}$$

- $\iota((s,s')) = \iota_1(s) \cdot \iota_2(s')$, *for all $s \in S_1, s' \in S_2$*
- $L((s,s')) = L_1(s) \cup L_2(s')$

We say that $\mathcal{M}_1 \parallel \mathcal{M}_2$ is the parallel composition *of $\mathcal{M}_1$ and $\mathcal{M}_2$.*

Let us write $\mathcal{M}[\alpha/\alpha']$ to denote the MDP where action α has been replaced by action α'. Formally: $(S, Act, Pr, \iota, L)[\alpha/\alpha'] = (S, Act', Pr', \iota, L)$ where:

- $Act' = (Act \setminus \{\alpha\}) \cup \{\alpha'\}$, and
- $Pr'(s, \beta, t) = \begin{cases} \Pr(s, \alpha', t) \text{ if } \beta = \alpha, \\ \Pr(s, \beta, t) \text{ } else \end{cases}$

To express properties of probabilistic systems, the probabilistic temporal logic $PCTL^\star$ [2, Chap. 10] can be used. We report its grammar:

$$\Phi ::= true \mid p \mid \Phi \wedge \Phi \mid \neg\Phi \mid \mathbb{P}_J(\varphi)$$
$$\varphi ::= \Phi \mid \varphi \wedge \varphi \mid \neg\varphi \mid X\varphi \mid G\varphi \mid F\varphi$$

where $p \in AP$ and $J \subseteq [0,1]$ is a rational interval. Terms of Φ are *state formulae*, while terms of φ are *path formulae*. Intuitively, formula $G\varphi$ (resp. $F\varphi$) holds w.r.t. some path iff every (resp. some) state visited along the path satisfies sub-formula φ. Given an MDP $\mathcal{M}$, we write $\mathcal{M} \models \Phi$ expressing that *all the initial states* of $\mathcal{M}$ satisfy Φ. Given a $PCTL^\star$ *path* formula φ and an MDP $\mathcal{M}$, we write $\mathcal{P}_{max}(\varphi, \mathcal{M})$ (resp. $\mathcal{P}_{min}(\varphi, \mathcal{M})$) denoting the maximum (resp. minimum) probability with which the specification φ is satisfied. Such a value can be computed in polynomial time w.r.t. its input [2, Chap. 10.5].

Given two MDPs $\mathcal{M}_1$ and $\mathcal{M}_2$, one can show that they are indistinguishable if (i) every transition to equivalent states on one system is mimicked on the other system, and (ii) equivalent states are reached with the same probability on the two systems. This is captured by *probabilistic bisimulation* [2, Chap. 10.5].

Definition 3 (Probabilistic bisimulation). *Given MDPs* $(S_1, Act_1, Pr_1,$
$\iota_1, L_1)$ *and* $(S_2, Act_2, Pr_2, \iota_2, L_2)$, *a probabilistic bisimulation is any relation*
$R \subseteq S \times S$ *such that* $R(s, s')$ *iff* $L(s) = L(s')$ *and* $Pr(s, \alpha, T) = Pr(s', \alpha, T)$,
for each action $\alpha \in Act$, *equivalence class* $T \in S/R$, *and* $s, s' \in S$, *where*
$(S, Act, Pr, \iota, L) = (S_1, Act_1, Pr_1, \iota_1, L_1) \parallel (S_2, Act_2, Pr_2, \iota_2, L_2)$.

The probabilistic bisimulation is an equivalence relation, thus given two states
s, t, let us write $s \approx_R t$ if R is a probabilistic bisimulation and $R(s, t)$ holds. When
R is clear from the context, we may omit it. It is known that bisimilar MDPs
satisfy the same $PCTL^\star$ formulae [2, Chap. 10.5].

Given two MDPs $\mathcal{M}_1$ and $\mathcal{M}_2$ and a sequence of action pairs $\Gamma = \alpha_1/\alpha_1', \ldots,$
α_n/α_n', let us write $\mathcal{M}_1 \approx_\Gamma \mathcal{M}_2$ to denote that $\mathcal{M}_1[\alpha_1/\alpha_1'] \ldots [\alpha_m/\alpha_m'] \approx$
$\mathcal{M}_2[\alpha_1/\alpha_1'] \ldots [\alpha_m/\alpha_m']$. We call the relation $\approx_\Gamma$ *probabilistic bisimulation up-*
to action replacement and intuitively denotes the fact that $\mathcal{M}_1$ and $\mathcal{M}_2$ are
bisimilar modulo a simple operation of renaming their actions.

Theorem 1. *Given two MDPs* $\mathcal{M}_1$, $\mathcal{M}_2$ *such that* $\mathcal{M}_1 \approx \mathcal{M}_2$, *then* $\mathcal{M}_1 \models \Phi$
iff $\mathcal{M}_2 \models \Phi$, *for any* $\Phi \in PCTL^\star$.

Corollary 1. *Given two MDPs* $\mathcal{M}_1$, $\mathcal{M}_2$ *s.t.* $\mathcal{M}_1 \approx \mathcal{M}_2$, *then* $\mathcal{P}_{max}(\varphi, \mathcal{M}_1) =$
$\mathcal{P}_{max}(\varphi, \mathcal{M}_2)$ *and* $\mathcal{P}_{min}(\varphi, \mathcal{M}_1) = \mathcal{P}_{min}(\varphi, \mathcal{M}_2)$ *for any* $\varphi \in PCTL^\star$.

Property 1 (Associativity). Given MDPs $\mathcal{M}_1 = (S_1, Act_1, Pr_1, \iota_1, L_1)$, $\mathcal{M}_2 =$
$(S_2, Act_2, Pr_2, \iota_2, L_2)$, and $\mathcal{M}_3 = (S_3, Act_3, Pr_3, \iota_3, L_3)$, then:

$$(\mathcal{M}_1 \parallel \mathcal{M}_2) \parallel \mathcal{M}_3 \approx \mathcal{M}_1 \parallel (\mathcal{M}_2 \parallel \mathcal{M}_3)$$

Proof (Sketched). Assume $s_1, s_1' \in S_1$, $s_2, s_2' \in S_2$, and $s_3, s_3' \in S_3$. The property is a consequence of the follogin facts: (1) $(S_1 \times S_2) \times S_3$ is isomorphic
to $S_1 \times (S_2 \times S_3)$. (2) By product associativity $(\iota_1(s_1) \cdot \iota_2(s_2)) \cdot \iota_3(s_3) =$
$\iota_1(s_1) \cdot (\iota_2(s_2) \cdot \iota_3(s_3))$. (3) By set union associativity $(L_1(s_1) \cup L_2(s_2)) \cup L_3(s_3) =$
$(L_1(s_1) \cup L_2(s_2)) \cup L_3(s_3)$. (4) Finally, for any $\alpha \in Act$ and $p \in [0, 1]$, one shows
by cases on the definition of Pr that $Pr(((s_1, s_2), s_3), \alpha, ((s_1', s_2'), s_3')) = p$ iff
$Pr((s_1, (s_2, s_3)), \alpha, (s_1', (s_2', s_3'))) = p$.

Property 2 (Commutativity). Given two MDPs $\mathcal{M}_1 = (S_1, Act_1, Pr_1, \iota_1, L_1)$ and
$\mathcal{M}_2 = (S_2, Act_2, Pr_2, \iota_2, L_2)$ then:

$$\mathcal{M}_1 \parallel \mathcal{M}_2 \approx \mathcal{M}_2 \parallel \mathcal{M}_1$$

Proof (Sketched). Assume $s_1 \in S_1$ and $s_2 \in S_2$. The property is a consequence
of the following facts: (1) $S_1 \times S_2$ is isomorphic to $S_2 \times S_1$. (2) By product
commutativity $\iota_1(s_1) \cdot \iota_2(s_2) = \iota_2(s_2) \cdot \iota_1(s_1)$. (3) By set union $L_1(s_1) \cup L_2(s_2) =$
$L_2(s_2) \cup L_1(s_1)$. (4) Finally, for any $\alpha \in Act$ and $p \in [0, 1]$, one shows by cases on
the definition of Pr that $Pr((s_1, s_2), \alpha, (s_1', s_2')) = p$ iff $Pr((s_2, s_1), \alpha, (s_2', s_1')) = p$.

6   Assessment Methodology

Here we describe the MDPs modeling user, links to storage nodes and attacker.
Later we show how the system can be verified for *any number* of links.

6.1   Modeling

At every modification, the AONT-RS schema encrypts the file with a fresh and random key. Then the data is encoded and dispersed. This means that a *write* operation invalidates the slices of previous versions of the same file. On the other side, a *read* operation gives a new opportunity to the attacker for collecting new slices up to the threshold k. We refer to this model as *write-once/read-many*.

At a first sight, the considered model might resemble the threshold public key encryption systems [12], where a private key is distributed among n decryption servers, so that at least k servers are needed for decryption. In reality, between the two systems there are important differences. In particular, in the AONT-based scheme we have only one user and the algorithm exploits symmetric ciphering. The only analogy is on the concept of threshold that, however, while in the case of threshold encryption is applied to the number of users that aim to decipher, in the present case is applied to the number of recovered slices.

In Figs. 3, 4 and 5 we represent the relevant MDPs using straight variable names for edge pre-conditions, while primed variable names are considered edge post-conditions. When the edge does not have a synchronization label (resp. a probability), we assume that the transition is asynchronous (resp. it has probability 1). When the boolean formula is omitted, we assume it is a tautology.

The user. Figure 3 shows the MDP $\text{USER}(d_1, \ldots, d_m)$ where m is the number of storage nodes. Its variables are: nw counts the number of written slices, c tracks the next node to write to, nra counts the number of remaining read attempts of the previously written message, $nr_1 \ldots nr_m$ count the slices to be read from the node i, in this attempt of reading the message. Note that the dispatch probabilities make a probability distribution, i.e. $\sum_{1 \leq i \leq m} d_i = 1$. USER also reads variables $ctr_1, \ldots, ctr_m$ from MDPs $\text{LINK}_1, \ldots, \text{LINK}_m$ to know the number of slices hosted by each node. USER starts by dispatching the n slices across the available m node links. When done, it goes to the reading stage, where it loops n_reads times reading back the previously written message.

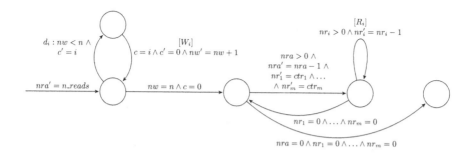

Fig. 3. The $\text{USER}(d_1, \ldots, d_m)$

The node links. MDP $\text{LINK}_i^c(a_i)$ depicted in Fig. 4 models any link to some storage node that could host up to c slices. It stores in ctr_i the amount of hosted slices *not known* by the attacker so far, and uses a binary flag $leak_i$ to remember whether the last written or read slice was intercepted by the intruder. Note that only one of the m copies of $LINK_i$ can have $leak_i = 1$ at any time. This forces the attacker to intercept the leaked slice before the user tries to write or read another one (possibly leaking it again). Probability a_i represents the likelihood of a slice being intercepted when traveling between the user and the storage node, while $1 - a_i$ is the probability of not being intercepted. The node link synchronizes with the user through actions R_i and W_i, and with the attacker using action L_i.

The Attacker. ATTACKER is the MDP modeling the intruder depicted in Fig. 5. It has a single local variable ctr_a counting the number of collected slices so far. The attack proceeds by collecting the slices leaked by the m copies of $LINK_i$ in the system. An attack is successful when $ctr_a \geq k$.

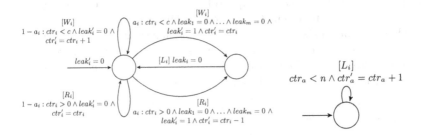

Fig. 4. The $LINK_i^c(a_i)$ **Fig. 5.** The ATTACKER

6.2   Security Assessment Analysis

It is evident that assessing the security of dispersal cloud storage algorithms is inherently a parameterized problem. Indeed, by allowing an arbitrarily large number of read operations by the user, the attacker has probability 1 of intercepting more than k slices (every read the attacker has one more chance of intercepting the missing slices, until it intercepts all of them). Similarly, assuming the secret is split into an arbitrarily large number of slices gives the attacker a negligible probability of succeeding in his/her attack. Between these ends lie all the parameters values of the actual implementations of AONT-based algorithms. Very often such values are not bound to a clearly stated security metric.

Our approach exploits bounded and probabilistic model checking to compute the likelihood of a successful attack, specified as a $PCTL^*$ formula, for several parameter configurations. The collected data allow us to draw a multi-dimensional graph relating the probability of a successful attack with the parameter values.

For any $m \in \mathbb{N}$, the dispersal cloud storage algorithms is modeled by:

$$\mathcal{M}_m^{AONT} := USER(d_1, \ldots, d_m) \parallel LINK_1^n(a_1) \parallel \ldots \parallel LINK_m^n(a_m) \parallel ATTACKER.$$

Finally, the probabilistic model checker is repeatedly invoked to solve the following problem varying parameter values:

$$\mathcal{P}_{max}(F(\mathit{ctr}_a \geq k), \mathcal{M}_m^{\text{AONT}}) \qquad (1)$$

6.3 Small Model Theorem for Node Links

A small model theorem allows to verify a class of infinite state systems by only checking a finite size system. The key observation is that, in a system where slices are intercepted when traveling between user and storage nodes, two or more node links with the same attack probability are indistinguishable from a single node link having the same attack probability, modulo some technicalities. In the case studies of Sect. 7 we will see that since the likelihood of intercepting a slice is determined by the physical properties of the employed LAN, the number of required node links for verifying real-world scenarios is usually very small.

Lemma 1 (Reduction). *For any natural numbers $c, d, i, j, k > 0$ such that $i \neq j$, any probability a. Given the MDPs $\text{LINK}_i^c(a)$, $\text{LINK}_j^d(a)$, and $\text{LINK}_k^{c+d}(a)$:*

$$\text{LINK}_i^c(a) \parallel \text{LINK}_j^d(a) \approx_\Gamma \text{LINK}_k^{c+d}(a)$$

where $\Gamma = R_i/R_k, R_j/R_k, W_i/W_k, W_j/W_k, L_i/L_k, L_j/L_i$.

Proof (Sketched). Fix $\text{LINK}_i^c(a) = (S_i, Act_i, \text{Pr}_i, \iota_i, L_i)$, $\text{LINK}_j^d(a) = (S_j, Act_j, \text{Pr}_j, \iota_j, L_j)$, $\text{LINK}_k^{c+d}(a) = (S_k, Act_k, \text{Pr}_k, \iota_k, L_k)$. One shows that there exists a relation $R \subseteq (S_i \times S_j) \times S_k$ that is indeed a probabilistic bisimulation up-to action replacing Γ. Take $R := \{((s,t), u) : s.ctr_i + t.ctr_j = u.ctr_k, s.leak_i = 1 \iff t.leak_j = 1 \vee u.leak_k = 1\}$.

Call Pr_{big} the probabilistic transition function of the composed MDP $\text{LINK}_i^c(a) \parallel \text{LINK}_j^d(a)$. Now one proves that the following commutative diagram holds:

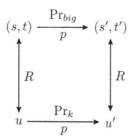

We first prove one direction. Fix any s, t, u such that $R((s,t), u)$ and then reason by cases on $\text{Pr}_{big}((s,t), \alpha, (s', t')) = p$.

- Case $\alpha = R_i$ (i.e. read on i): if $p = a$ (i.e. a leaking happened) then $s.ctr_i > 0$, $s'.ctr_i = s.ctr_i - 1$, $s.leak_i = t.leak_j = t'.leak_j = 0$, $s'.leak_i = 1$. By $R((s,t), u)$ we know that $u.ctr_k = s.ctr_i + t.ctr_j$ and $u.leak_k = \min(s.leak_i + t.leak_j, 1) = 0$.

Let us name u' the (unique) state satisfying the following: $u'.ctr_k = u.ctr_k - 1$ and $u'.leak_k = 1$. It is evident that $R((s', t'), u')$, concluding this branch of the case. If $p = 1 - a$ (i.e. no leaking happened) one similarly observes that $s.ctr_i > 0$, $s'.ctr_i = s.ctr_i$, and $s.leak_i = t.leak_j = s'.leak = t'.leak = 0$. Now, under our assumptions, let us define u' to be the (unique) state where $u'.ctr_k = u.ctr_k = s'.ctr_i + t'.ctr_j$ and $u'.leak_k = 0 = \min(s'.leak + t'.leak, 1)$. Observing that $R((s', t'), u')$ ends this case.

- Case $\alpha = R_j$ (i.e. read on j): it is symmetric to the previous one.
- Cases $\alpha \in \{W_i, W_j, L_i, L_j\}$ are straightforward to check following the reasoning for the case $\alpha = R_i$.

Finally, fix any s, t, u such that $R((s, t), u)$ and reason by cases on $\Pr_k(u, \alpha, u') = p$ to show that the opposite direction holds. $\qquad\square$

Given a sorted list of numbers $a_1, \ldots, a_m$ s.t. $a_1 \leq \ldots \leq a_m$, let us call its *distinction* the list of indices $i_1, \ldots, i_{q+1}$ satisfying the following:

- $i_1 = 1$, $i_{q+1} = m$, and $i_1 < \ldots < i_{q+1}$,
- $\forall j \in [1, q].\forall k \in [i_j, i_{j+1} - 1].\ a_{i_j} = a_k$, and
- $\forall j \in [1, q].\ a_{i_j} < a_{i_{j+1}}$.

Such constraints mean that the list $a_1, \ldots, a_m$ can be partitioned into q sublists, each containing identical values, and each pair of lists containing distinct values. For example, the distinction of the sorted list of probabilities $0.00, 0.00, 0.05, 0.10, 0.10, 0.10, 0.15$, is the list of indices $1, 3, 4, 7$.

The core theoretical contribution of this work shows that one can do parameterized probabilistic model checking of systems with any number of LINKs, by considering only a finite number of them. Such number is often called *cutoff*.

Theorem 2 (Small Model Theorem). *For any naturals $m, c_1, \ldots, c_m > 0$ and probabilities $a_1, \ldots, a_m$. Given the MDPs $\text{LINK}_1^{c_1}(a_1), \ldots, \text{LINK}_m^{c_m}(a_m)$. For any MDP $\mathcal{M}$ and formula $\Phi \in PCTL^*$ the following holds:*

$$\mathcal{M} \parallel \text{LINK}_1^{c_1}(a_1) \parallel \ldots \parallel \text{LINK}_m^{c_m}(a_m) \models \Phi \Leftrightarrow$$
$$\mathcal{M} \parallel \text{LINK}_1^{c_{i_1}}(a_{i_1}) \parallel \ldots \parallel \text{LINK}_q^{c_{i_q}}(a_{i_q}) \models \Phi$$

where, for some $0 < q \leq m$, the list of indices $i_1, \ldots, i_q$ is a distinction of the list $a_1, \ldots, a_m$ (assume w.l.o.g. that the latter is sorted), the dispatch probabilities are given by $d_{i_j} = \sum_{k=i_j}^{i_{j+1}-1} d_k$ while the capacities are defined as $c_{i_j} = \sum_{k=i_j}^{i_{j+1}-1} c_k$.

Proof. Let us recursively apply Lemma 1. The latter reduces any pair of LINKs with identical attack probabilities to a single LINK. The procedure ends when all the LINKs have distinct attack probabilities. By Lemma 1, for every $j \in [1, q]$, the LINK having attack probability a_{i_j} has capacity c_{i_j} (resp. dispatch probability d_{i_j}) defined as the sum of the capacities (resp. of the dispatch probabilities) of all the original LINKs with identical attack probability. By Lemma 1 and Theorem 1 they satisfy the same $PCTL^*$ formulae. $\qquad\square$

7   Case Studies

We analyse here two typical scenarios of real-world implementations of dispersal cloud storage systems: (a) the user is connected to a wireless LAN, and (b) the user is connected to two LANs, one wired and one wireless (combined scenario).

In our experiments we choose the number of slices n to range between 10 and 100, and the number of read events n_reads between 1 and 31. As explained in Sect. 4, the probability of an attack in the wireless network depends on the distance of the attacker from the user. Here we set such a distance to 20 m which, according to Fig. 2, corresponds to a probability of intercepting a slice of 0.009 (resp. 0.148) for a network operating at 54 Mbps (resp. 48 Mbps). The successful attack probability of the wired LAN is instead assumed to be zero. Consequently, we can apply the small model theorem discussed in Sect. 6.3, and reduce the actual number m of LINKs in the model checked system to a fixed value, i.e. the number of different intercept probabilities in the system. Thus, m equals the number of considered LANs. Finally, in the presented case studies we consider a threshold $k = 0.7 \cdot n$ and for the combined scenario that slices are routed more likely to the wired LAN (75%) than to the wireless LAN (25%).

To assess confidentiality in both scenarios we used SecMC[1], an open-source modular tool allowing to define model checking workflows. The tool repeatedly invokes the PRISM model checker [18]. Each invocation instantiates a parametric MDP and returns the probability of a successful attack (as defined by the security metric (1)). Figure 6 plots the obtained results. In it, each line corresponds to a given number of slices used to split the message, and every line relates the likelihood of a successful attack to the number of read attempts by the client.

Our analysis reveals that while the considered cloud dispersal protocols protect against untrusted storage nodes, they do not ensure a high level of confidentiality against an eavesdropper in the same wireless LAN of the user, in the case that some storage nodes do not use end-to-end cryptography, or the implementation of the latter is broken. Fixing the same parameters n and n_reads, the measure of confidentiality may be several orders of magnitude bigger in the case of 54 Mbps networks w.r.t. 48 Mbps. But, especially in the wireless scenario, the probability of an attack grows too fast with the number of file reads. Furthermore, in networks with 48 Mbps rate or lower, even using 100 or more slices, it is enough to force the user to read the file 6 or more times to reach a probability of reconstructing the file close to 100%.

The methodology can assess security metrics while designing cloud dispersal algorithms. Assume the designer fixes this reference scenario: 54 Mbps wireless LAN, an eavesdropper 20 m far from the user. Let assume that the probability of an attack should be bounded by 10^{-21}. This provided, the methodology suggests to limit the number of reads before overwriting and redistributing the file between 11 and 15 (resp. between 26 and 30) in case of 51 slices (resp. 91 slices).

The methodology proved to be quite feasible: the presented case studies were run on a machine Intel Xeon E5520 2.27 Ghz Quad-core 48 GB RAM and

---

[1] https://bitbucket.org/fcloseunivpm/secmc.

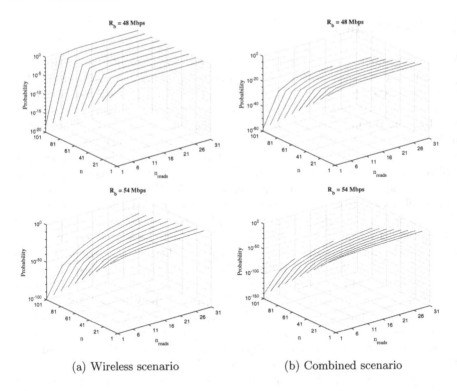

(a) Wireless scenario   (b) Combined scenario

Fig. 6. Probabilities of a successful attack to confidentiality

required 52 h for the two wireless scenarios, and 151 h for the two combined scenarios. Let us remark that the tool can certainly be improved to take advantage of multi-core architectures, since every invocation of the model checker is independent from the others.

8   Conclusions

We have introduced a novel formal probabilistic model to verify security properties of online storage clouds based on data dispersal algorithms. In our model we have considered a client that can read and write in the storage nodes and a passive intruder that can steal individual slices from heterogeneous communication channels without compromising the storage nodes.

From a theoretical point of view, we also provided a novel abstraction technique enabling us to use a relatively small probabilistic model with a fixed number of communication channels and storage nodes while the assessed security metric scale to any arbitrary number of storage nodes.

Our methodology can be applied to (1) formally specify a custom security metric for the system under consideration, and (2) certify at design time the parameter assignments ensuring a given level of security, i.e., for which the computed likelihood of a successful attack is below a given threshold value.

Acknowledgements. This work is part of the project FCloSe (Federated Cloud Security) funded by the RSA-B 2015 programme of the Università Politecnica delle Marche. Ezio Bartocci is supported by the Austrian National Research Network (nr. S 11405-N23) SHiNE funded by the Austrian Science Fund (FWF).

References

1. Aminof, B., Kotek, T., Rubin, S., Spegni, F., Veith, H.: Parameterized model checking of rendezvous systems. In: Baldan, P., Gorla, D. (eds.) CONCUR 2014. LNCS, vol. 8704, pp. 109–124. Springer, Heidelberg (2014). doi:10.1007/978-3-662-44584-6_9

2. Baier, C., Katoen, J.P.: Principles of Model Checking. Springer, Heidelberg (2008)

3. Baldi, M., Cucchiarelli, A., Senigagliesi, L., Spalazzi, L., Spegni, F.: Parametric and probabilistic model checking of confidentiality in data dispersal algorithms. In: Proceedings of HPCS 2016: International Conference on High Performance Computing and Simulation, pp. 476–483 (2016)

4. Baldi, M., Maturo, N., Montali, E., Chiaraluce, F.: AONT-LT: a data protection scheme for cloud and cooperative storage systems. In: Proceedings of HPCS 2014: International Conference on High Performance Computing and Simulation, pp. 566–571 (2014)

5. Bartocci, E., Grosu, R., Katsaros, P., Ramakrishnan, C.R., Smolka, S.A.: Model repair for probabilistic systems. In: Abdulla, P.A., Leino, K.R.M. (eds.) TACAS 2011. LNCS, vol. 6605, pp. 326–340. Springer, Heidelberg (2011). doi:10.1007/978-3-642-19835-9_30

6. Basin, D.A., Cremers, C., Meadows, C.A.: Model checking security protocols. In: Handbook of Model Checking. Springer, Heidelberg (2017)

7. Bertrand, N., Fournier, P.: Parameterized verification of many identical probabilistic timed processes. In: Proceedings of FSTTCS 2013: The IARCS Annual Conference on Foundations of Software Technology and Theoretical Computer Science, LIPIcs, vol. 24, pp. 501–513 (2013)

8. Bloem, R., Jacobs, S., Khalimov, A., Konnov, I., Rubin, S., Veith, H., Widder, J.: Decidability in parameterized verification. SIGACT News **47**(2), 53–64 (2016)

9. Bowers, K.D., Juels, A., Oprea, A.: Hail: a high-availability and integrity layer for cloud storage. In: Proceedings of the 16th ACM Conference on Computer and Communications Security, pp. 187–198. ACM (2009)

10. Chung, J.Y., Joe-Wong, C., Ha, S., Hong, J.W.K., Chiang, M.: Cyrus: towards client-defined cloud storage. In: Proceedings of the 10th European Conference on Computer Systems, p. 17. ACM (2015)

11. Clarke, E.M., Grumberg, O., Long, D.E.: Model checking and abstraction. ACM Trans. Program. Lang. Syst. **16**(5), 1512–1542 (1994)

12. Desmedt, Y.: Threshold cryptosystems. In: Seberry, J., Zheng, Y. (eds.) AUSCRYPT 1992. LNCS, vol. 718, pp. 1–14. Springer, Heidelberg (1993). doi:10.1007/3-540-57220-1_47

13. Dolev, D., Yao, A.C.: On the security of public key protocols. IEEE Trans. Inf. Theory **29**(2), 198–207 (1983)

14. Durumeric, Z., Kasten, J., Adrian, D., Halderman, J.A., Bailey, M., Li, F., Weaver, N., Amann, J., Beekman, J., Payer, M., Paxson, V.: The matter of heartbleed. In: Proceedings of the 2014 Internet Measurement Conference, pp. 475–488. ACM (2014)

15. Escobar, S., Meadows, C.A., Meseguer, J.: A rewriting-based inference system for the NRL protocol analyzer and its meta-logical properties. Theor. Comput. Sci. **367**(1–2), 162–202 (2006)

16. Georgiev, M., Iyengar, S., Jana, S., Anubhai, R., Boneh, D., Shmatikov, V.: The most dangerous code in the world: validating SSL certificates in non-browser software. In: Proceedings of the ACM Conference on Computer and Communications Security, pp. 38–49 (2012)

17. Johnson, T.T., Mitra, S.: A small model theorem for rectangular hybrid automata networks. In: Giese, H., Rosu, G. (eds.) FMOODS/FORTE-2012. LNCS, vol. 7273, pp. 18–34. Springer, Heidelberg (2012). doi:10.1007/978-3-642-30793-5_2

18. Kwiatkowska, M., Norman, G., Parker, D.: PRISM 4.0: verification of probabilistic real-time systems. In: Gopalakrishnan, G., Qadeer, S. (eds.) CAV 2011. LNCS, vol. 6806, pp. 585–591. Springer, Heidelberg (2011). doi:10.1007/978-3-642-22110-1_47

19. Legay, A., Delahaye, B., Bensalem, S.: Statistical model checking: an overview. In: Barringer, H., et al. (eds.) RV 2010. LNCS, vol. 6418, pp. 122–135. Springer, Heidelberg (2010). doi:10.1007/978-3-642-16612-9_11

20. Lengál, O., Lin, A.W., Majumdar, R., Rümmer, P.: Fair termination for parameterized probabilistic concurrent systems. In: Legay, A., Margaria, T. (eds.) TACAS 2017. LNCS, vol. 10205, pp. 499–517. Springer, Heidelberg (2017). doi:10.1007/978-3-662-54577-5_29

21. Lenzini, G., Mauw, S., Ouchani, S.: Security analysis of socio-technical physical systems. Comput. Electr. Eng. **47**, 258–274 (2015)

22. Li, M., Qin, C., Li, J., Lee, P.P.: CDstore: toward reliable, secure, and cost-efficient cloud storage via convergent dispersal. IEEE Internet Comp. **20**(3), 45–53 (2016)

23. Li, Y., Pang, J.: Formalizing provable anonymity in Isabelle/HOL. Formal Aspects Comput. **27**(2), 255–282 (2015)

24. Merani, M.L., Barcellona, C., Tinnirello, I.: Multi-cloud privacy preserving schemes for linear data mining. In: Proceedings of ICC 2015: IEEE International Conference on Communications, pp. 7095–7101 (2015)

25. Ouchani, S., Debbabi, M.: Specification, verification, and quantification of security in model-based systems. Computing **97**(7), 691–711 (2015)

26. Pagliarecci, F., Spalazzi, L., Spegni, F.: Model checking grid security. Fut. Gener. Comput. Syst. **29**(3), 811–827 (2013)

27. Panti, M., Spalazzi, L., Tacconi, S., Valenti, S.: Automatic verification of security in payment protocols for electronic commerce. In: ICEIS 2002, Proceedings of the 4th International Conference on Enterprise Information Systems, pp. 968–974 (2002)

28. Pei, G., Henderson, T.: Validation of OFDM error rate model in ns-3 (2010). www.nsnam.org/pei/80211ofdm.pdf

29. Resch, J., Plank, J.: AONT-RS: blending security and performance in dispersed storage systems. In: Proceedings 9th FAST Conference (2011)

30. Seidel, S.Y., Rappaport, T.S.: 914 MHz path loss prediction models for indoor wireless communications in multifloored buildings. IEEE Trans. Microwave Theory Tech. **40**(2), 202–217 (1992)

31. Shamir, A.: How to share a secret. Commun. ACM **22**(11), 612–613 (1979)

32. Shen, L., Feng, S., Sun, J., Li, Z., Wang, G., Liu, X.: CloudS: a multi-cloud storage system with multi-level security. In: Wang, G., Zomaya, A., Perez, G.M., Li, K. (eds.) ICA3PP 2015. LNCS, vol. 9530, pp. 703–716. Springer, Cham (2015). doi:10.1007/978-3-319-27137-8_51

33. Shmatikov, V.: Probabilistic analysis of an anonymity system. J. Comput. Secur. **12**(3–4), 355–377 (2004)

34. Spalazzi, L., Spegni, F.: Parameterized model-checking of timed systems with conjunctive guards. In: Giannakopoulou, D., Kroening, D. (eds.) VSTTE 2014. LNCS, vol. 8471, pp. 235–251. Springer, Cham (2014). doi:10.1007/978-3-319-12154-3_15

35. Strunk, A., Mosch, M., Groß, S., Thoß, Y., Schill, A.: Building a flexible service architecture for user controlled hybrid clouds. In: Proceedings of the 2012 Seventh International Conference on Availability, Reliability and Security (ARES), pp. 149–154. IEEE (2012)

36. Tang, H., Liu, F., Shen, G., Jin, Y., Guo, C.: Unidrive: synergize multiple consumer cloud storage services. In: Proceedings of the 16th Annual Middleware Conference, pp. 137–148. ACM (2015)

37. Yang, F., Yang, G., Hao, Y.: The modeling library of eavesdropping methods in quantum cryptography protocols by model checking. Int. J. Theor. Phys. **55**(7), 3414–3427 (2016)

On the Cost of Diagnosis with Disambiguation

Loïc Hélouët[(✉)] and Hervé Marchand

INRIA Rennes, Campus de Beaulieu, 35042 Rennes Cedex, France
{loic.helouet,herve.marchand}@inria.fr

Abstract. Diagnosis consists in deciding from a partial observation of a system whether a fault has occurred. A system is *diagnosable* if there exists a mechanism (a *diagnoser*) that accurately detects faults a finite number of steps after their occurrence. In a regular setting, a *diagnoser* builds an estimation of possible states of the system after an observation to decide if a fault has occurred. This paper addresses diagnosability (deciding whether a system is diagnosable) and its cost for safe Petri nets. We define an energy-like cost model for Petri nets: transitions can consume or restore energy of the system. We then give a partial order representation for state estimation, and extend the cost model and the capacities of diagnosers. Diagnosers are allowed to use additional energy to refine their estimations. In this setting, diagnosability is an energy game, and checking diagnosability under energy constraints is in 2-EXPTIME.

1 Introduction

This paper addresses diagnosability of partially observable systems with disambiguation mechanisms, under energy constraints. We consider systems that can consume and produce energy, and which energy consumption can be modeled as weights attached to actions. In the standard diagnosis setting [17], faults are occurrences of particular faulty events (complex fault patterns have also been proposed in [14]). Systems under diagnosis are equipped with sensors (software probes or physical equipments) that can signal *some* state changes or occurrences of *some* actions, yielding partial observation of the system. The objective of diagnosis is to build monitors that receive observations from sensors and raise alarms when a fault occurrence is certain. Cost of diagnosis has mainly been defined as the cost needed to exploit sensors [5,18] in order to guarantee diagnosability. However, real life systems need not be passive: when a fault is suspected, a monitor can perform tests (read the status of a variable, use a calculator) to leverage ambiguity. It is hence natural to consider *active* diagnosis, where monitors can perform additional costly actions to get information. The question in this active setting is then whether a non-diagnosable system is diagnosable with the help of additional tests, while satisfying energy constraints.

The model used in this paper is *finite safe Petri nets*, with observable and unobservable transitions, equipped with a cost model. The system starts with an initial energy provision, and actions produce or consume energy. Optional

© Springer International Publishing AG 2017
N. Bertrand and L. Bortolussi (Eds.): QEST 2017, LNCS 10503, pp. 140–156, 2017.
DOI: 10.1007/978-3-319-66335-7_9

energy consuming tests can be used to reduce ambiguity on the current state of the system. Faults are a subset of transitions of the net, and they are considered permanent: once a faulty transition has been fired, the system remains faulty. Observations are sequences of observable events. As we are working with *safe* Petri nets, diagnosis can obviously be recast in a labeled transition system setting. However, using Petri nets provides a compact way to represent the *state estimate* that a diagnoser can build after an observation. We use *observation guided unfolding* to find processes that may have produced an observation. Upon sensible restrictions, unfolding always terminates. One can also maintain finite state estimates along arbitrary long observations. We then define when a diagnosis can be produced by looking at properties of its processes. The exact current state of a system is not always precisely known, as several processes can correspond to the same observation. Additional ambiguity comes from the fact that diagnosers do not observe what has effectively occurred since the last observable event. Even with uncertainty on current state of a system, faults can be detected: a fault has occurred with certainty iff all processes of its observation guided unfolding contain a fault. No fault has occurred if all processes of the unfolding are fault free. If the unfolding contains both kind of processes, then it is said *ambiguous*. With this structure, a system is *diagnosable* iff it cannot remain ambiguous for an arbitrary long time. Diagnosability of a net is a PSPACE-complete problem.

A natural question is whether increasing the power of diagnosers can make a system diagnosable. This setting is useful when a system is built from proprietary components that can not be modified, and one has to rely on every possible observation to detect faults. We define disambiguation functions, that bring additional information on the current state of the observed system, and can be used to reduce the set of possible states the system can be in. The disambiguation functions proposed in this work are simple: they provide information on the contents of a place (this models access to boolean variables), but disambiguation is not limited to this simple setting. We assume that tests cost energy, and hence have to be used parsimoniously. The cost of disambiguation can be easily integrated in the original cost model by assigning a negative weight to each test. This situation can occur in automotive systems, where one cannot modify a component to avoid ambiguity, and where power management is a concern. For instance, in some ABS systems, a warning sign alights when the ABS is not operational. This can be caused by dirty sensors, hydraulic problems, or if the battery level is too low to initialize the ABS calculator (which consumes a lot of power). The latter case is a minor problem that is solved when the engine is started, but if the battery level is really low, launching tests to know the exact status of the ABS can exhaust the battery and prevent starting the car. Our *active diagnosis* setting is the following: immediately after an observation, diagnosers are allowed to use *one* disambiguation function to refine their state estimate. In addition, use of tests must not exhaust the system's energy. Diagnosability can hence be redefined as the possibility to make a system diagnosable with the help of tests without exhausting its energy. This question can be answered in terms of energy games with partial information. We build an arena that represents the behavior of the diagnosed system, the possible sets of state estimates that a diagnoser might build, and the remaining energy. Diagnosability with disambiguation and an upper bounded energy budget (ULWUB diagnosability) is equivalent to a partial observation co-Büchi game [7] between a system and a diagnoser: a system with energy constraints is not diagnosable iff it can impose arbitrary long ambiguity or energy exhaustion. As a consequence, ULWUB diagnosability is in 2-EXPTIME. Testing whether disambiguation can make a system diagnosable without energy consideration is already in 2-EXPTIME.

This paper is organized as follows: Sect. 2 introduces the main notations of the paper, and costs. Section 3 defines a partial order setting for diagnosis. Section 4 defines disambiguation, and shows that diagnosability with disambiguation is in 2-EXPTIME, and that diagnosability with costly disambiguation and energy constraints is also in 2-EXPTIME, before conclusion. Due to lack of space, proofs are only sketched, and provided in a research report [13].

2   Models and Definition of the Diagnosis Problem

Definition 1. *A Petri net is a tuple* $\mathcal{N} = (P, T, {}^\bullet(), (){}^\bullet, \lambda, m_0)$, *where* P *is a set of places,* T *is a set of transitions,* ${}^\bullet() : T \to 2^P$ *is a preset relation, and* $(){}^\bullet : T \to 2^P$ *is a postset relation, and* $\lambda : T \to \Sigma$ *associates labels to transitions.*

A *marking* of net $\mathcal{N}$ is a function $M : P \to \mathbb{N}$. We only consider *finite* and *safe* Petri nets, i.e. such that for every reachable marking M and for every place $p \in P$, $M(p) \leq 1$. Hence, markings can be seen as subsets of marked places, and we write $M \subseteq X$ when the set of marked places in M is contained in X. The *size* of $\mathcal{N}$ is simply $|P|$. We partition alphabet Σ into observable and unobservable actions, i.e. $\Sigma = \Sigma_o \uplus \Sigma_{uo}$. A transition $t \in T$ is *observable* iff $\lambda(t) \in \Sigma_o$, and *unobservable* otherwise. Intuitively, only a part of the events that occur during a run of the system is monitored and logged. This assumption is sensible: some parts of a system usually have to be considered as black boxes, and remain unobserved. This partial observation setting can also be a design choice to maintain logs of reasonable sizes, use a small number of sensors,... We also consider a subset $\Sigma_f \subseteq \Sigma_{uo}$ of faulty actions, and say that $t \in T$ is *faulty* iff $\lambda(t) \in \Sigma_f$.

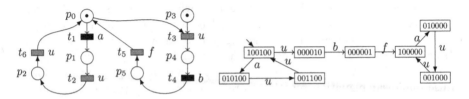

Fig. 1. A Petri net (left) and the associated LTS (right). Circles are places, marked places contain a token. Black rectangles are observable transitions and grey ones unobservable transitions. $\Sigma = \{a, b, f, u\}, \Sigma_o = \{a, b\}, \Sigma_f = \{f\}$.

Petri nets have an interleaved semantics, defined as follows: a transition $t \in T$ is *enabled* (denoted by $M[t\rangle$) from marking $M \subseteq P$ iff ${}^{\bullet}t \subseteq M$. Firing t from M results in a marking $M' = (M \setminus {}^{\bullet}t) \uplus t^{\bullet}$. This is denoted $M[t\rangle M'$. We also write $M \xrightarrow{t} M'$ when $M[t\rangle M'$, and $M \xrightarrow{*} M'$ when there exists a sequence of transitions $M[t_1.t_2...t_n\rangle M'$. A *run* of $\mathcal{N}$ is a sequence $\rho = m_0 \xrightarrow{t_1} M_1 \ldots M_n$. The set of *reachable configurations* from marking m_0 is the set $Reach(\mathcal{N}, m_0) = \{M \mid m_0 \xrightarrow{*} M\}$. The *labeling* of a run $\rho = m_0 \xrightarrow{t_1} M_1 \ldots M_n$ is the word $w_\rho = \lambda(t_1).\lambda(t_2) \ldots \lambda(t_n)$, and its *observation* is the word $\Pi_{\Sigma_o}(w_\rho)$, where $\Pi_{\Sigma_o}(.)$ is the usual *projection* erasing all letters of $\Sigma \setminus \Sigma_o$. We denote by $|\rho|_{\Sigma_o}$ the size of $\Pi_{\Sigma_o}(w_\rho)$. We write $M \xRightarrow{a} M'$ iff there exists a run ρ from M to M' and $\Pi_{\Sigma_o}(w_\rho) = a$. Obviously, a net $\mathcal{N}$ defines a finite labeled transition system $LTS_\mathcal{N} = (Q = Reach(\mathcal{N}, m_0), T, \delta, q_0 = m_0)$, where $\delta \subseteq Q \times T \times Q$ is a transition relation such that $(q, t, q') \in \delta$ iff $q \xrightarrow{t} q'$. The size of $LTS_\mathcal{N}$ is in $O(2^{|P|})$.

Definition 2. *A k-diagnoser is a mechanism $\mathcal{D}$ that accepts observations and returns a value from $\{0, 1\}$, such that:*

- *for each non-faulty run ρ, $\mathcal{D}(\Pi_{\Sigma_o}(\rho)) = 0$, and*
- *for each faulty run such that at least k observable transitions have occurred since the first occurrence of a fault, $\mathcal{D}(\Pi_{\Sigma_o}(\rho)) = 1$.*

We say that a system is k-*diagnosable* if there exists no run ρ with k observations after a fault such that the observation of ρ has one faulty explanation and one non-faulty. A system is *diagnosable* if it is k-diagnosable for some $k \in \mathbb{N}$. A slightly different definition of diagnosability is proposed by [4]: it says that for every faulty run ρ, there exists a bound k_ρ such that a fault is detected at most k_ρ steps after it occurs. In a regular setting, both definitions coincide, but in a non-regular setting (e.g. for unbounded Petri nets) this is not always the case (as shown in [10]). It is frequently assumed that diagnosers are regular, i.e. one can compute an automaton that reads observations, and accepts all words such that $\mathcal{D}(w) = 1$. Upon acceptance, one can claim a fault occurred.

Definition 3. *A net is* diagnosable *if it is k-diagnosable for some $k \in \mathbb{N}$.*

In a regular setting, a necessary and sufficient condition [17] for diagnosability is existence of a bound K such that for every faulty run ρ, for every $\rho' = \rho.\rho_1$ with $|\rho_1|_{\Sigma_o} \geq K$, every run ρ'' such that $\Pi_{\Sigma_o}(\rho') = \Pi_{\Sigma_o}(\rho'')$ is faulty.

Proposition 1. *Deciding whether a Petri net $\mathcal{N}$ is diagnosable is PSPACE-complete. Furthermore, this can be decided in $O(2^{2 \cdot |P|})$.*

Proof (Sketch). One can search in PSPACE a pair of faulty/non-faulty runs with equivalent observation, that end on a loop (yielding infinite non-diagnosable executions). For the hardness part, one can reduce the problem of emptiness of the intersection of regular languages to a diagnosability problem. The exponential bound comes from the quadratic diagnosability decision procedure of [15] for automata, that can be applied to $LTS_\mathcal{N}$.   □

We assume that systems follow some energy consumption schemes: performing some actions consume energy, some others restore energy. The system starts with a known initial provision B_0. We also assume that systems have limited energy storage capacities (for instances batteries) and hence set an upper bound B_{max} for the amount of energy that can be stored by the system. A first question to consider is whether the system is properly designed and does not exhaust its energy provision, i.e. the remaining energy provision when the system does not use tests is always maintained above 0. For this, we define a cost model, and check that the system does not a priori consume more energy than it produces.

Definition 4. *A* cost model *for a Petri net $\mathcal{N}$ is a map $C : T \to \mathbb{N}$ that associates integers to transitions of T.*

Definition 5. *Let $\mathcal{N}$ be a net, $B_0 \in \mathbb{N}$ be an initial energy budget, $B_{max} \in \mathbb{N}$ be an upper energy bound, and let C be a cost function. The* accumulated weight *with initial provision B_0 under weak upper bound B_{max} along a path ρ is denoted $W_{B_0 \downarrow B_{max}}(\rho)$, and is defined inductively as $W_{B_0 \downarrow B_{max}}(\rho) = r_{|\rho|}$, with $r_0 = B_0$, $r_{i+1} = \min(r_i + C(t_{i+1}), B_{max})$. $\mathcal{N}$ satisfies the* universal lower-weak-upper-bound *problem (ULWUB) iff, $\forall \rho$ starting from m_0 we have $0 \leq W_{B_0 \downarrow B_{max}}(\rho) \leq B_{max}$.*

Informally, the accumulated weight along a run is the initial energy provision increased or decreased by the cost of each action at every step in ρ, and bounded at each step by the maximal amount of energy the system can accumulate. If a system is well designed, and abstracting away external energy losses such as batteries aging, it should be able to run forever without exhausting its energy budget. This property can be easily verified. It was already shown that the ULWUB problem is polynomial for weighted automata [3], and consists in detecting lassos ending with cycles of negative weight. This result applies to $LTS_{\mathcal{N}}$. Assuming that the upper bound for the energy budget is an integer smaller than $2^{|P|}$ we have:

Proposition 2. *The ULWUB problem is in PSPACE.*

Proof (Sketch). We can reuse the techniques proposed by [3], i.e. detect paths of length $< 2^{|P|}$ that end with a negative energy provision, or lassos that end with a cycle of negative cumulated weight. One does not need to explore paths of length greater than $2^{|P|}$, nor lassos of length $2^{|P|+1}$. This nondeterministic exploration memorizes at most 2 markings of the system and maintains energy budgets with $O(|P|)$ bits, i.e., it is in PSPACE (w.r.t. the number of places in $\mathcal{N}$). $\qquad\square$

3   Diagnosis with Processes of a Net

The semantics of Petri nets can also be given in a non-interleaved setting with *processes*, a partially ordered representation of transitions firings.

Definition 6. *A process of a net* $\mathcal{N} = (P, T, {}^{\bullet}(), (){}^{\bullet}, \lambda_{\mathcal{N}}, m_0)$ *is a tuple* $\varepsilon = (E, B, \lambda)$, *where E is a set of events, B is a set of conditions, and $\lambda : E \rightarrow \Sigma$ is a labeling of events. An* event *is a pair of the form $e = (X, t)$, where $X \subseteq B$ is a set of conditions needed to execute e, and t a particular transition. Given an event $e = (X, t)$, we denote by $t(e) = t$ the transition e refers to. To be consistent with $\mathcal{N}$, we have $\lambda(e) = \lambda_{\mathcal{N}}(t(e))$. A* condition *is a pair of the form $b = (e, p)$, where $e \in E$ and $p \in P$ is a reference to a place of $\mathcal{N}$. We denote by $Place(b)$ the place referred to in b.*

In processes, events are occurrences of transitions firing, and conditions places which contents is consumed/produced when firing a transition. Processes define a partial ordering (antisymmetric, transitive, reflexive relation) among their elements. We write $x \prec y$ when $x = (e, p)$ is a condition and $y = (X, t)$ an event such that $x \in X$, or when $x = (X, t)$ is an event and y a condition of the form $y = (x, p)$. This way, a process also defines a partial order among its events: we write $e \leq_{\varepsilon} e'$ when $e \prec^* e'$. Given an element x in a process, the set of its predecessors is denoted $\downarrow x$ and is the set $\downarrow x = \{y \in E \cup B \mid y \prec^* x\}$. Similarly, the set of successors of x is denoted $\uparrow x$ and is the set $\uparrow x = \{y \in E \cup B \mid x \prec^* y\}$. An element x is *minimal* (resp. *maximal*) in a process iff $\downarrow x = \{x\}$ (resp. $\uparrow x = \{x\}$). Minimal and maximal elements in processes are conditions. We denote by $min(\varepsilon)$ the set of minimal places in ε and by $max(\varepsilon)$ the set of maximal places in ε. Processes are *conflict-free*: for every pair of distinct events $e, e' \in E$, we have ${}^{\bullet}e \cap {}^{\bullet}e' = \emptyset$. Processes are also *join-free*: $\forall e \neq e' \in E$, $e^{\bullet} \cap e'^{\bullet} = \emptyset$.

Processes have been well studied (see for instance [8]) but for completeness, we give an inductive construction technique for processes of a net $\mathcal{N}$, starting from marking m_0. We assume a dummy event $\perp$, and create a set of conditions $B_0 = \{(\perp, p) \mid p \in m_0\}$. The initial process is $\varepsilon_0 = (\emptyset, B_0, \lambda_0)$, where λ_0 is the empty map. Then, we iterate the following construction for each process $\varepsilon_i = (E_i, B_i, \lambda_i)$: we compute the set Max_i of maximal conditions in ε_i. Max_i corresponds to the state (marking) of the system once all transitions appearing in process ε_i have been executed. An occurrence of transition t can be appended to ε_i as soon as ${}^{\bullet}t \subseteq Place(Max_i)$. Intuitively, after executing ε_i, the places needed by t to fire are filled. Note that more than one transition can satisfy this condition. When t can be appended to ε_i we can define $e_{i+1} = (Max_i \cap Place^{-1}({}^{\bullet}t), t)$, the event consuming conditions from Max_i that are instances of places in ${}^{\bullet}t$, and build the process $\varepsilon_{i+1} = (E_i \cup e_{i+1}, B_i \cup \{(e_{i+1}, p) \mid p \in t^{\bullet}\}, \lambda_i \cup (e_{i+1} \rightarrow \lambda(t)))$.

A *linear extension* of a process $\varepsilon = (E, B, \lambda)$ is a sequence of events $e_1 \ldots e_n$ such that $n = |E|$ and for every $i < j$, $e_j \not\prec^* e_i$. A *linearization* of ε is a word $w = a_1.a_2 \ldots a_n$ such that there exists a linear extension u of ε with $\lambda(u) = w$. Intuitively, linearizations of ε are words that could be logged sequentially during execution of ε. Processes are a compact way to represent executions of Petri nets. For every run $\rho = m_0 \xrightarrow{t_1} m_1 \ldots \xrightarrow{t_n} m_n$ of $\mathcal{N}$, there exists a unique process ε_{ρ} obtained by appending successively $t_1, \ldots t_n$. Considering conditions as places, and events as transitions, processes are occurrence nets, i.e. an acyclic, join-free and conflict free type of net. We can hence safely talk about runs of a process and denote by $Confs(\varepsilon) \subseteq 2^P$ the configurations that can be reached during executions of $\mathcal{N}$ represented by ε. Figure 2 shows two processes for the net of Fig. 1, with $\Sigma_o = \{a, b\}, \Sigma_{uo} = \{u, f\}$ and $\Sigma_f = \{f\}$. Conditions are represented as circles, and events as rectangle. Black rectangles represent occurrences of observable transitions, and grey rectangles of unobservable ones. The left process corresponds to an observation $a.b$, and contains no fault, the right one to an observation $b.a$, and contains a fault.

Fig. 2. Two processes for the net of Fig. 1.

Definition 7. *Let $\Sigma = \Sigma_o \uplus \Sigma_{uo}$ be a finite alphabet, $\mathcal{N}$ be a Petri net labeled by Σ, and $w \in \Sigma_o^*$ be an observation. A process ε of $\mathcal{N}$ is an* explanation *of w iff $\varepsilon = \varepsilon_\rho$ for some run $\rho = m_0 \xrightarrow{t_1} m_1 \ldots \xrightarrow{t_n} m_n$ of $\mathcal{N}$ such that $\Pi_{\Sigma_o}(w_\rho) = w$.*

Hence, a process ε is an explanation of an observation w iff w is the projection of a linearization $a_1 \ldots a_{|w|}$ of ε on Σ_o. Note that explanations can contain an arbitrary number of occurrences of unobservable transitions occurring after or concurrently with the last observable transition. Our objective is to define mechanisms that detect faults from partial observations after a finite number of steps of the system. Moreover, we want these mechanisms to run with finite memory. This is not always possible for general Petri nets [4], but for safe Petri nets, sensible restrictions allow diagnosers to memorize finite suffixes of processes.

Let $w \in \Sigma_o^*$ be the observation of some run of $\mathcal{N}$. We can build inductively a set $\mathbb{E}_w$ of processes that "explain" w as follows. We start from a set $\mathbb{E}_w^0$ that contains process $\varepsilon_0 = (\emptyset, B_0, \lambda_0)$. At each step, starting from a set of processes $\mathbb{E}_w^i$, one can select a process ε, and append to it either an unobservable transition, or a transition labeled by the next letter to explain in w. The construction stops when reaching a set of processes $\mathbb{E}_w^n$ that cannot be extended without adding more observable events than in w.

As w is an observation of $\mathcal{N}$, $\mathbb{E}_w$ contains at least one process. Due to concurrency and choices, several transitions can usually be appended to a process. Hence the observation guided construction above is non-deterministic. Every unobservable transition can be appended to $\varepsilon \in \mathbb{E}_w^i$ if it is allowed from configuration $Max(\varepsilon)$, and an observable transition can be appended if it carries the label of the next unexplained action in w. There can be several occurrences of such transitions in a labeled net. As the set of appendable transitions may contain conflicting transitions $\mathbb{E}_w$ can contain more than one process. Interested readers can find an algorithm for the construction of $\mathbb{E}_w$ in [13].

Addition of an unobservable transition to a process $\varepsilon \in \mathbb{E}_w^i$ is not conditioned by w. Hence, in general, process construction (w.r.t w) needs not stop. Indeed, one can append successively an arbitrary number of unobservable transitions, and without restriction, the set of explanations $\mathbb{E}_w$ for observation $w \in \Sigma_o^*$ may not be finite. We hence define the following restriction:

Definition 8. *A Petri net* $\mathcal{N} = (P, T, {}^{\bullet}(), ()^{\bullet}, \lambda)$ *with observable alphabet* $\Sigma_o \subseteq \Sigma$ *is* boundedly silent *iff there exists a bound* $K \in \mathbb{N}$ *such that for every marking* $M \in Reach(\mathcal{N})$, *there exists no process* $\varepsilon = (E, B)$ *that starts at* M *and such that* $\lambda(t(E)) \subseteq (\Sigma \setminus \Sigma_o)$ *and* $|\varepsilon| > K$.

Requiring boundedly silent nets is a sensible assumption asking that systems fire an observable event regularly. A similar notion exists for diagnosis of systems described with automata: it requires that systems have no unobservable cycle.

Proposition 3. *Let* $\mathcal{N}$ *be a boundedly silent Petri net, with bound* K, *and let* $w \in \Sigma_o^*$ *be an observation. Then,* $\mathbb{E}_w$ *is finite, and contains processes built in at most* $(|w| + 1).K$ *steps.*

Proof (sketch). By induction on the length of w.

In [13], we give an effective procedure $Unfold(\mathcal{N}, w, m_0)$ that unfolds a boundedly silent net $\mathcal{N}$ starting from m_0 to compute $\mathbb{E}_w$. The algorithm proceeds inductively and returns all explanations of w provided by $\mathcal{N}$. Every process ε_w^i in $\mathbb{E}_w$ has exactly $|w|$ observable transitions, and every observable transition of ε_w^i can be associated a letter of w that it explains. Note however that processes are "saturated" by appending all unobservable transitions that may have occurred without generating observations. Hence, some processes built by our algorithm contain unobservable transitions that are not *needed* to explain observation w. We can use the same algorithm to build a set of silent processes depicting maximal unobservable behaviors, starting from any marking by calling $Unfold(\mathcal{N}, \epsilon, M)$.

If *all* processes in $\mathbb{E}_w$ contain faulty processes, then one can claim without error that a fault has occurred. Similarly, if all processes in $\mathbb{E}_w$ are non-faulty, then one can claim that the observed behavior is non-faulty. If $\mathbb{E}_w$ is composed of faulty and non-faulty processes, then ambiguity remains on whether a fault occurred. This is a standard setting in diagnosis. Another frequent assumption in diagnosis is that systems are stopped immediately after occurrence of an observable action. Our setting slightly differs, as saturating processes means that one considers executions that have not yet produced an additional observation after w. Though this does not make a huge difference in decidability or algorithms, this setting seems more natural, especially in a context where unobserved additional transitions might be concurrent with the occurrence of the last observable ones. Let us comment this situation: let $\varepsilon \sqsubseteq \varepsilon'$ be two explanations of w. Supposing that ε is the actual behavior of $\mathcal{N}$ that led to observation w, then one cannot decide whether some events in $\varepsilon' \setminus \varepsilon$ have occurred or not. This is a new source of ambiguity: if ε contains no faulty transition, and ε' is faulty, then ambiguity arises from the fact that one cannot yet decide whether this faulty transition has already occurred or not. Note also that future occurrence of a fault after ε is not mandatory either, as after ε, $\mathcal{N}$ can still execute a non-faulty process ε'' such that $\varepsilon \sqsubseteq \varepsilon''$ instead of ε'.

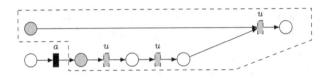

Fig. 3. A process for the net of Fig. 1 wrt observation a. The frontier is denoted by greyed conditions. The summary is the set of events and conditions in the dashed zone.

Building explanations online with observations results in ever growing processes. However, remembering a whole process is not needed for fault detection. Indeed, the important information to keep from a process is whether a fault has occurred, and the maximal conditions after observable events that *must* have happened in this process. It suffices to find the possible configurations of a system after an observation. This information can be memorized as a finite *summary.*

Definition 9. *Let* $\varepsilon = (E, B, \lambda)$ *be a process explaining* $w \in \Sigma_o^*$. *An event* e *in* ε *must have occurred if* $\lambda(e) \in \Sigma_o$, *or* $\exists f, e \le f$ *and* f *must have occurred. We denote by* $Must(\varepsilon)$ *the set of events that must have occurred in* ε. *The silent part of* ε *is the restriction of* ε *to events in* $E' = E \setminus {\downarrow} Must(\varepsilon)$ *and conditions in* ${}^\bullet E' \cup E'^\bullet$. *The execution frontier* $Frontier(\varepsilon)$ *of process* ε *is the set of minimal places in the silent part of* ε. *The summary of* ε *is denoted* $Sum(\varepsilon)$, *and is the restriction of* ε *to* ${\uparrow} Frontier(\varepsilon)$.

Let us give the intuition behind the notions of frontier and summary of a process. A frontier is a possible configuration that the system reaches when executing the smallest possible subset of events containing all observable events needed to explain w. This does not mean that after observing w the system is necessarily in this state, as unobservable transitions may have occurred concurrently with the last observable actions of w. Summaries capture parts of executions that *may* have occurred. A summary is also a process, starting from $Frontier(\varepsilon)$. Considering processes as standard nets, we can prove that $Frontier(\varepsilon)$ is a marking of ε and hence also that $Place(Frontier(\varepsilon))$ is a marking of $\mathcal{N}$ (see [13] for complete proof). Hence, up to a relabeling of minimal conditions in the silent summary $\varepsilon_S = Sum(\varepsilon)$, we have that ε_S is equivalent to some process in $Unfold(\mathcal{N}, \epsilon, Frontier(\varepsilon))$. Moreover, remembering silent parts of processes suffices to describe the set of possible configurations a system might be in after an observation. Figure 3 illustrates the notions of frontier and summary on a process of the net given in Fig. 1.

Proposition 4. *Let* ε *be an explanation of observation* $w \in \Sigma_o^*$, *and* ε_S *be its summary. Then for every marking* M *of* ε_S, $Place(M)$ *is a marking that is reachable via a run* ρ *of* $\mathcal{N}$ *such that* $\lambda(\rho) = w$.

Proposition 5. *Let* ε *be an explanation of word* $w \in \Sigma_o^*$, *and* M *be its frontier. Then* $\varepsilon' \sqsupseteq \varepsilon$ *is an explanation of* $w.a \in \Sigma_o^*$ *iff there exists* $\varepsilon_a \in Unfold(\mathcal{N}, a, M)$ *such that* $\varepsilon' = \varepsilon \cup \varepsilon_a$ *(where union of processes means union component wise).*

When performing diagnosis, the important information to memorize in processes is whether a fault has occurred, or may have occurred, and the configurations the system might be in after some observation. Note that the set of summaries obtained by pruning explanations for an observation is always finite. Note also that $Frontier(\varepsilon)$ is the minimal set of places in $\varepsilon_S = Sum(\varepsilon)$. In the rest of the paper, we will write $\varepsilon_S \xrightarrow{a} \varepsilon'$ if ε_S is a summary with frontier M, and $\varepsilon' \in Unfold(\mathcal{N}, a, M)$ (one can execute observable action a from some configuration in the state estimation contained in ε_S and obtain a process ε'. Note that ε' is not a summary, but can be projected to obtain $\varepsilon'_S = Sum(\varepsilon')$. Note also that in general $\varepsilon' \not\sqsupseteq \varepsilon$, as the execution chosen from M may differ from the possible execution depicted by ε_S. The common part between ε_S and ε'_S is $\varepsilon_S \cap \varepsilon'$, and the part of ε' that does not appear in ε_S is denoted $\varepsilon' \setminus \varepsilon_S$.

A *possible state* after an observation $w \in \Sigma_o^*$ is a pair (V, ε_S) where ε_S is a summary of some explanation ε of w, and V is a tag from $\{F, N, A\}$ called a *verdict*. Tag F stands for faulty, N stands for non-faulty, and A stands for ambiguous. Tags are set as follows:

- $V = N$ iff $\lambda(\varepsilon) \cap \Sigma_f = \emptyset$: no fault have occurred in ε (even if occurrence of some events in ε_S is uncertain, none of them is faulty),
- $V = F$ iff $\lambda(\varepsilon \setminus \varepsilon_S) \cap \Sigma_f \neq \emptyset$: a fault occurred *before* any of the uncertain events,
- $V = A$ iff $\lambda(\varepsilon \setminus \varepsilon_S) \cap \Sigma_f = \emptyset$ and $\lambda(\varepsilon_S) \cap \Sigma_f \neq \emptyset$: one event in ε is a fault in the uncertain part of ε. There is no guarantee that this event was executed.

A *state estimate* after $w \in \Sigma_o^*$ is a set of possible states $SE = \{(V_i, \varepsilon_S^i)\}$. State estimate SE is said *faulty* if all possible states in SE carry tag F, and *normal* if all verdicts in SE are N. Last, it is said *ambiguous* if at least one verdict V_i is equal to A, or there exists two contradictory verdicts $V_i = F$ and $V_j = N$. So, ambiguity appears both when different verdicts originate from different explanations, and when a possible state contains a faulty event which execution is uncertain. As for summaries, state estimates can be maintained online with observations. Let $SE = \{(V_1, \varepsilon_S^1), \ldots (V_k, \varepsilon_S^k)\}$ be the state estimate obtained after observation w, and let $a \in \Sigma_o$. The update of SE after observation $a \in \Sigma_o$ is the set $SE' = \{(V_1', \varepsilon_S^{1'}), \ldots (V_1', \varepsilon_S^{k'})\}$ such that for every $(V_i', \varepsilon_S^{i'})$ of SE' there exists a possible state (V_i, ε_S^i) in SE such that $\varepsilon_S^i \xrightarrow{a} \varepsilon_a$, $\varepsilon_S^{i'} = Sum(\varepsilon_a)$ and:

- $V_i = F$ implies $V_i' = F$,
- $V_i \in \{N, A\}$ and $\lambda(\varepsilon_a) \cap \Sigma_f = \emptyset$ implies $V_i' = N$,
- $V_i \in \{N, A\}$ and $\lambda(\varepsilon_a \setminus \varepsilon_S^{i'}) \cap \Sigma_f \neq \emptyset$ implies $V_i' = F$,
- $V_i \in \{N, A\}$ and $\lambda(\varepsilon_a \setminus \varepsilon_S^{i'}) \cap \Sigma_f = \emptyset$ and $\lambda(\varepsilon_S^{i'}) \cap \Sigma_f \neq \emptyset$ implies $V_i' = A$.

We write $SE \xrightarrow{a} SE'$ when SE' is the state estimate obtained from SE after observation a. One can notice that if ε is an explanation for w, with a frontier F and a summary ε_S, then as ε_S is the part of ε obtained by saturation with unobservable transitions, any process of the form $\varepsilon \setminus \varepsilon_S \cup \varepsilon'_S$ replacing ε_S by an unobservable summary $\varepsilon'_S \in Unfold(\mathcal{N}, \epsilon, F)$ is also an explanation of w. Hence, state estimates should contain all processes that differ only via their summaries, which allows a compact representation (CSE) for state estimates. Let $SE = \{(V_i, \varepsilon_S^i)\}$. SE can be represented as sets of pairs of the form $CSE = \{(V_i, M_i)\}$ where each M_i is the set of places in a frontier F_i appearing in some summary of SE, and V_i is a verdict attached to some summary with frontier F_i. Obviously, one can recover a state estimate from a compact representation that is isomorphic to the original estimate (isomorphism is obtained by relabeling minimal conditions) and diagnosis can safely resume from each M_i for any extension of w, as shown in Proposition 5. A diagnoser that maintains state estimates can claim that an observation corresponds to a faulty (resp. non-faulty) behavior iff all verdicts are set to F (resp. N). A transition relation $CSE \xrightarrow{a} CSE'$ among CSEs can be designed identically to the transition relation among state estimates.

Proposition 6. *Given a safe boundedly silent Petri net $\mathcal{N}$, there exists only a finite number (in $O(2^{3.2^{|P|}})$) of compact representations of state estimates reachable after an observation of an execution of $\mathcal{N}$.*

A *diagnoser automaton* for a safe boundedly silent net $\mathcal{N}$ is an automaton $\mathcal{D}_\mathcal{N} = (S_{\mathcal{D}_\mathcal{N}}, \rightarrow_{\mathcal{D}_\mathcal{N}}, F_{\mathcal{D}_\mathcal{N}})$ such that $S_{\mathcal{D}_\mathcal{N}}$ is the set of reachable compact state estimates for $\mathcal{N}$, $\rightarrow_{\mathcal{D}_\mathcal{N}} \subseteq S_{\mathcal{D}_\mathcal{N}} \times \Sigma_o \times S_{\mathcal{D}_\mathcal{N}}$ is the transition relation among CSEs. $\mathcal{D}_\mathcal{N}$ is deterministic and of size in $O(2^{2^{|P|}})$. It reads observable labels appearing during executions of $\mathcal{N}$, and raises an alarm when the reached state estimate is faulty. Note that $\mathcal{D}_\mathcal{N}$ needs not be built a priori to perform diagnosis: one can maintain a single state estimate online, which is updated at each occurrence of a new observation, with a memory in $O(2^{|P|})$. Building this doubly exponential $\mathcal{D}_\mathcal{N}$ is not needed either to check diagnosability (Propostion 1). However, we show in Sect. 4 that states of $\mathcal{D}_\mathcal{N}$ are needed to address disambiguation with costs.

4   Disambiguation Under Budget Constraints

The main reason for non-diagnosability is unbounded ambiguity about the actual run executed by a system. However, in safety critical systems, long uncertainty about major failures is not acceptable. If the actual status of a system is not known and a major failure is suspected, then safety checks must be performed, even if this implies consuming more energy. We model these additional checks as follows: after each observed action, diagnosers have the possibility to perform one test to reduce ambiguity on the actual status of the system. For safe Petri nets, where places can be seen as boolean variables, it seems natural to model tests as mechanisms that provide information on the possible status of places. We first consider whether disambiguation suffices to avoid ambiguity, and then whether it can be performed without exhausting the system's resources.

Definition 10. *A* disambiguation function *is a partial map* $d : 2^P \to 2^{2^P}$ *that, given an* actual *marking, returns a* set of *possible* markings *of the system.*

Disambiguation functions return a set of markings that correspond to a partial observation of an actual (unknown) marking. The knowledge returned by the disambiguation map is not necessarily a state estimation that a diagnoser might have built. So, in general, disambiguation functions provide additional information, that can be used to make a system diagnosable. A first example is map id, that takes a marking M as argument, and returns set $\{M\}$. Another example is the map d_p that associates to every marking M all configurations where $p \in P$ is marked if $M(p) = 1$ and all configurations where p is not marked otherwise. Such a function can be used to inform users on the status of a boolean variable. The result returned by a disambiguation function is not only related to the contents of a place and can encode a constraint on possible markings (e.g. $M(p_1) = 0 \Rightarrow M(p_2) = 1$). In the sequel, we will assume that systems are provided with a finite set $Dis = \{d_1, \ldots, d_n\}$ of (costly) disambiguation functions, and that a diagnoser uses at most one of them after every observation.

Let ε_S be a summary, and let d be a disambiguation function. We say that ε_S is *compatible* with d in marking M if at least one configuration of ε_S is made of places that belongs to $d(M)$. Let ε_S be a summary, d be a disambiguation function, and M be a marking such that ε_S is compatible with d in M. Then, getting information using disambiguation provided by function d can help users refine their estimation of the state of the system. The *refinement* of ε_S by $d(M)$ is denoted by $\varepsilon_{S \setminus d(M)}$, and is the projection of ε_S on successors of configurations of ε_S that are compatible with information provided by $d(M)$. It is computed as follows: $\varepsilon_{S \setminus d(M)} = (E_d, B_d, \lambda_d)$ where $E_d = E \cap \uparrow Cond(\varepsilon_S, d, M)$, $B_d = B \cap \uparrow Cond(\varepsilon_S, d, M)$, with $Cond(\varepsilon_S, d, M) = \{b \in B \mid \exists C \in Conf(\varepsilon_S), b \in C \wedge Place(C) \in d(M)\}$, and λ_d is the restriction of λ to E_d.

Of course, when a system is diagnosable, one does not need disambiguation mechanisms to perform diagnosis. However, if a system is not diagnosable, the questions of diagnosability with the help of tests makes sense. Recall that diagnosis can only be performed after some observation. If ambiguity remains after an observation, then one can perform a test from Dis, or wait for a better moment to reduce ambiguity. This situation is well captured as a strategy.

Definition 11. *A* strategy *is a function* $\sigma : \Sigma_o^* \to Dis \cup \{d_\bot\}$ *that indicates, for every observation* w *whether a disambiguation from* Dis *should occur* $(\sigma(w) \in Dis)$ *or if no disambiguation shall be applied* $(\sigma(w) = d_\bot)$.

The set of processes compatible with w and pruned by strategy σ is the set of processes built inductively as follows: we first set $\mathbb{E}_{w,\sigma}^0 = \{Unfold(m_0, \epsilon, \mathcal{N})\}$ Then, we compute $\mathbb{E}_{w,\sigma}^{i+1}$ from $\mathbb{E}_{w,\sigma}^i$ as follows: we use disambiguation $d = \sigma(w_{[1..i+1]})$ prescribed by σ. Then, for every process $\varepsilon_S \in \mathbb{E}_{w,\sigma}^i$, every $\varepsilon' \in Unfold(\varepsilon_S, w_i, \mathcal{N})$ if there exists $M \in Confs(sum(\varepsilon'))$ such that $sum(\varepsilon')$ is compatible with $d(M)$ we add ε' to $\mathbb{E}_{w,\sigma}^{i+1}$. As the application of disambiguation only restricts the set of processes, the notions of summaries, possible states, state estimates and their compact representations (see Definition 9) can be used. One can maintain finite sets of state estimates that are compatible with w and pruned by a strategy σ. Similarly, as disambiguation applies after a new observation, a finite set of representations of possible states reached by a net $\mathcal{N}$ together with a verdict can be maintained online with occurrences observable actions with finite memory. **Diagnosability with disambiguation** can now be formalized:

Given a Petri net $\mathcal{N}$ and a set of disambiguation functions $\mathcal{D}is = \{d_1, \ldots, d_n\}$, is there a strategy σ and a constant $K \in \mathbb{N}$ such that for every faulty run ρ of $\mathcal{N}$, and every run $\rho' = \rho.\rho_1$ such that $|\rho'_1|_{\Sigma_o} > K$ every process in $\mathbb{E}_{w,\sigma}^{|w|}$ (i.e. with $w = \Pi_{\Sigma_o}(\rho')$ and pruned by strategy σ) is faulty?

Theorem 1. *Given a Petri net $\mathcal{N}$ and a set of disambiguation functions $\mathcal{D}is$, one can decide in 2-EXPTIME whether $\mathcal{N}$ is diagnosable with the help of $\mathcal{D}is$.*

Proof (Sketch). We can define diagnosis with disambiguation as a partial information co-Büchi game. A two players co-Büchi game with a winning condition C is won by player 0 iff it has strategy such that only states in C are visited infinitely often. We first build an arena which nodes contain: the actual marking of the net, the fault status ($\{N, F\}$) of the system, and state estimates. The game is turn-based. Nodes of this arena are either nodes of the diagnoser (player 1), or nodes of the system (player 0). From its node, a diagnoser can choose one disambiguation function to refine its state estimation. From its nodes, the system can choose any sequence of action containing a single observation, and hence move to another node of the diagnoser. The partial information in the game comes from the fact that the real state of the system is not known. In this game, the system wins if status of faults can remain ambiguous forever, i.e. it can produce a fault, and force the game to remain in nodes with ambiguous state estimates. The complexity comes from the doubly exponential size of the arena. □

The translation to a partial information co-Büchi game is interesting for reasons to go beyond the simple complexity characterization. A partial information co-Büchi game on an arena G can be translated to a perfect information parity game over an arena G^K that represents knowledge of players, and as parity games are positional [11], it means that bounded memory strategies are sufficient to make a system diagnosable with disambiguation. Usually, an additional exponential cost has to be paid to solve partial information games [6]. However, state estimates already contain knowledge of players. An additional exponential blowup needs not be paid, and solving the game is "only" doubly exponential. At first sight, diagnosis with disambiguation looks close to the diagnosis with adaptive observable alphabet proposed by [5], and to the active diagnosis (see for instance [12]) where some transitions can be disabled to make a system diagnosable. Adaptive observations and disambiguation are orthogonal techniques, and our framework differs from active diagnosis as it does not modify the overall behavior of the monitored system. We discuss these issues in conclusion.

Information on the current state of a system can be obtained by running some tests, activating sensors, checking for the status of some component, etc.

Such operations usually have a cost, for instance in terms of time or energy. Fortunately, knowing precisely at every instant the current state of a system is not needed to make it diagnosable. So, one does not have to perform the most expensive tests at every step to ensure diagnosability. In the rest of the paper, we assume that tests follow our energy consumption scheme, and consume energy. We hence extend the cost function to consider the cost of tests. From now, we will consider that a cost functions is a map $C : T \cup \mathcal{D}is \to \mathbb{N}$ that associates integers to transitions of T and negative integers to elements of $\mathcal{D}is$. As tests cost energy, they cannot be used systematically to disambiguate systems states. The next question to consider is whether one can use tests to disambiguate state estimates of a system without exhausting its energy.

The **ULWUB diagnosability** question is defined as follows: Given a non-diagnosable Petri net $\mathcal{N}$, $\mathcal{D}is$, C, B_0 and B_{max}, is there a strategy that makes $\mathcal{N}$ diagnosable while maintaining the energy provision above 0 ? We assume that a diagnoser does not know precisely the remaining energy provision of the system. This assumption seems realistic to model examples such as the car ABS given in introduction. It can however be relaxed, in which case information on remaining energy can help reducing ambiguity. Hence, a diagnoser uses only its belief, i.e., its estimation of the possible state of the system and bounds for the remaining energy. A diagnoser should take the same disambiguation decision for all states with the same belief. In the rest of the paper, we show that this question can be answered as an energy game, that can again be solved as a particular instance of co-Büchi game. In a setting where tests have a cost, one may have to wait before launching a test that could exhaust the energy budget of the system. Again, the right moment to perform a test can be defined as a strategy.

Theorem 2. *The ULWUB diagnosability is decidable in 2-EXPTIME.*

Proof (sketch). We build an arena where nodes are nodes of the system, or nodes of the diagnoser. As for disambiguation without budget constraints, the diagoser can choose a test to disambiguate its state estimates. Nodes contain the current state of the system, its fault status, the remaining energy provision. They also contain state estimates, and for each summary in state estimates, an upper and lower bound for remaining energy. In system nodes, the system can play any sequence of actions with a single observable event. The current state, fault status, and budget are adapted accordingly. The state and budget estimate are also changed according to the possible executions and their costs. If remaining energy is known, then it is represented in nodes as a single rational value, and is used to reduce ambiguity on state estimates. From diagnoser nodes, the diagnoser can choose a particular disambiguation function to refine its state estimate, which decreases the actual energy budget and the upper/lower estimations by the cost of this test. The diagnoser can also choose to avoid tests, and give its turn to the system without refining its state estimate. If either the system or the diagnoser exhaust the energy budget of the system, the game reaches a particular exhaustion sink node that cannot be left. Then, in this arena, the diagnoser cannot disambiguate faults with energy constraints iff the system has a strategy to remain in ambiguous or exhaustion nodes forever. This is again a partial knowledge co-Büchi game over a doubly exponential size arena. As for disambiguation, building the knowledge of players in our arena amounts to building copies of state and budget estimates. Hence, an additional exponential blowup needs not be paid, and solving the game is doubly exponential.   □

5   Conclusion

This work is not the first one addressing diagnosis with *Petri nets* (see the survey in [16]). An offline algorithm to build all explanations for an observation using Petri net unfoldings is proposed in [9]. Fault detection is a simpler problem than explanation reconstruction, that can be addressed online, as a part of each possible run can be forgotten when observations grow. A diagnosis framework for general Petri nets is proposed in [1]. The authors exhibit examples of unbounded Petri nets that are diagnosable, i.e., in which every fault can be detected within a finite number of steps even if no general upper bound on the number of observations needed before detection exists. In this setting, diagnosers are no longer regular. [10] represents systems as general Petri nets, and diagnosers as Petri nets too, but do not address diagnosability. Diagnoser nets maintain sets of *extended markings*, i.e. reachable markings augmented with finite vectors of fault status. Extended markings are close to our state estimates, and their construction is close to observation guided unfolding. However, the approach of [10] systematically enumerates all possible markings.

Our framework can be seen as *active diagnosis*, as diagnosers can perform tests. In active diagnosis, diagnosers guide the observed system. They play the role of a controller that prevents actions to avoid ambiguity in otherwise non-diagnosable systems. The active diagnosability problem then consists in deciding whether such a controller exists. It is EXPTIME-complete for automata [12]. In [2], active diagnosis is recast in a probabilistic setting. The main objective is then to ensure *safe active diagnosability*, i.e. show existence of controller that guarantees diagnosability while ensuring that non-faulty runs still have a strictly positive probability. Active probabilistic diagnosability is EXPTIME-complete, but safe active diagnosis is undecidable. Active diagnosis and disambiguation are orthognal techniques. Active diagnosis forbids some controllable transitions to recover diagnosability, which changes the behavior of the system. Disambiguation brings additional information to diagnosers, but does not change the behavior of the disambiguated system: for every run ρ of the arena over actions in $\Sigma \uplus \mathcal{D}is$, $\Pi_\Sigma(\rho)$ is a run of the original net $\mathcal{N}$. The *dynamic masks* of [5] are also a form of active diagnosis. In this setting, the observable alphabet can be changed dynamically to ensure diagnosability. Dynamic masks and disambiguation are also uncomparable, as we do not change the set of observable actions but rather introduce new actions to refine state estimation.

Cost of diagnosis was considered in [5,18]. In [18], a cost is computed for each state of a finite transition system. It represents the effort needed to diagnose a fault. Non-diagnosable systems have diverging costs, and a system is diagnosable if one can associate a finite cost to each state. The cost model considered assigns a weight to each observable action, and the cost associated to a state is a sum of discounted costs of observations for runs leaving this state with a particular dynamic observation strategy (that may change the observable alphabet after each observation). Similarly, [5] defines the cost of diagnosis in terms of weights attached to sets of observable actions. The diagnosers and observers considered are dynamic: the set of observed actions may change depending on the observation. Cost of diagnosis is then the maximal mean cost needed to perform the observations required by the diagnoser. [5] show that checking existence of a k-diagnoser with an optimal cost can be solved in $O(|\Sigma| \times m \times 2^{n^2} \times 2^k \times 2^{2^{|\Sigma|}})$, where n is the number of states and m the number of transitions in the considered automaton. Our framework for diagnosis with costs can be seen as a quantitative game, as defined in [3]. However, in addition to quantitative considerations, in diagnosis players have to meet/prevent ω-regular objectives. We believe that the setting proposed in this paper easily adapts to most of the cost constraints appearing in [3], such as strict upper and lower bounds on energy level.

Using pieces of processes is a compact way to represent possible states of a system. Summaries can be computed efficiently with unfolding techniques (see [13] for details). However, in the worst cases, one may still need to enumerate all possible markings of a net. We believe that in practice this situation does not occur for systems with concurrency, but this has to be demonstrated on case studies, to measure the efficiency gain when working with a Petri net $\mathcal{N}$ and its unfolding rather than directly from its LTS $LTS_{\mathcal{N}}$. In a similar way, in our approach processes and summaries are partial order models, but observations are linearizations. An immediate question is whether our setting can be extended to a fully concurrent one where observations are partial orders too. The natural notion of explanation is when an observed order O *embeds* into the ordering on events depicted in a process ε of $\mathcal{N}$. However, in this setting, diagnosers may not work with finite memory. A possible extension of this work is then to find sensible restrictions on nets that allow diagnosers that only need to memorize summaries of processes and observations of bounded sizes.

Diagnosability with disambiguation and with ULWUB constraints are solved as partial information co-Büchi games in 2-EXPTIME. As for active diagnosis [2,12] partial observation leads to an exponential blowup that does not have to be paid twice when solving games. A future work is to find lower bounds. We also plan to consider the impact of several hypotheses of this work on the complexity of ULWUB diagnosability. For instance, if the diagnoser has access to the remaining energy of the system, exhaustion state is not needed, and one does not have to maintain energy estimation. This should not change the complexity of the ULWUB diagnosability game, but at least reduces the size of the arena.

References

1. Basile, F.: Overview of fault diagnosis methods based on petri net models. In: Procedings of IEEE European Control Conference (ECC) (2014)
2. Bertrand, N., Fabre, É., Haar, S., Haddad, S., Hélouët, L.: Active diagnosis for probabilistic systems. In: Muscholl, A. (ed.) FoSSaCS 2014. LNCS, vol. 8412, pp. 29–42. Springer, Heidelberg (2014). doi:10.1007/978-3-642-54830-7_2
3. Bouyer, P., Fahrenberg, U., Larsen, K.G., Markey, N., Srba, J.: Infinite runs in weighted timed automata with energy constraints. In: Cassez, F., Jard, C. (eds.) FORMATS 2008. LNCS, vol. 5215, pp. 33–47. Springer, Heidelberg (2008). doi:10.1007/978-3-540-85778-5_4
4. Cabasino, M.P., Giua, A., Lafortune, S., Seatzu, C.: Diagnosability analysis of unbounded petri nets. In: Proceedings of CDC 2009, pp. 1267–1272. IEEE (2009)
5. Cassez, F., Tripakis, S.: Fault diagnosis with static and dynamic observers. Fundam. Informaticae 88(4), 497–540 (2008)
6. Chatterjee, K., Doyen, L.: The complexity of partial-observation parity games. In: Fermüller, C.G., Voronkov, A. (eds.) LPAR 2010. LNCS, vol. 6397, pp. 1–14. Springer, Heidelberg (2010). doi:10.1007/978-3-642-16242-8_1
7. Chatterjee, K., Doyen, L., Henzinger, T.A., Raskin, J.-F.: Algorithms for omega-regular games with imperfect information. In: Ésik, Z. (ed.) CSL 2006. LNCS, vol. 4207, pp. 287–302. Springer, Heidelberg (2006). doi:10.1007/11874683_19
8. Esparza, J., Römer, S., Vogler, W.: An improvement of Mc Millan's unfolding algorithm. Formal Methods Syst. Des. 20(3), 285–310 (2002)
9. Fabre, E., Benveniste, A., Haar, S., Jard, C.: Distributed monitoring of concurrent and asynchronous systems. Discret. Event Dyn. Syst. 15(1), 33–84 (2005)
10. Genc, S., Lafortune, S.: Distributed diagnosis of discrete-event systems using petri nets. In: van der Aalst, W.M.P., Best, E. (eds.) ICATPN 2003. LNCS, vol. 2679, pp. 316–336. Springer, Heidelberg (2003). doi:10.1007/3-540-44919-1_21
11. Grädel, E., Thomas, W., Wilke, T. (eds.): Automata Logics, and Infinite Games: A Guide to Current Research. LNCS, vol. 2500. Springer, Heidelberg (2002). doi:10.1007/3-540-36387-4
12. Haar, S., Haddad, S., Melliti, T., Schwoon, S.: Optimal constructions for active diagnosis. In: Proceedings of FSTTCS 2013. LIPICS, vol. 24, pp. 527–539 (2013)
13. Hélouët, L., Marchand, H.: On the cost of diagnosis with disambiguation. Technical report, INRIA Rennes (2017). https://hal.inria.fr/hal-01537796
14. Jéron, T., Marchand, H., Pinchinat, S., Cordier, M-O.: Supervision patterns in discrete event systems diagnosis. In: Proceedings of WODES, pp. 262–268 (2006)
15. Jiang, S., Huang, Z., Chandra, V., Kumar, R.: A polynomial time algorithm for diagnosability of discrete event systems. IEEE TAC 46(8), 1318–1321 (2001)
16. Pocci, M.: Test and diagnosis of discrete event systems using Petri nets. Ph.D. thesis, Univ. Aix-Marseille and Univ. Cagliari (2013)
17. Sampath, M., Sengupta, R., Lafortune, S., Sinaamohideen, K., Teneketzis, D.: Diagnosability of discrete event systems. IEEE TAC 40(9), 1555–1575 (1995)
18. Thorsley, D., Teneketzis, D.: Diagnosis of cyclic discrete-event systems using active acquisition of information. In: Proceedings of WODES 2006, pp. 248–255 (2006)

Multi-class Resource Sharing with Batch Arrivals and Complete Blocking

Paul Ezhilchelvan[(⊠)] and Isi Mitrani

School of Computing Science, Newcastle University,
Newcastle upon Tyne NE1 7RU, UK
{paul.ezhilchelvan,isi.mitrani}@ncl.ac.uk

Abstract. A cloud provider hosts virtual machines of different types, with different resource requirements. There are bounds on the total amounts of each kind of resource that are available. Requests arrive in batches of different sizes, and are accepted if all the VMs in the batch can be accommodated; otherwise the request is blocked, with an associated loss of revenue. The trade-offs between costs and benefits are evaluated by means of an appropriate model, for which a novel solution is proposed. The applicability of that solution is extended, by means of a simplification, to very large-scale systems. Numerical examples and comparisons with simulations are presented.

1 Introduction

A cloud provider may offer services of different types, with different patterns of demand, resource requirements and charges. A job of a given type is run by instantiating an appropriate Virtual Machine (VM), provided that the resources it requires are available. There are bounds on the total amounts of different resources, so that whether a VM can be instantiated or not, depends both on the type of the new job and on the numbers and types of the other jobs already running.

We are concerned with systems where user demands arrive in batches whose sizes may be fixed or random, and may depend on type. Moreover, the nature of the applications is such that users are not interested in partial acceptance: either all VMs in a batch must be instantiated, or none. This is known as the *complete blocking* policy, in contrast to the *partial blocking* policy whereby a part of a batch may be accepted and the rest rejected.

There are many applications which require a batch of VMs in order to complete a certain task within a certain period of time. These are often concerned with the analysis of large volumes of data, such as those arising in the fields of sociology, biology or high energy physics. In particular, the 'MapReduce' framework (e.g., see [3]), allowing the deployment of batches of VMs, is widely used.

The trade-offs in this context are between the costs incurred by providing resources, and the revenues obtained by running jobs. The problem is to decide

© Springer International Publishing AG 2017
N. Bertrand and L. Bortolussi (Eds.): QEST 2017, LNCS 10503, pp. 157–169, 2017.
DOI: 10.1007/978-3-319-66335-7_10

what amounts of the various resources to provide, so as to maximize the average long-term profit (revenues minus costs) per unit time, or to achieve certain quality-of-service objectives. To that end, we analyze and solve an Erlang-type loss model with multi-class batch arrivals and complete blocking.

We assume that the demand parameters are given, and the system reaches steady state during a period where those parameters remain fixed. In practice, the resource provisioning policies would have to be supplemented by some monitoring and parameter estimation technique that would detect when the traffic parameters change. Such techniques exist (see below). It is also worth pointing out that batch arrivals can, and have been, used to model bursty arrival streams.

The model that is addressed here has been considered before, but has not been solved. In their 1995 paper [2], Choudhury, Cheung and Whitt claimed to provide a product-form solution for both the partial blocking and the complete blocking policies. That solution agreed with the results obtained by Kaufman and Regge [9], for general distribution of batch sizes, and by van Doorn and Planken [4], for geometrically distributed batches. However, both of those papers had analyzed only the partial blocking case.

We agree that a product-form solution holds in systems with partial blocking of batches, but will show that it does not hold in the case of complete blocking. Therefore, that problem is still open. The purpose of the present paper is to fill some of the gap by proposing and evaluating an accurate approximate solution based on fixed-point iterations.

The rest of the existing literature on multi-class resource sharing deals mainly with demands arriving one at a time in Poisson streams (i.e., no batches). Much of the work is in the context of circuit-switched networks, e.g., Kelly [10], Hampshire et al. [7], Kaufman [8], Roberts [14] and Ross [15]. In the telephony field, the resources are the circuits available on various links, and the job types are indexed by the set of links that can be reserved for a call. The optimal allocation of VMs on servers hired from a cloud was explored in Ezhilchelvan and Mitrani [6], and in Tan et al. [17]. Again, one-at-a-time Poisson arrivals were assumed.

More distantly related is quite a large body of work on server allocation with a single job type. Ezhilchelvan and Mitrani [5] showed that dynamic allocation policies do not bring significant benefits over static ones. The trade-off between performance and energy consumption was examined by Mazzucco et al. [11,12], using models and empirical observations. Their focus, and also that of Bodík et al. [1], was on estimating the traffic and reacting to changes in the parameters.

The model, and the profit maximization problem, are described in Sect. 2. The solution presented by Choudhury et al. is shown to be erroneous in Sect. 3, while Sect. 4 describes and evaluates the proposed fixed-point approximation. Section 5 is concerned with very large-scale systems. A simplified version of the fixed-point approximation is shown to be accurate and numerically stable. Some conclusions and directions for future work are summarized in Sect. 6.

2   The Model

The hosting infrastructure provides R different types of resources, such as CPUs, memory, interconnection bandwidth, etc. The total amount of available resource of type r is D_r, referred to as the 'resource capacity' of type r $(r = 1, 2, \ldots, R)$. Those resources are shared by VMs, or jobs, of T different types. A job of type j requires an amount $d_{j,r}$ of type r resource $(j = 1, 2, \ldots, T; r = 1, 2, \ldots, R)$. In order to run a job, all its resource requirements must be satisfied. For every j, at least one of the requirements $d_{j,r}$ is greater than 0, which imposes a limit on the maximum number, m_j, of type j jobs that can run in parallel:

$$m_j = \min_r \left\{ \left\lfloor \frac{D_r}{d_{j,r}} \right\rfloor \right\}, \tag{1}$$

where $\lfloor x \rfloor$ is the largest integer less than or equal to x.

Moreover, if D_r is replaced by the type r resource capacity *currently available* (determined by the numbers and types of jobs currently running), then (1) provides a limit on the number of type j jobs that can be admitted at a given moment.

Requests of type j arrive in an independent Poisson stream with rate λ_j. Each such request consists of a batch of jobs, all of type j, whose size has an arbitrary distribution dependant on j: there are k jobs in the batch with probability $q_{j,k}$. If there is at least one job in the batch that cannot be run because at least one of the resources it requires cannot be provided, then the whole batch is rejected. That is the *complete blocking* policy.

The probabilities $q_{j,k}$ can be arbitrary. However, a sensible batch size distribution should be consistent with (1). In other words, there should be a limit, K_j, on batch sizes of type j, such that $K_j \leq m_j$. Otherwise, some batches would be rejected even if there are no other jobs present.

Service times for jobs of type j are assumed to be i.i.d. random variables distributed exponentially with mean $1/\mu_j$ $(j = 1, 2, \ldots, T)$. The insensitivity property of the Erlang model does not hold when jobs arrive in batches (see [8]).

Suppose that each unit of resource of type r costs c_r to provide, and each job of type j that is run brings in a revenue of v_j. Denote by $\alpha_{j,k}$ the steady-state probability that an incoming batch of type j and size k is accepted. Then the total steady-state average profit, V, that the system generates per unit time, can be expressed as

$$V = \sum_{j=1}^{T} \lambda_j \sum_{k=1}^{K_j} q_{j,k} \alpha_{j,k} k v_j - \sum_{r=1}^{R} c_r D_r. \tag{2}$$

Clearly, increasing the resource capacities D_r leads to higher revenues, but also higher costs. The profit optimization problem is to choose D_r so as to maximize V. One could also consider a Quality-of-Service problem, which is to find the minimum values of D_r for which the acceptance probabilities $\alpha_{j,k}$ exceed certain pre-defined targets. In both cases, it is necessary to solve the model in order to determine $\alpha_{j,k}$.

3   An Erroneous Solution

The technique used in Choudhury et al. [2] is to replace an incoming batch of type j and size k (when $q_{j,k} > 0$), by a single *macro job* that goes through a series of k queues in tandem: in the first queue it uses $kd_{j,r}$ units of type r resource and is served at rate $k\mu_j$, regardless of how many other such macro jobs are present; in the second queue it uses $(k-1)d_{j,r}$ units of resource and is served at rate $(k-1)\mu_j$; this goes on until queue k, where it uses $d_{j,r}$ units of resource and is served at rate μ_j. After that, the macro job departs.

In this formulation, which is equivalent to the one in terms of batches of ordinary jobs, the system state is a vector of integers $[n_{j,k,s}]$, specifying the numbers of macro jobs of type j and size k that are now in queue s of their series of queues (i.e., they have $k+1-s$ ordinary jobs remaining). The authors find that, for both partial and complete blocking, the probabilities of those vectors satisfy local balance, and therefore the steady-state distribution has product form:

$$\pi(\mathbf{n}) = G \prod_{j=1}^{T} \prod_{k=1}^{K_j} \prod_{s=1}^{k} \frac{(\lambda_j q_{j,k})^{n_{j,k,s}}}{((k+1-s)\mu_j)^{n_{j,k,s}} n_{j,k,s}!}, \tag{3}$$

where $\mathbf{n}$ is the state vector $[n_{j,k,s}]$ and G is a normalization constant.

To demonstrate that this product form does not hold in the case of complete blocking, we offer the following simple counter-example. Take a system with a single resource type, a single job type, and a single batch size. The resource capacity is 4, the resource requirement per job is 1 and the incoming batch size is 3. The arrival and service rates are both 1.

There are now 3 queues in series, so the system state is a triple (n_1, n_2, n_3). The feasible states are $(0,0,0)$, $(1, 0, 0)$, $(0, 1, 0)$, $(0, 0, 1)$, $(1, 0, 1)$, $(0, 1, 1)$ and $(0, 0, 2)$. The state $(0, 2, 0)$, for instance, is not feasible because if there was a macro job in queue 2, consuming 2 resource units, then a new macro job requiring 3 units would not have been admitted.

Consider the two states $(0, 0, 1)$ and $(1, 0, 1)$. If (3) is correct, then their stationary probabilities are $\pi(0,0,1) = G$ and $\pi(1,0,1) = G/3$. Hence, $\pi(0,0,1) = 3\pi(1,0,1)$. On the other hand, the only way of leaving state $(1,0,1)$ is by a service completion, either at queue 1, at rate 3, or at queue 3, at rate 1. The total completion rate is 4. The only way of entering state $(1, 0, 1)$ is by an arrival of a new batch, at rate 1, when the system is in state $(0, 0, 1)$. Therefore, $\pi(0,0,1) = 4\pi(1,0,1)$. This contradiction demonstrates that the solution (3) is not correct.

The failure of the product form is due to the fact that, contrary to the assertion in [2], local balance does not hold in the case of complete blocking. In the above example, in states $(1, 0, 1)$ and $(0, 1, 1)$, a service completion at queue 3 (leading to states $(1, 0, 0)$ and $(0, 1, 0)$ respectively), cannot be balanced by an arrival because in either case the new batch would be rejected. For that reason, although in principle there might exist a different, as yet undiscovered product form, we believe that this is very unlikely.

In the absence of a tractable exact solution, we now turn to the task of finding an accurate approximation.

4   A Fixed-Point Approximation

We propose treating each job type as if it was an isolated, one-dimensional Markov process taking place within a static environment determined by the other job types. More precisely, when considering jobs of type j, assume that all jobs of other types in the system are consuming a fixed total amount, $Z_{j,r}$, of type r resource ($r = 1, 2, \ldots, R$). In other words, type j operates in an environment where the available resource of type r is $D_r - Z_{j,r}$. Hence, the maximum number of type j jobs that can be admitted, m_j, is given by (1), with D_r replaced by $D_r - Z_{j,r}$.

Under the above assumption, the type j Markov process evolves on the state space $\{0, 1, \ldots, m_j\}$. Let $\pi_j(n)$ be the stationary probability of state n, i.e., the probability that there are n type j jobs present. We can write a set of recurrence relations for these probabilities by examining the flows across a cut separating states $0, 1, \ldots, n-1$ from states $n, n+1, \ldots, m_j$. The relevant balance diagram is illustrated in Fig. 1.

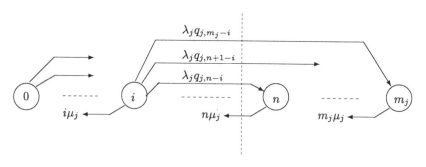

Fig. 1. Balance diagram for type j

The downward (i.e., to the left) flow across the cut contains just a single transition, from state n, due to a departure; the corresponding rate is $n\mu_j$. On the other hand, if an arrival occurs in any state, $i(i < n)$, and the size of the incoming batch is at least $n - i$ but does not exceed $m_j - i$, then it contributes to the upward (to the right) flow across the cut. Equating the two flows provides a balance equation for each n:

$$n\mu_j \pi_j(n) = \lambda_j \sum_{i=0}^{n-1} \pi_j(i) \sum_{k=n-i}^{m_j-i} q_{j,k}; \quad n = 1, 2, \ldots, m_j. \tag{4}$$

The simplest way to solve these equations is to set $\pi_j(0) = 1$, evaluate $\pi_j(n)$ for $n = 1, 2, \ldots, m_j$ according to (4), and then re-normalize by dividing each of them by their sum.

Having computed the probabilities $\pi_j(n)$, the probability that an incoming batch of type j and size k is accepted, is given by

$$\alpha_{j,k} = \sum_{n=0}^{m_j-k} \pi_j(n). \tag{5}$$

The average number, L_j, of type j jobs present is equal to

$$L_j = \sum_{n=1}^{m_j} n\pi_j(n). \tag{6}$$

Consequently, the average amount, $z_{j,r}$, of type r resource consumed by jobs of type j, is given by

$$z_{j,r} = d_{j,r} L_j. \tag{7}$$

In common with most models of this type, the above solution depends only on the ratio $\rho_j = \lambda_j/\mu_j$, and not on the individual values of those parameters.

We can now approximate the effect that type j has on the other job types by treating its *average* consumption of resource r, $z_{j,r}$ as a *fixed* consumption of resource r. This will form part of the environment in which another given job type operates.

Suppose that we have somehow obtained an estimated vector, $\mathbf{L} = (L_1, L_2, \ldots, L_T)$, of the average numbers of jobs of different types in the system. Carry out the following procedure.

1. For every j, compute the type r resource, $Z_{j,r}$, consumed by job types other than j:

$$Z_{j,r} = \sum_{i=1, i\neq j}^{T} z_{i,r}, \tag{8}$$

with $z_{i,r}$ being given by (7).
2. For every j, solve the isolated Markov process of type j and compute a new value for L_j (and hence new values for $z_{j,r}$ and the acceptance probabilities $\alpha_{j,k}$).

This procedure implements a mapping, $f(\cdot)$, from one vector of averages, call it $\mathbf{L}^{old}$, to another vector of averages, $\mathbf{L}^{new}$. Our approximate solution consists in finding a vector, $\mathbf{L}^*$, whose 'new' image is the same as the 'old' one. That is, $\mathbf{L}^*$ is a fixed point of the mapping $f(\cdot)$:

$$\mathbf{L}^* = f(\mathbf{L}^*). \tag{9}$$

At the fixed point $\mathbf{L}^*$, the environments in which the different job types operate are consistent with each other. That is, for every job type j, the resources it consumes, $z_{j,r}$, do not alter the resources consumed by the other job types, $Z_{j,r}$. In that sense, this fixed point is like a Nash equilibrium in multi-person games.

The acceptance probabilities corresponding to $\mathbf{L}^*$ are substituted into (2) in order to compute the profit V, or are used to see whether the QoS targets have been met.

To compute $\mathbf{L}^*$, we use an iterative schema of the form

$$\mathbf{L}^{(i+1)} = f(\mathbf{L}^{(i)}). \tag{10}$$

These iterations start with some initial vector such as $\mathbf{L}^{(0)} = (0, 0, \ldots, 0)$, and stop when two consecutive vectors are sufficiently close to each other. To reduce the number of iterations, it is advisable that the evaluations are of the Gauss-Seidel type, i.e. as soon as a new value for some L_j is obtained, the corresponding value of $z_{j,r}$ is used in determining the environment for other job types.

Note. Fixed-point approximations of queuing systems have been used in the past, mainly in the context of open or closed networks. There, the decomposition is in terms of nodes and the fixed point equations attempt to capture the interactions between them (e.g., see Sadre et al. [16], Whitt [18]). Kelly's fixed-point approximation for multi-class circuit-switched networks [10] is concerned with shared resources, but does not model batch arrivals. The decomposition is with respect to offered loads for individual units of resource. Kelly was able to prove the existence of a fixed point by appealing to Brouwer's theorem, and he demonstrated convergence to it.

We have been unable to prove either the existence or the uniqueness of a fixed point for our mapping $f(\cdot)$. This is due to the fact that Eq. (1) involves the 'floor' function $\lfloor x \rfloor$, which is not continuous. However, $f(\cdot)$ is uniformly bounded and our experience with many examples has been that the iterations (10) always converge.

As far as we are aware, a decomposition by job type, where the fixed point equations capture different contributions to a shared environment, has not been used before.

To examine the quality of the proposed approximation, consider an example system with four job types, 1, 2, 3 and 4, or 'small', 'medium', 'large' and 'very large'. There is a single resource type and the individual resource requirements of the four job types are $d_1 = 1$, $d_2 = 2$, $d_3 = 4$ and $d_4 = 8$. These numbers are motivated by similarities with the T2 family of VM instances offered by the Amazon EC2 (Elastic Computing Cloud) service (see [19]). The resource that is being shared in this context is vCPUs (virtual CPUs).

Type 1 jobs arrive singly, at a rate of $\lambda_1 = 6$ jobs per unit time. Their average service times are $1/\mu_1 = 1$. For type 2, the possible batch sizes are 1 or 2, with equal probability ($q_{2,1} = q_{2,2} = 0.5$). The traffic parameters are $\lambda_2 = 4$ and $1/\mu_2 = 1$. Jobs of type 3 and type 4 arrive in batches of size 1, 2, or 3, with probabilities 0.4, 0.3 and 0.3, respectively. Their arrival and service parameters are $\lambda_3 = 2$, $1/\mu_3 = 0.5$, $\lambda_4 = 1$, $1/\mu_4 = 0.5$.

The cost incurred per unit of resource is $c = 0.2$, and the revenues brought in by the different job types increase with the resource consumed: $v_1 = 1$, $v_2 = 3$, $v_3 = 6$, $v_4 = 10$.

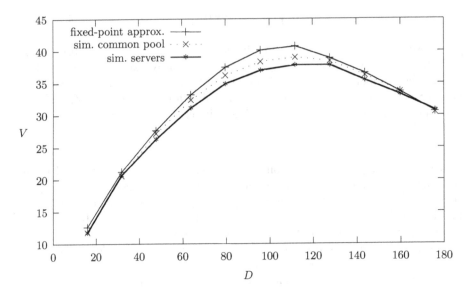

Fig. 2. Estimated and simulated average profit

In Fig. 2, the estimated average profit is plotted against the offered resource capacity, D. The three curves correspond to (a) the fixed-point approximation, (b) a simulation of the model as described in Sect. 2 and (c) a simulation of a system where the resource is not provided as a common pool, but is contained within servers. In this last case, each server has 16 vCPUs. Thus, in order to offer 128 vCPUs, one has to provide 8 servers. The discrete server system is less efficient than the common pool, because, in order to instantiate a VM on a server, all its resource requirement must be available on that server. It is not enough to have some of the requirement available on one server and some on another.

The two simulation graphs may be considered exact. Their confidence intervals are too narrow to plot. The figure shows that the fixed-point approximation tends to overestimate the average profit slightly. This is not surprising, since replacing a random environment with a fixed one reduces the variance of the process. The relative differences between approximated and simulated (common pool) points are on the order of 5% or less.

The discrete server simulation confirms the intuition that system is less efficient, but again the differences from the common pool are on the order of 5% or less. In particular, all three graphs agree that the optimal resource capacity to provide is 112 vCPUs, or 7 servers.

The convergence of iterations (10) to the fixed point is very fast. The termination criterion in this example was that $|L_j^{(i+1)} - L_j^{(i)}| < 10^{-6}$ for all job types j. The entire approximation graph took less than tenth of a second to compute, and no point required more than 4 iterations.

We have also simulated a discrete server system where each server contains 8 vCPUs instead of 16. The differences between that system and the one shown in the figure are negligible.

In the next example, all arrival rates are scaled up by a factor of 100: $(\lambda_1, \lambda_2, \lambda_3, \lambda_4) = (600, 400, 200, 100)$. The resource is again provided in a common pool, with scaled capacities ranging from 7000 to 10000. The other parameters, revenues and costs are kept as before. The aim of this exercise is to see whether the fixed-point approximation is sufficiently robust to cope with the much larger state space and offered loads. The estimated and simulated average profits are displayed in Fig. 3.

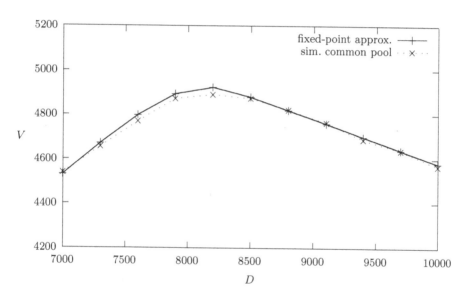

Fig. 3. System scaled up by a factor of 100

The accuracy of the approximation has increased. Not only does the model predict the optimal resource allocation correctly, but the relative differences between estimated and simulated profits are now less than 1%. This is not really surprising, since the flows of traffic of various job types through the system behave more and more like deterministic fluids when capacities and offered loads increase.

We conclude that, for systems of this and similar sizes, it would be preferable to use the fixed-point approximation even if an exact product-form solution was available. The reason is that the complexity and numerical problems associated with computing the relevant normalization constants increase very quickly with the number of job types, the incoming batch sizes and the offered resource capacities.

It should be acknowledged, however, that the fixed-point approximation also has its limits of applicability. For example, if the system in Fig. 3 is scaled up by another factor of 10, bringing the arrival rates to $(\lambda_1, \lambda_2, \lambda_3, \lambda_4) = (6000, 4000, 2000, 1000)$ and resource capacities in the region of 100000, our solution breaks down. The failure is not in the fixed-point iterations, but in the one-dimensional solutions (4) which start experiencing numerical overflows.

It is thus desirable to develop another approximation which can be applied to very large systems and produce reasonable estimates, albeit with some loss of accuracy.

5   Very Large-Scale Systems

We have observed that the solution of an isolated job type ceases to work when the resource capacities and the offered loads are on the order of tens of thousands or more. For such large systems, a simpler and more robust approximation is required.

With this in mind, we propose to represent the various batch arrivals of type j by single 'macro' jobs with appropriately chosen resource requirements. Then, for the purpose of the fixed-point solution, the isolated type j model becomes a classic Erlang loss process. The benefit of this simplification is that the Erlang B function, which provides the rejection probability, can be computed in a stable manner for large values of the parameters.

The arrival rate and average service time of type j macro jobs are λ_j and $1/\mu_j$, respectively. To define the resource requirement of type r for a macro job of type j, $\delta_{j,r}$, we take the average over the possible type j batch sizes:

$$\delta_{j,r} = d_{j,r} \sum_{k=1}^{K_j} k q_{j,k}. \tag{11}$$

Hence, if all other job types consume a fixed amount, $Z_{j,r}$, of type r resource, then the maximum number of type j macro jobs that can be admitted into the system is

$$m_j = \min_r \left\{ \left\lfloor \frac{D_r - Z_{j,r}}{\delta_{j,r}} \right\rfloor \right\}. \tag{12}$$

The probability, β_j, that an incoming macro job of type j will be rejected, is given by the Erlang-B function (e.g., see [13])

$$\beta_j = B(m_j, \rho_j) = \frac{\rho_j^{m_j}}{m_j!} \left[\sum_{i=0}^{m_j} \frac{\rho^i}{i!} \right]^{-1}. \tag{13}$$

A numerically stable procedure for computing the function $B(m, \rho)$ is provided by the recurrence relation

$$B(m, \rho) = \frac{\frac{\rho}{m} B(m-1, \rho)}{1 + \frac{\rho}{m} B(m-1, \rho)}, \tag{14}$$

starting with $B(0, \rho) = 1$ (e.g., see [6]).

The average number, L_j, of type j macro jobs in the system is then given by Little's result:

$$L_j = \rho_j(1 - \beta_j). \tag{15}$$

The average amount of type r resource that those jobs consume is $z_{j,r} = L_j \delta_{j,r}$.

We now have the necessary elements for carrying out the iterations described in the previous section and finding the fixed-point vectors of average numbers of macro jobs present, L_j^*, and corresponding rejection probabilities, β_j^*. The average profit achieved per unit time is given by

$$V = \sum_{j=1}^{T} \lambda_j v_j (1 - \beta_j) \sum_{k=1}^{K_j} k q_{j,k} - \sum_{r=1}^{R} c_r D_r. \tag{16}$$

To illustrate the efficacy and accuracy of this approximation, we have scaled up the example with four job types introduced in the previous section. The arrival rates are now $(\lambda_1, \lambda_2, \lambda_3, \lambda_4) = (6000, 4000, 2000, 1000)$, and the single resource capacity is varied in the range $D \in (70000, 90000)$. All other parameters are as before.

In Fig. 4, the average profits predicted by the macro fixed-point approximation are compared with those produced by simulation runs in each of which a total of ten million batches of all types and sizes arrived into the system. Computing the fixed-point was very fast and free from numerical problems. The simulation runs were several orders of magnitude slower, because we wanted narrow confidence intervals.

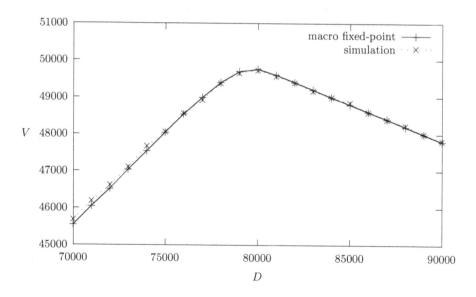

Fig. 4. System scaled up by a factor of 1000

The figure shows that the simplified approximation is remarkably accurate at this scale. The two plots are almost indistinguishable. This confirms the tendency observed earlier, that as the scale of the system increases, its behaviour agrees more closely with the deterministic assumptions that underlie the fixed-point approach.

6   Conclusions

We have addressed a practically relevant problem concerned with service provisioning in public clouds. The multi-class model with batch arrivals and complete blocking appeared to be solved, but we have shown that the existing solution is incorrect. An alternative solution based on fixed-point iterations is introduced. This replaces the multi-dimensional stochastic process with a number of single-dimensional ones, using averages to model the interactions between them. The accuracy of the fixed-point solution is good for small systems, and gets better when the system size increases. A simplified version of that solution is shown to apply to very large-scale systems.

The exact solution of the complete blocking model is still an open problem, as is also the solution of the model where resources are provided in discrete servers, rather than in common pools. Those problems are interesting and worthy of further study. However, we feel that even if the exact solutions were available, it would be better to tackle the task of optimizing a real-life system by using the proposed approximations. They are easily implementable and sufficiently accurate.

Other, more general models may be tackled by the methods described here. For example, one may wish to operate a complete blocking policy for some job types, and partial blocking for others. The one-dimensional solution for partial blocking is different, but the fixed-point approach would still apply.

References

1. Bodík, P., Griffith, R., Sutton, C., Fox, A., Jordan, M., Patterson, D.: Statistical machine learning makes automatic control practical for internet datacenters. In: Conference on Hot Topics in Cloud Computing (HotCloud 2009), Berkeley, CA, USA (2009)
2. Choudhury, G.L., Leung, K.K., Whitt, W.: Resource-sharing models with state-dependent arrivals of batches. In: Stewart, W.J. (ed.) Computations with Markov Chains, pp. 255–282. Springer, Boston (1995). doi:10.1007/978-1-4615-2241-6_16
3. Dean, J., Ghemawat, S.: MapReduce: simplified data processing on large clusters. Commun. ACM **51**(1), 107–113 (2008)
4. van Doorn, E.A., Planken, F.J.M.: Blocking probabilities in a loss system with arrivals in geometrically distributed batches and heterogeneous service requirements. ACM/IEEE Trans. Netw. 1, 664–667 (1993)
5. Ezhilchelvan, P., Mitrani, I.: Static and dynamic hosting of cloud servers. In: Beltrán, M., Knottenbelt, W., Bradley, J. (eds.) EPEW 2015. LNCS, vol. 9272, pp. 19–31. Springer, Cham (2015). doi:10.1007/978-3-319-23267-6_2

6. Ezhilchelvan, P., Mitrani, I.: Optimal provisioning of servers for hosting services of multiple types. Simul. Model. Pract. Theory **75**, 17–28 (2017)
7. Hampshire, R.C., Massey, W.A., Mitra, D., Wang, Q.: Provisioning for bandwidth sharing and exchange. In: Anandalingam, G., Raghavan, S. (eds.) Telecommunications Network Design and Management. ORCS, vol. 23, pp. 207–225. Springer, Boston (2003)
8. Kaufman, J.S.: Blocking in a shared resource environment. IEEE Trans. Commun. **29**, 1474–1481 (1981)
9. Kaufman, J.S., Rege, K.M.: Blocking in a shared resource environment with batched Poisson arrival processes. Perform. Eval. **24**, 249–263 (1996)
10. Kelly, F.: Blocking probabilities in large circuit switched networks. Adv. Appl. Probab. **18**, 473–505 (1986)
11. Mazzucco, M., Dyachuk, D., Dikaiakos, M.: Profit-aware server allocation for green internet services. In: IEEE International Symposium on Modeling, Analysis and Simulation of Computer and Telecommunication Systems (MASCOTS), pp. 277–284 (2010)
12. Mazzucco, M., Vasar, M., Dumas, M.: Squeezing out the cloud via profit-maximizing resource allocation policies. In: IEEE International Symposium on Modeling, Analysis and Simulation of Computer and Telecommunication Systems (MASCOTS), pp. 19–28 (2012)
13. Mitrani, I.: Probabilistic Modelling. Cambridge University Press, Cambridge (1998)
14. Roberts, J.W.: A service system with heterogeneous user requirement. In: Pujolle, G. (ed.) Performance of Data Communications Systems and Their Application, North-Holland, pp. 423–431 (1981)
15. Ross, K.W.: Multiservice Loss Models for Broadband Telecommunication Networks. Springer, Heidelberg (1995)
16. Sadre, R., Haverkort, B.R., Ost, A.: An efficient and accurate decomposition method for open finite- and infinite-buffer queueing networks. In: Stewart, W., Plateau, B. (eds.) Proceedings of 3rd International Workshop on Numerical Solution of Markov Chains, p. 120 (1999)
17. Tan, Y., Lu, Y., Xia, C.H.: Provisioning for large scale loss network systems with applications in cloud computing. ACM SIGMETRICS Perform. Eval. Rev. **40**(3), 83–85 (2012)
18. Whitt, W.: The queueing network analyzer. Bell Syst. Tech. J. **62**(9), 2779–2815 (1983)
19. Amazon Web Services (2016). https://aws.amazon.com/ec2/instance-types/

Parametric Verification

Reachability in Parametric Interval Markov Chains Using Constraints

Anicet Bart[1]([⊠]), Benoît Delahaye[2], Didier Lime[3], Éric Monfroy[2], and Charlotte Truchet[2]

[1] Institut Mines-Télécom Atlantique - LS2N, UMR 6004, Nantes, France
anicet.bart@ls2n.fr
[2] Université de Nantes - LS2N, UMR 6004, Nantes, France
{benoit.delahaye,eric.monfroy,charlotte.truchet}@ls2n.fr
[3] École Centrale de Nantes - LS2N, UMR 6004, Nantes, France
didier.lime@ls2n.fr

Abstract. Parametric Interval Markov Chains (pIMCs) are a specification formalism that extend Markov Chains (MCs) and Interval Markov Chains (IMCs) by taking into account imprecision in the transition probability values: transitions in pIMCs are labeled with parametric intervals of probabilities. In this work, we study the difference between pIMCs and other Markov Chain abstractions models and investigate the two usual semantics for IMCs: once-and-for-all and at-every-step. In particular, we prove that both semantics agree on the maximal/minimal reachability probabilities of a given IMC. We then investigate solutions to several parameter synthesis problems in the context of pIMCs – consistency, qualitative reachability and quantitative reachability – that rely on constraint encodings. Finally, we propose a prototype implementation of our constraint encodings with promising results.

1 Introduction

Discrete time Markov chains (MCs for short) are a standard probabilistic modeling formalism that has been extensively used in the litterature to reason about software [7] and real-life systems [14]. However, when modeling real-life systems, the exact value of transition probabilities may not be known precisely. Several formalisms abstracting MCs have therefore been developed. Parametric Markov chains [1] (pMCs for short) extend MCs by allowing parameters to appear in transition probabilities. In this formalism, parameters are variables and transition probabilities may be expressed as polynomials over these variables. A given pMC therefore represents a potentially infinite set of MCs, obtained by replacing each parameter by a given value. pMCs are particularly useful to represent systems where dependencies between transition probabilities are required. Indeed,

This work is partially supported by the ANR national research programs PACS (ANR-14-CE28-0002) and Coverif (ANR-15-CE25-0002), and the regional programme Atlantic2020 funded by the French Region Pays de la Loire and the European Regional Development Fund.

© Springer International Publishing AG 2017

N. Bertrand and L. Bortolussi (Eds.): QEST 2017, LNCS 10503, pp. 173–189, 2017.
DOI: 10.1007/978-3-319-66335-7_11

a given parameter may appear in several distinct transition probabilities, therefore requiring that the same value is given to all its occurences. Interval Markov chains [15] (IMC for short) extend MCs by allowing precise transition probabilities to be replaced by intervals, but cannot represent dependencies between distinct transitions. IMCs have mainly been studied with two distinct semantics interpretation. Under the *once-and-for-all* semantics, a given IMC represents a potentially infinite number of MCs where transition probabilities are chosen inside the specified intervals while keeping the same underlying graph structure. The *at-every-step* semantics, which was the original semantics given to IMCs in [15], does not require MCs to preserve the underlying graph structure of the original IMC but instead allows an "unfolding" of the original graph structure where different probability values may be chosen (inside the specified interval) at each occurence of the given transition.

Model-checking algorithms and tools have been developed in the context of pMCs [9,13,16] and IMCs with the once-and-for-all semantics [5,12]. State of the art tools [9] for pMC verification compute a rational function on the parameters that characterizes the probability of satisfying a given property, and then use external tools such as SMT solving [9] for computing the satisfying parameter values. For these methods to be viable in practice, the number of parameters used is quite limited. On the other hand, the model-checking procedure for IMCs presented in [5] is adapted from machine learning and builds successive refinements of the original IMCs that optimize the probability of satisfying the given property. This algorithm converges, but not necessarilly to a global optimum. It is worth noticing that existing model checking procedures for pMCs and IMCs strongly rely on their underlying graph structure. As a consequence, to the best of our knowledge, no solutions for model-checking IMCs with the at-every-step semantics have been proposed yet.

In this paper, we focus on Parametric interval Markov chains [11] (pIMCs for short), that generalize both IMCs and pMCs by allowing parameters to appear in the endpoints of the intervals specifying transition probabilities, and we provide four main contributions.

First, we formally compare abstraction formalisms for MCs in terms of succinctness: we show in particular that pIMCs are *strictly more succinct* than both pMCs and pIMCs when equipped with the right semantics. In other words, everything that can be expressed using pMCs or IMCs can also be expressed using pIMCs while the reverse does not hold. Second, we prove that the once-and-for-all and the at-every-step semantics are equivalent w.r.t. rechability properties, both in the IMC and in the pIMC settings. Notably, this result gives theoretical backing to the generalization of existing works on the verification of IMCs to the at-every-step semantics. Third, we study the parametric verification of fundamental properties at the pIMC level: consistency, qualitative reachability, and quantitative reachability. Given the expressivity of the pIMC formalism, the risk of producing a pIMC specification that is incoherent and therefore does not model any concrete MC is high. We therefore propose constraint encodings for deciding whether a given pIMC is consistent and, if so, synthesizing parameter values ensuring consistency. We then extend these encodings to qualitative reach-ability, *i.e.*, ensuring that given state labels are reachable in *all* (resp. *none*) of the MCs modeled by a given pIMC. Finally, we focus on the quantitative reach-ability problem, *i.e.*, synthesizing parameter values such that the probability of reaching given state labels satisfies fixed bounds in *at least one* (resp. *all*) MCs modeled by a given pIMC. While consistency and qualitative reachability for pIMCs have already been studied in [11], the constraint encodings we propose in this paper are significantly smaller (linear instead of exponential). To the best of our knowledge, our results provide the first solution to the quantitative reacha-bility problem for pIMCs. Our last contribution is the implementation of all our verification algorithms in a prototype tool that generates the required constraint encodings and can be plugged to any SMT solver for their resolution. Due to space limitation, proofs and detailed examples are given in [4].

2 Background

In this section we introduce notions and notations that will be used throughout the paper. Given a finite set of variables $X = \{x_1, \ldots, x_k\}$, we write D_x for the domain of the variable $x \in X$ and D_X for the set of domains associated to the variables in X. A valuation v over X is a set $v = \{(x, d) | x \in X, d \in D_x\}$ of elementary valuations (x, d) where for each $x \in X$ there exists a unique pair of the form (x, d) in v. When clear from the context, we write $v(x) = d$ for the value given to variable x according to valuation v. A rational function f over X is a division of two (multivariate) polynomials g_1 and g_2 over X with rational coefficients, *i.e.*, $f = g_1/g_2$. We write $\mathbb{Q}$ the set of rational numbers and $\mathbb{Q}_X$ the set of rational functions over X. The evaluation $v(g)$ of a polynomial g under the valuation v replaces each variable $x \in X$ by its value $v(x)$.

An *atomic constraint* over X is a Boolean expression of the form $f(X) \bowtie g(X)$, with $\bowtie \in \{\leq, \geq, <, >, =\}$ and f and g two functions over variables in X and constants. A constraint is *linear* if the functions f and g are linear. A *constraint* over X is a Boolean combination of atomic constraints over X.

Given a finite set of states S, we write $\mathsf{Dist}(S)$ for the set of probability distri-butions over S, *i.e.*, the set of functions $\mu : S \to [0, 1]$ such that $\sum_{s \in S} \mu(s) = 1$. We write $\mathbb{I}$ for the set containing all the interval subsets of $[0, 1]$. In the follow-ing, we consider a universal set of symbols A that we use for labelling the states of our structures. We call these symbols *atomic propositions*. We will use Latin alphabet in state context and Greek alphabet in atomic proposition context.

Constraints. Constraints are first order logic predicates used to model and solve combinatorial problems [19]. A problem is described with a list of variables, each in a given domain of possible values, together with a list of constraints over these variables. Such problems are then sent to solvers which decide whether the problem is satisfiable, *i.e.*, if there exists a valuation of the variables satisfying all the constraints, and in this case computes a solution. Checking satisfiability of constraint problems is difficult in general, as the space of all possible valuations has a size exponential in the number of variables.

Formally, a Constraint Satisfaction Problem (CSP) is a tuple $\Omega = (X, D, C)$ where X is a finite set of variables, $D = D_X$ is the set of all the domains associated to the variables from X, and C is a set of constraints over X. We say that a valuation over X satisfies Ω if and only if it satisfies all the constraints in C. We write $v(C)$ for the satisfaction result of the valuation of the constraints C according to v (*i.e.*, true or false). In the following we call CSP *encoding* a scheme for formulating a given problem into a CSP. The size of a CSP corresponds to the number of variables and atomic constraints appearing in the problem. Note that, in constraint programming, having less variables or less constraints during the encoding does not necessarily imply faster solving time of the problems.

Discrete Time Markov Chains. A Discrete Time Markov Chain (DTMC or MC for short) is a tuple $\mathcal{M} = (S, s_0, p, V)$, where S is a finite set of states containing the initial state s_0, $V : S \to 2^A$ is a labelling function, and $p : S \to \text{Dist}(S)$ is a probabilistic transition function. We write MC for the set containing all the discrete time Markov chains.

A Markov Chain can be seen as a directed graph where the nodes correspond to the states of the MC and the edges are labelled with the probabilities given by the transition function of the MC. In this representation, a missing transition between two states represents a transition probability of zero. As usual, given a MC $\mathcal{M}$, we call a *path* of $\mathcal{M}$ a sequence of states obtained from executing $\mathcal{M}$, *i.e.*, a sequence $\omega = s_1, s_2, \ldots$ s.t. the probability of taking the transition from s_i to s_{i+1} is strictly positive, $p(s_i)(s_i + 1) > 0$, for all i. A path ω is finite iff it belongs to S^*, *i.e.*, it represents a finite sequence of transitions from $\mathcal{M}$.

Example 1. Figure 1 illustrates the Markov chain $\mathcal{M}_1 = (S, s_0, p, V) \in$ MC where $S = \{s_0, s_1, s_2, s_3, s_4\}$, the atomic proposition are restricted to $\{\alpha, \beta\}$, the initial state is s_0, and the labelling function V corresponds to $\{(s_0, \emptyset), (s_1, \alpha), (s_2, \beta), (s_3, \{\alpha, \beta\}), (s_4, \alpha)\}$. The sequences of states (s_0, s_1, s_2), (s_0, s_2), and (s_0, s_2, s_2, s_2), are three (finite) paths from the initial state s_0 to the state s_2.

Reachability. A Markov chain $\mathcal{M}$ defines a unique probability measure $\mathbb{P}^{\mathcal{M}}$ over the paths from $\mathcal{M}$. According to this measure, the probability of a finite path $\omega = s_0, s_1, \ldots, s_n$ in $\mathcal{M}$ is the product of the probabilities of the transitions executed along this path, *i.e.*, $\mathbb{P}^{\mathcal{M}}(\omega) = p(s_0)(s_1) \cdot p(s_1)(s_2) \cdot \ldots \cdot p(s_{n-1})(s_n)$. This distribution naturally extends to infinite paths (see [2]) and to sequences of states over S that are not paths of $\mathcal{M}$ by giving them a zero probability.

Given a MC $\mathcal{M}$, the overall probability of reaching a given state s from the initial state s_0 is called the *reachability probability* and written $\mathbb{P}_{s_0}^{\mathcal{M}}(\lozenge s)$ or $\mathbb{P}^{\mathcal{M}}(\lozenge s)$ when clear from the context. This probability is computed as the sum of the probabilities of all finite paths starting in the initial state and reaching this state for the first time. Formally, let $\text{reach}_{s_0}(s) = \{\omega \in S^* \mid \omega = s_0, \ldots s_n \text{ with } s_n = s \text{ and } s_i \neq s \; \forall 0 \leq i < n\}$ be the set of such paths. We then define $\mathbb{P}^{\mathcal{M}}(\lozenge s) = \sum_{\omega \in \text{reach}_{s_0}(s)} \mathbb{P}^{\mathcal{M}}(\omega)$ if $s \neq s_0$ and 1 otherwise. This notation naturally extends to the reachability probability of a state s from a state t that is not s_0, written $\mathbb{P}_t^{\mathcal{M}}(\lozenge s)$ and to the probability of reaching a label $\alpha \subseteq A$ written $\mathbb{P}_{s_0}^{\mathcal{M}}(\lozenge \alpha)$. In the following, we say that a state s (resp. a label $\alpha \subseteq A$) is reachable in $\mathcal{M}$ iff the reachability probability of this state (resp. label) from the initial state is strictly positive.

Example 2 (Example 1 continued). In Fig. 1 the probability of the path $(s_0, s_2, s_1, s_1, s_3)$ is $0.3 \cdot 0.5 \cdot 0.5 \cdot 0.5 = 0.0375$ and the probability of reaching the state s_1 is $\mathbb{P}^{\mathcal{M}_1}(\lozenge s_1) = p(s_0)(s_1) + \Sigma_{i=0}^{+\infty} p(s_0)(s_2) \cdot p(s_2)(s_2)^i \cdot p(s_2)(s_1) = p(s_0)(s_1) + p(s_0)(s_2) \cdot p(s_2)(s_1) \cdot (1/(1 - p(s_2)(s_2))) = 1$. Furthermore, the probability of reaching β corresponds to the probability of reaching the state s_2.

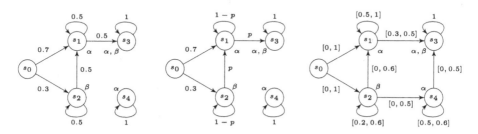

Fig. 1. MC $\mathcal{M}_1$ **Fig. 2.** pMC $\mathcal{I}'$ **Fig. 3.** IMC $\mathcal{I}$

3   Markov Chains Abstractions

Modelling an application as a Markov Chain requires knowing the exact probability for each possible transition of the system. However, this can be difficult to compute or to measure in the case of a real-life application (*e.g.*, precision errors, limited knowledge). In this section, we start with a generic definition of Markov chain abstraction models. Then we recall three abstraction models from the literature, respectively pMC, IMC, and pIMC, and finally we present a comparison of these existing models in terms of succinctness.

Definition 1 (Markov chain Abstraction Model). *A Markov chain abstraction model (an abstraction model for short) is a pair* $(\mathrm{L}, \models)$ *where* L *is a nonempty set and* $\models$ *is a relation between* MC *and* L. *Let* $\mathcal{P}$ *be in* L *and* $\mathcal{M}$ *be in* MC *we say that* $\mathcal{M}$ *implements* $\mathcal{P}$ *iff* $(\mathcal{M}, \mathcal{P})$ *belongs to* $\models$ *(i.e.,* $\mathcal{M} \models \mathcal{P}$*). When the context is clear, we do not mention the satisfaction relation* $\models$ *and only use* L *to refer to the abstraction model* $(\mathrm{L}, \models)$.

A *Markov chain Abstraction Model* is a specification theory for MCs. It consists in a set of abstract objects, called *specifications*, each of which representing a (potentially infinite) set of MCs – *implementations* – together with a satisfaction relation defining the link between implementations and specifications. As an example, consider the powerset of MC (*i.e.*, the set containing all the possible sets of Markov chains). Clearly, $(2^{\mathrm{MC}}, \in)$ is a Markov chain abstraction model, which we call the *canonical abstraction model*. This abstraction model has the advantage of representing all the possible sets of Markov chains but it also has the disadvantage that some Markov chain abstractions are only representable by an infinite extension representation. Indeed, recall that there exists subsets of $[0, 1] \subseteq \mathbb{R}$ which cannot be represented in a finite space (e.g., the Cantor set [6]). We now present existing MC abstraction models from the literature.

3.1   Existing MC Abstraction Models

Parametric Markov Chain is a MC abstraction model from [1] where a transition can be annotated by a rational function over *parameters*. We write pMC for the set containing all the parametric Markov chains.

Definition 2 (Parametric Markov Chain). *A Parametric Markov Chain (pMC for short) is a tuple $\mathcal{I} = (S, s_0, P, V, Y)$ where S, s_0, and V are defined as for MCs, Y is a set of variables (parameters), and $P : S \times S \to \mathbb{Q}_Y$ associates with each potential transition a parameterized probability.*

Let $\mathcal{M} = (S, s_0, p, V)$ be a MC and $\mathcal{I} = (S, s_0, P, V, Y)$ be a pMC. The satisfaction relation $\models_p$ between MC and pMC is defined by $\mathcal{M} \models_p \mathcal{I}$ iff there exists a valuation v of Y s.t. $p(s)(s')$ equals $v(P(s, s'))$ for all s, s' in S.

Example 3. Figure 2 shows a pMC $\mathcal{I}' = (S, s_0, P, V, Y)$ where S, s_0, and V are similar to the same entities in the MC $\mathcal{M}$ from Fig. 1, the set of variable Y contains only one variable p, and the parametric transitions in P are given by the edge labelling (*e.g.*, $P(s_0, s_1) = 0.7$, $P(s_1, s_3) = p$, and $P(s_2, s_2) = 1 - p$). Note that the pMC $\mathcal{I}'$ is a specification containing the MC $\mathcal{M}$ from Fig. 1.

Interval Markov Chains extend MCs by allowing to label transitions with intervals of possible probabilities instead of precise probabilities. We write IMC for the set containing all the interval Markov chains.

Definition 3 (Interval Markov Chain [15]). *An Interval Markov Chain (IMC for short) is a tuple $\mathcal{I} = (S, s_0, P, V)$, where S, s_0, and V are defined as for MCs, and $P : S \times S \to \mathbb{I}$ associates with each potential transition an interval of probabilities.*

Example 4. Figure 3 illustrates IMC $\mathcal{I} = (S, s_0, P, V)$ where S, s_0, and V are similar to the MC given in Fig. 1. By observing the edge labelling we see that $P(s_0, s_1) = [0, 1]$, $P(s_1, s_1) = [0.5, 1]$, and $P(s_3, s_3) = [1, 1]$. On the other hand, the intervals of probability for missing transitions are reduced to $[0, 0]$, e.g., $P(s_0, s_0) = [0, 0]$, $P(s_0, s_3) = [0, 0]$, $P(s_1, s_4) = [0, 0]$.

In the literature, IMCs have been mainly used with two distinct semantics: *at-every-step* and *once-and-for-all*. Both semantics are associated with distinct satisfaction relations which we now introduce.

The *once-and-for-all* IMC semantics ([9,18,20]) is alike to the semantics for pMC, as introduced above. The associated satisfaction relation $\models_I^o$ is defined as follows: A MC $\mathcal{M} = (T, t_0, p, V^M)$ satisfies an IMC $\mathcal{I} = (S, s_0, P, V^I)$ iff $(T, t_0, V^M) = (S, s_0, V^I)$ and for all reachable state s and all state $s' \in S$, $p(s)(s') \in P(s, s')$. In this sense, we say that MC implementations using the once-and-for-all semantics need to have the same structure as the IMC specification.

On the other hand, the *at-every-step* IMC semantics, first introduced in [15], operates as a simulation relation based on the transition probabilities and state labels, and therefore allows MC implementations to have a different structure than the IMC specification. The associated satisfaction relation $\models_I^a$ is defined as follows: A MC $\mathcal{M} = (T, t_0, p, V^M)$ satisfies an IMC $\mathcal{I} = (S, s_0, P, V^I)$ iff there exists a relation $\mathcal{R} \subseteq T \times S$ such that $(t_0, s) \in \mathcal{R}$ and whenever $(t, s) \in \mathcal{R}$, we have 1. the labels of s and t correspond: $V^M(t) = V^I(s)$, 2. there exists a correspondence function $\delta : T \to (S \to [0, 1])$ s.t. (a) $\forall t' \in T$ if $p(t)(t') > 0$ then $\delta(t')$ is a distribution on S (b) $\forall s' \in S : (\Sigma_{t' \in T} p(t)(t') \cdot \delta(t')(s')) \in P(s, s')$, and (c) $\forall (t', s') \in T \times S$, if $\delta(t')(s') > 0$, then $(t', s') \in \mathcal{R}$. By construction, it is clear that $\models_I^a$ is more general than $\models_I^o$, i.e., that whenever $\mathcal{M} \models_I^o \mathcal{I}$, we also have $\mathcal{M} \models_I^a \mathcal{I}$. The reverse is obviously not true in general, even when the underlying graphs of $\mathcal{M}$ and $\mathcal{I}$ are isomorphic (see [4] for details).

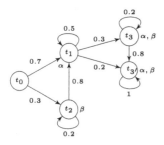

Fig. 4. MC $\mathcal{M}_2$ satisfying the IMC $\mathcal{I}$ from Fig. 3 with a different structure

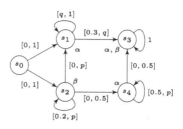

Fig. 5. pIMC $\mathcal{P}$

Example 5 (Example 4 continued). Consider the MC $\mathcal{M}_1$ with state space S from Fig. 1 and the MC $\mathcal{M}_2$ with state space T from Fig. 4. They both satisfy the IMC $\mathcal{I}$ with state space S given in Fig. 3. Furthermore, $\mathcal{M}_1$ satisfies $\mathcal{I}$ with the same structure. On the other hand, for the MC $\mathcal{M}_2$ given in Fig. 4, the state s_3 from $\mathcal{I}$ has been "split" into two states t_3 and $t_{3'}$ in $\mathcal{M}_2$ and the state t_1 from $\mathcal{M}_2$ "aggregates" states s_1 and s_4 in $\mathcal{I}$. The relation $\mathcal{R} \subseteq T \times S$ containing the pairs (t_0, s_0), (t_1, s_1), (t_1, s_4), (t_2, s_2), (t_3, s_3), and $(t_{3'}, s_3)$ is a satisfaction relation between $\mathcal{M}_2$ and $\mathcal{I}$.

Parametric Interval Markov Chains, as introduced in [11], abstract IMCs by allowing (combinations of) parameters to be used as interval endpoints in IMCs. Under a given parameter valuation the pIMC yields an IMC as introduced above. pIMCs therefore allow the representation, in a compact way and with a finite structure, of a potentially infinite number of IMCs. Note that one parameter can appear in several transitions at once, requiring the associated transition probabilities to depend on one another. Let Y be a finite set of parameters and v be a valuation over Y. By combining notations used for IMCs and pMCs the set $\mathbb{I}(\mathbb{Q}_Y)$ contains all parametrized intervals over $[0, 1]$, and for all $I = [f_1, f_2] \in \mathbb{I}(\mathbb{Q}_Y)$, $v(I)$ denotes the interval $[v(f_1), v(f_2)]$ if $0 \leq v(f_1) \leq v(f_2) \leq 1$ and the empty set otherwise[1]. We write pIMC for the set containing all the parametric interval Markov chains.

Definition 4 (Parametric Interval Markov Chain [11]). *A Parametric Interval Markov Chain (*pIMC *for short) is a tuple* $\mathcal{P} = (S, s_0, P, V, Y)$, *where* S, s_0, V *and* Y *are defined as for* pMCs, *and* $P : S \times S \to \mathbb{I}(\mathbb{Q}_Y)$ *associates with each potential transition a (parametric) interval.*

In [11] the authors introduced pIMCs where parametric interval endpoints are limited to linear combination of parameters. In this paper we extend the pIMC model by allowing rational functions over parameters as endpoints of parametric intervals. Given a pIMC $\mathcal{P} = (S, s_0, P, V, Y)$ and a valuation v, we write $v(\mathcal{P})$ for the IMC (S, s_0, P_v, V) obtained by replacing the transition function P from $\mathcal{P}$ with the function $P_v : S \times S \to \mathbb{I}$ defined by $P_v(s, s') = v(P(s, s'))$ for all $s, s' \in S$. The IMC $v(\mathcal{P})$ is called an *instance* of pIMC $\mathcal{P}$. Finally, depending on the semantics chosen for IMCs, two satisfaction relations can be defined between MCs and pIMCs. They are written $\models_{\text{pI}}^a$ and $\models_{\text{pI}}^o$ and defined as follows: $\mathcal{M} \models_{\text{pI}}^a \mathcal{P}$ (resp. $\models_{\text{pI}}^o$) iff there exists an IMC $\mathcal{I}$ instance of $\mathcal{P}$ s.t. $\mathcal{M} \models_{\text{I}}^a \mathcal{I}$ (resp. $\models_{\text{I}}^o$).

Example 6. Consider the pIMC $\mathcal{P} = (S, 0, P, V, Y)$ given in Fig. 5. The set of states S and the labelling function are the same as in the MC and the IMC presented in Figs. 1 and 3 respectively. The set of parameters Y has two elements p and q. Finally, the parametric intervals from the transition function P are given by the edge labelling (*e.g.*, $P(s_1, s_3) = [0.3, q]$, $P(s_2, s_4) = [0, 0.5]$, and $P(s_3, s_3) = [1, 1]$). Note that the IMC $\mathcal{I}$ from Fig. 3 is an instance of $\mathcal{P}$ (by assigning the value 0.6 to the parameter p and 0.5 to q). Furthermore, as said in Example 5, the Markov Chains $\mathcal{M}_1$ and $\mathcal{M}_2$ (from Figs. 1 and 4 respectively) satisfy $\mathcal{I}$, therefore $\mathcal{M}_1$ and $\mathcal{M}_2$ satisfy $\mathcal{P}$.

In the following, we consider that the size of a pMC, IMC, or pIMC corresponds to its number of states plus its number of transitions not reduced to 0, $[0, 0]$ or $\emptyset$. We will also often need to consider the predecessors (Pred), and the successors (Succ) of some given states. Given a pIMC with a set of states S, a state s in S, and a subset S' of S, we write:

- $\text{Pred}(s) = \{s' \in S \mid P(s', s) \notin \{\emptyset, [0, 0]\}\}$ – $\text{Pred}(S') = \bigcup_{s' \in S'} \text{Pred}(s')$
- $\text{Succ}(s) = \{s' \in S \mid P(s, s') \notin \{\emptyset, [0, 0]\}\}$ – $\text{Succ}(S') = \bigcup_{s' \in S'} \text{Succ}(s')$

---

[1] Indeed, when $0 \leq v(f_1) \leq v(f_2) \leq 1$ is not respected, the interval is inconsistent and therefore empty.

3.2   Abstraction Model Comparisons

IMC, pMC, and pIMC are three Markov chain Abstraction Models. In order to compare their expressiveness and compactness, we introduce the comparison operators $\sqsubseteq$ and $\equiv$. Let $(L_1, \models_1)$ and $(L_2, \models_2)$ be two Markov chain abstraction models containing respectively $\mathcal{L}_1$ and $\mathcal{L}_2$. We say that $\mathcal{L}_1$ is entailed by $\mathcal{L}_2$, written $\mathcal{L}_1 \sqsubseteq \mathcal{L}_2$, iff all the MCs satisfying $\mathcal{L}_1$ satisfy $\mathcal{L}_2$ modulo bisimilarity. (*i.e.*, $\forall \mathcal{M} \models_1 \mathcal{L}_1, \exists \mathcal{M}' \models_2 \mathcal{L}_2$ s.t. $\mathcal{M}$ is bisimilar to $\mathcal{M}'$). We say that $\mathcal{L}_1$ is (semantically) equivalent to $\mathcal{L}_2$, written $\mathcal{L}_1 \equiv \mathcal{L}_2$, iff $\mathcal{L}_1 \sqsubseteq \mathcal{L}_2$ and $\mathcal{L}_2 \sqsubseteq \mathcal{L}_1$. Definition 5 introduces succinctness based on the sizes of the abstractions.

Definition 5 (Succinctness). *Let* $(L_1, \models_1)$ *and* $(L_2, \models_2)$ *be two Markov chain abstraction models.* L_1 *is at least as succinct as* L_2, *written* $L_1 \leq L_2$, *iff there exists a polynomial p such that for every* $\mathcal{L}_2 \in L_2$, *there exists* $\mathcal{L}_1 \in L_1$ *s.t.* $\mathcal{L}_1 \equiv \mathcal{L}_2$ *and* $|\mathcal{L}_1| \leq p(|\mathcal{L}_2|)$.[2] *Thus,* L_1 *is strictly more succinct than* L_2, *written* $L_1 < L_2$, *iff* $L_1 \leq L_2$ *and* $L_2 \nleq L_1$.

We start with a comparison of the succinctness of the pMC and IMC abstractions. Since pMCs allow the expression of dependencies between the probabilities assigned to distinct transitions while IMCs allow all transitions to be independant, it is clear that there are pMCs without any equivalent IMCs (regardless of the IMC semantics used), therefore $(\texttt{IMC}, \models_I^o) \nleq \texttt{pMC}$ and $(\texttt{IMC}, \models_I^a) \nleq \texttt{pMC}$ (see [4] for details). On the other hand, IMCs imply that transition probabilities need to satisfy linear inequalities in order to fit given intervals. However, these types of constraints are not allowed in pMCs. It is therefore easy to exhibit IMCs that, regardless of the semantics considered, do not have any equivalent pMC specification. As a consequence, $\texttt{pMC} \nleq (\texttt{IMC}, \models_I^o)$ and $\texttt{pMC} \nleq (\texttt{IMC}, \models_I^a)$.

We now compare pMCs and IMCs to pIMCs. Recall that the pIMC model is a Markov chain abstraction model allowing to declare parametric interval transitions, while the pMC model allows only parametric transitions (without intervals), and the IMC model allows interval transitions without parameters. Clearly, any pMC and any IMC can be translated into a pIMC with the right semantics (once-and-for-all for pMCs and the chosen IMC semantics for IMCs). This means that $(\texttt{pIMC}, \models_{pI}^o)$ is more succinct than pMC and pIMC is more succinct than IMC for both semantics. Furthermore, since pMC and IMC are not comparable due to the above results, we have that the pIMC abstraction model is strictly more succinct than the pMC abstraction model and than the IMC abstraction model with the right semantics. Our comparison results are presented in Proposition 1. Further explanations and examples are given in [4].

Proposition 1. *The Markov chain abstraction models can be ordered as follows w.r.t. succinctness:* $(\texttt{pIMC}, \models_{pI}^o) < (\texttt{pMC}, \models_p)$, $(\texttt{pIMC}, \models_{pI}^o) < (\texttt{IMC}, \models_I^o)$ *and* $(\texttt{pIMC}, \models_{pI}^a) < (\texttt{IMC}, \models_I^a)$.

Note that $(\texttt{pMC}, \models_p) \leq (\texttt{IMC}, \models_I^o)$ could be achieved by adding unary constraints on the parameters of a pMC, which is not allowed here. However, this would not have any impact on our other results.

---

[2] $|\mathcal{L}_1|$ and $|\mathcal{L}_2|$ are the sizes of $\mathcal{L}_1$ and $\mathcal{L}_2$, respectively.

4   Qualitative Properties

As seen above, pIMCs are a succinct abstraction formalism for MCs. The aim
of this section is to investigate qualitative properties for pIMCs, *i.e.*, properties
that can be evaluated at the specification (pIMC) level, but that entail properties on its MC implementations. pIMC specifications are very expressive as
they allow the abstraction of transition probabilities using both intervals and
parameters. Unfortunately, as it is the case for IMCs, this allows the expression
of incorrect specifications. In the IMC setting, this is the case either when some
intervals are ill-formed or when there is no probability distribution matching the
interval constraints of the outgoing transitions of some reachable state. In this
case, no MC implementation exists that satisfies the IMC specification. Deciding whether an implementation that satisfies a given specification exists is called
the consistency problem. In the pIMC setting, the consistency problem is made
more complex because of the parameters which can also induce inconsistencies
in some cases. One could also be interested in verifying whether there exists
an implementation that reaches some target states/labels, and if so, propose a
parameter valuation ensuring this property. Both the consistency and the consistent reachability problems have already been investigated in the IMC and pIMC
setting [10,11]. In this section, we briefly recall these problems and propose new
solutions based on CSP encodings. Our encodings are linear in the size of the
original pIMCs whereas the algorithms from [10,11] are exponential.

4.1   Existential Consistency

A pIMC $\mathcal{P}$ is existential consistent iff there exists a MC $\mathcal{M}$ satisfying $\mathcal{P}$ (*i.e.*,
there exists a MC $\mathcal{M}$ satisfying an IMC $\mathcal{I}$ instance of $\mathcal{P}$). As seen in Sect. 2,
pIMCs are equipped with two semantics: once-and-for-all ($\models_{\text{pI}}^{o}$) and at-every-
step ($\models_{\text{pI}}^{a}$). Recall that $\models_{\text{pI}}^{o}$ imposes that the underlying graph structure of
implementations needs to be isomorphic to the graph structure of the corre-
sponding specification. In contrast, $\models_{\text{pI}}^{a}$ allows implementations to have a different graph structure. It therefore seems that some pIMCs could be inconsistent
w.r.t $\models_{\text{pI}}^{o}$ while being consistent w.r.t $\models_{\text{pI}}^{a}$. On the other hand, checking the
consistency w.r.t $\models_{\text{pI}}^{o}$ seems easier because of the fixed graph structure.

In [10], the author firstly proved that both semantics are equivalent w.r.t.
existential consistency, and proposed a CSP encoding for verifying this property
which is exponential in the size of the pIMC. Based on this result of semantics
equivalence w.r.t. existential consistency from [10] we propose a new CSP encod-
ing, written $\mathbf{C}_{\exists\mathbf{c}}$, for verifying the existential consistency property for pIMCs.

Let $\mathcal{P} = (S, s_0, P, V, Y)$ be a pIMC, we write $\mathbf{C}_{\exists\mathbf{c}}(\mathcal{P})$ for the CSP produced by
$\mathbf{C}_{\exists\mathbf{c}}$ according to $\mathcal{P}$. Any solution of $\mathbf{C}_{\exists\mathbf{c}}(\mathcal{P})$ will correspond to a MC satisfying
$\mathcal{P}$. In $\mathbf{C}_{\exists\mathbf{c}}(\mathcal{P})$, we use one variable π_p with domain $[0, 1]$ per parameter p in Y; one
variable $\theta_s^{s'}$ with domain $[0, 1]$ per transition (s, s') in $\{\{s\} \times \text{Succ}(s) \mid s \in S\}$;
and one Boolean variable ρ_s per state s in S. These Boolean variables will
indicate for each state whether it appears in the MC solution of the CSP (*i.e.*,

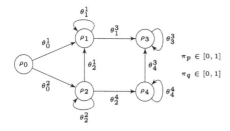

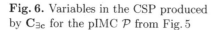

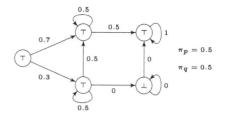

Fig. 6. Variables in the CSP produced by $\mathbf{C}_{\exists c}$ for the pIMC $\mathcal{P}$ from Fig. 5

Fig. 7. A solution to the CSP $\mathbf{C}_{\exists c}(\mathcal{P})$ for the pIMC $\mathcal{P}$ from Fig. 5

in the MC satisfying the pIMC $\mathcal{P}$). For each state $s \in S$, Constraints are as follows:

(1) ρ_s, if $s = s_0$

(2) $\rho_s \Leftrightarrow \Sigma_{s' \in \text{Succ}(s)} \theta_s^{s'} = 1$

(3) $\neg\rho_s \Leftrightarrow \Sigma_{s' \in \text{Pred}(s) \backslash \{s\}} \theta_{s'}^s = 0$, if $s \neq s_0$

(4) $\neg\rho_s \Leftrightarrow \Sigma_{s' \in \text{Succ}(s)} \theta_s^{s'} = 0$

(5) $\rho_s \Rightarrow \theta_s^{s'} \in P(s, s')$, for all $s' \in \text{Succ}(s)$

Recall that given a pIMC $\mathcal{P}$ the objective of the CSP $\mathbf{C}_{\exists c}(\mathcal{P})$ is to construct a MC $\mathcal{M}$ satisfying $\mathcal{P}$. Constraint (1) states that the initial state s_0 appears in $\mathcal{M}$. Constraint (3) ensures that for each non-initial state s, variable ρ_s is set to false iff s is not reachable from its predecessors. Constraint (2) ensures that if a state s appears in $\mathcal{M}$, then its outgoing transitions form a probability distribution. On the contrary, Constraint (4) propagates non-appearing states (*i.e.*, if a state s does not appear in $\mathcal{M}$ then all its outgoing transitions are set to zero). Finally, Constraint (5) states that, for all appearing states, the outgoing transition probabilities must be selected inside the specified intervals.

Example 7. Consider the pIMC $\mathcal{P}$ given in Fig. 5. Figure 6 describes the variables in $\mathbf{C}_{\exists c}(\mathcal{P})$: one variable per transition (*e.g.*, θ_0^1, θ_0^2, θ_1^1), one Boolean variable per state (*e.g.*, ρ_0, ρ_1), and one variable per parameter (π_p and π_q). The following constraints correspond to the Constraints (2), (3), (4), and (5) generated by our encoding $\mathbf{C}_{\exists c}$ for the state 2 of $\mathcal{P}$:

$$\neg\rho_2 \Leftrightarrow \theta_0^2 = 0 \qquad \rho_2 \Leftrightarrow \theta_2^1 + \theta_2^2 + \theta_2^4 = 1 \qquad \rho_2 \Rightarrow 0.2 \leq \theta_2^2 \leq \pi_p$$
$$\neg\rho_2 \Leftrightarrow \theta_2^1 + \theta_2^2 + \theta_2^4 = 0 \qquad \rho_2 \Rightarrow 0 \leq \theta_2^1 \leq \pi_p \qquad \rho_2 \Rightarrow 0 \leq \theta_2^4 \leq 0.5$$

Finally, Fig. 7 describes a solution for the CSP $\mathbf{C}_{\exists c}(\mathcal{P})$. Note that given a solution of a pIMC encoded by $\mathbf{C}_{\exists c}$, one can construct a MC satisfying the given pIMC by keeping all the states s s.t. ρ_s is equal to true and considering the transition function given by the probabilities in the $\theta_s^{s'}$ variables. We now show that our encoding works as expected.

Proposition 2. *A pIMC $\mathcal{P}$ is existential consistent iff $\mathbf{C}_{\exists c}(\mathcal{P})$ is satisfiable.*

Our existential consistency encoding is linear in the size of the pIMC instead of exponential for the encoding from [11] which enumerates the powerset of the states in the pIMC resulting in deep nesting of conjunctions and disjunctions.

4.2   Qualitative Reachability

Let $\mathcal{P} = (S, s_0, P, V, Y)$ be a pIMC and $\alpha \subseteq A$ be a state label. We say that α is *existential reachable* in $\mathcal{P}$ iff there exists an implementation $\mathcal{M}$ of $\mathcal{P}$ where α is reachable (*i.e.*, $\mathbb{P}^{\mathcal{M}}(\Diamond\alpha) > 0$). In a dual way, we say that α is *universal reachable* in $\mathcal{P}$ iff α is reachable in any implementation $\mathcal{M}$ of $\mathcal{P}$. As for existential consistency, we use a result from [10] that states that both pIMC semantics are equivalent w.r.t. existential (and universal) reachability. We therefore propose a new CSP encoding, written $\mathbf{C}_{\exists r}$, that extends $\mathbf{C}_{\exists c}$, for verifying these properties. Formally, CSP $\mathbf{C}_{\exists r}(\mathcal{P}) = (X \cup X', D \cup D', C \cup C')$ is such that $(X, D, C) = \mathbf{C}_{\exists c}(\mathcal{P})$, X' contains one integer variable ω_s with domain $[0, |S|]$ per state s in S, D' contains the domains of these variables, and C' is composed of the following constraints for each state $s \in S$:

(6) $\omega_s = 1$, if $s = s_0$ **(7)** $\omega_s \neq 1$, if $s \neq s_0$ **(8)** $\rho_s \Leftrightarrow (\omega_s \neq 0)$

(9) $\omega_s > 1 \Rightarrow \bigvee_{s' \in \mathtt{Pred}(s)\setminus\{s\}}(\omega_s = \omega_{s'} + 1) \wedge (\theta_s^{s'} > 0)$, if $s \neq s_0$

(10) $\omega_s = 0 \Leftrightarrow \bigwedge_{s' \in \mathtt{Pred}(s)\setminus\{s\}}(\omega_{s'} = 0) \vee (\theta_s^{s'} = 0)$, if $s \neq s_0$

Recall first that CSP $\mathbf{C}_{\exists c}(P)$ constructs a Markov chain $\mathcal{M}$ satisfying $\mathcal{P}$. Informally, for each state s in $\mathcal{M}$ the Constraints **(6)**, **(7)**, **(9)** and **(10)** in $\mathbf{C}_{\exists r}$ ensure that $\omega_s = k$ iff there exists in $\mathcal{M}$ a path from the initial state to s of length $k - 1$ with non zero probability; and state s is not reachable in $\mathcal{M}$ from the initial state s_0 iff ω_s equals to 0. Finally, Constraint **(8)** enforces the Boolean reachability indicator variable ρ_s to bet set to true iff there exists a path with non zero probability in $\mathcal{M}$ from the initial state s_0 to s (*i.e.*, $\omega_s \neq 0$).

Let S_α be the set of states from $\mathcal{P}$ labeled with α. $\mathbf{C}_{\exists r}(\mathcal{P})$ therefore produces a Markov chain satisfying $\mathcal{P}$ where reachable states s are such that $\rho_s = \mathrm{true}$. As a consequence, α is existential reachable in $\mathcal{P}$ iff $\mathbf{C}_{\exists r}(\mathcal{P})$ admits a solution such that $\bigvee_{s \in S_\alpha} \rho_s$; and α is universal reachable in $\mathcal{P}$ iff $\mathbf{C}_{\exists r}(\mathcal{P})$ admits no solution such that $\bigwedge_{s \in S_\alpha} \neg\rho_s$. This is formalised in the following proposition.

Proposition 3. *Let $\mathcal{P} = (S, s_0, P, V, Y)$ be a pIMC, $\alpha \subseteq A$ be a state label, $S_\alpha = \{s \mid V(s) = \alpha\}$, and (X, D, C) be the CSP $\mathbf{C}_{\exists r}(\mathcal{P})$.*

- *CSP $(X, D, C \cup \bigvee_{s \in S_\alpha} \rho_s)$ is satisfiable iff α is existential reachable in $\mathcal{P}$*
- *CSP $(X, D, C \cup \bigwedge_{s \in S_\alpha} \neg\rho_s)$ is unsatisfiable iff α is universal reachable in $\mathcal{P}$*

As for the existential consistency problem, we have an exponential gain in terms of size of the encoding compared to [11]: the number of constraints and variables in $\mathbf{C}_{\exists r}$ is linear in terms of the size of the encoded pIMC.

Remark. In $\mathbf{C}_{\exists r}$ Constraints **(3)** inherited from $\mathbf{C}_{\exists c}$ are entailed by Constraints **(8)** and **(10)** added to $\mathbf{C}_{\exists r}$. Thus, in a practical approach one may ignore Constraints **(3)** from $\mathbf{C}_{\exists c}$ if they do not improve the solver performances.

5   Quantitative Properties

We now move to the verification of quantitative reachability properties in pIMCs. Quantitative reachability has already been investigated in the context of pMCs and IMCs with the once-and-for-all semantics. Due to the complexity of allowing implementation structures to differ from the structure of the specifications, quantitative reachability in IMCs with the at-every-step semantics has, to the best of our knowledge, never been studied. In this section, we propose our main theoretical contribution: a theorem showing that both IMC semantics are equivalent with respect to quantitative reachability, which allows the extension of all results from [5,20] to the at-every-step semantics. Based on this result, we also extend the CSP encodings introduced in Sect. 4 in order to solve quantitative reachability properties on pIMCs regardless of their semantics.

5.1   Equivalence of $\models_I^o$ and $\models_I^a$ w.r.t Quantitative Reachability

Given an IMC $\mathcal{I} = (S, s_0, P, V)$ and a state label $\alpha \subseteq A$, a quantitative reachability property on $\mathcal{I}$ is a property of the type $\mathbb{P}^{\mathcal{I}}(\Diamond \alpha) \sim p$, where $0 < p < 1$ and $\sim \in \{\leq, <, >, \geq\}$. Such a property is verified iff there exists an MC $\mathcal{M}$ satisfying $\mathcal{I}$ (with the chosen semantics) such that $\mathbb{P}^{\mathcal{M}}(\Diamond \alpha) \sim p$.

As explained above, all existing techniques and tools for verifying quantitative reachability properties on IMCs only focus on the once-and-for-all semantics. Indeed, in this setting, quantitative reachability properties are easier to compute because the underlying graph structure of all implementations is known. However, to the best of our knowledge, there are no works addressing the same problem with the at-every-step semantics or showing that addressing the problem in the once-and-for-all setting is sufficiently general. The following theorem fills this theoretical gap by proving that both semantics are equivalent w.r.t quantitative reachability. In other words, for all MC $\mathcal{M}$ such that $\mathcal{M} \models_I^a \mathcal{I}$ and all state label α, there exist MCs $\mathcal{M}_<$ and $\mathcal{M}_>$ such that $\mathcal{M}_< \models_I^o \mathcal{I}$, $\mathcal{M}_> \models_I^o \mathcal{I}$ and $\mathbb{P}^{\mathcal{M}_<}(\Diamond \alpha) \leq \mathbb{P}^{\mathcal{M}}(\Diamond \alpha) \leq \mathbb{P}^{\mathcal{M}_>}(\Diamond \alpha)$. This is formalized in the following theorem.

Theorem 1. *Let $\mathcal{I} = (S, s_0, P, V)$ be an IMC, $\alpha \subseteq A$ be a state label, $\sim \in \{\leq, <, >, \geq\}$ and $0 < p < 1$. $\mathcal{I}$ satisfies $\mathbb{P}^{\mathcal{I}}(\Diamond \alpha) \sim p$ with the once-and-for-all semantics iff $\mathcal{I}$ satisfies $\mathbb{P}^{\mathcal{I}}(\Diamond \alpha) \sim p$ with the at-every-step semantics.*

The proof is constructive (see [4]): we use the structure of the relation $\mathcal{R}$ from the definition of $\models_I^a$ in order to build the MCs $\mathcal{M}_<$ and $\mathcal{M}_>$.

5.2   Constraint Encodings

Note that the result from Theorem 1 naturally extends to pIMCs. We therefore exploit this result to construct a CSP encoding for verifying quantitative reachability properties in pIMCs. As in Sect. 4, we extend the CSP $\mathbf{C}_{\exists c}$, that produces a correct MC implementation for the given pIMC, by imposing that this MC implementation satisfies the given quantitative reachability property.

In order to compute the probability of reaching state label α at the MC level, we use standard techniques from [2] that require the partitioning of the state space into three sets $S_\top$, $S_\perp$, and $S_?$ that correspond to states reaching α with probability 1, states from which α cannot be reached, and the remaining states, respectively. Once this partition is chosen, the reachability probabilities of all states in $S_?$ are computed as the unique solution of a linear equation system (see [2], Theorem 10.19, p. 766). We now explain how we identify states from $S_\perp, S_\top$ and $S_?$ and how we encode the linear equation system, which leads to the resolution of quantitative reachability.

Let $\mathcal{P} = (S, s_0, P, V, Y)$ be a pIMC and $\alpha \subseteq A$ be a state label. We start by setting $S_\top = \{s \mid V(s) = \alpha\}$. We then extend $\mathbf{C}_{\exists\mathbf{r}}(\mathcal{P})$ in order to identify the set $S_\perp$. Let $\mathbf{C}'_{\exists\mathbf{r}}(\mathcal{P}, \alpha) = (X \cup X', D \cup D', C \cup C')$ be such that $(X, D, C) = \mathbf{C}_{\exists\mathbf{r}}(\mathcal{P})$, X' contains one Boolean variable λ_s and one integer variable α_s with domain $[0, |S|]$ per state s in S, D' contains the domains of these variables, and C' is composed of the following constraints for each state $s \in S$:

(11) $\alpha_s = 1$, if $\alpha = V(s)$ **(12)** $\alpha_s \neq 1$, if $\alpha \neq V(s)$ **(13)** $\lambda_s \Leftrightarrow (\rho_s \wedge (\alpha_s \neq 0))$

(14) $\alpha_s > 1 \Rightarrow \bigvee_{s' \in \mathrm{Succ}(s) \setminus \{s\}} (\alpha_s = \alpha_{s'} + 1) \wedge (\theta_s^{s'} > 0)$, if $\alpha \neq V(s)$

(15) $\alpha_s = 0 \Leftrightarrow \bigwedge_{s' \in \mathrm{Succ}(s) \setminus \{s\}} (\alpha_{s'} = 0) \vee (\theta_s^{s'} = 0)$, if $\alpha \neq V(s)$

Note that variables α_s play a symmetric role to variables ω_s from $\mathbf{C}_{\exists\mathbf{r}}$: instead of indicating the existence of a path from s_0 to s, they characterize the existence of a path from s to a state labeled with α. In addition, due to Constraint **(13)**, variables λ_s are set to true iff there exists a path with non zero probability from the initial state s_0 to a state labeled with α passing by s. Thus, α cannot be reached from states s.t. $\lambda_s = \mathsf{false}$. Therefore, $S_\perp = \{s \mid \lambda_s = \mathsf{false}\}$.

Finally, we encode the equation system from [2] in a last CSP encoding that extends $\mathbf{C}'_{\exists\mathbf{r}}$. Let $\mathbf{C}_{\exists\overline{\mathbf{r}}}(\mathcal{P}, \alpha) = (X \cup X', D \cup D', C \cup C')$ be such that $(X, D, C) = \mathbf{C}'_{\exists\mathbf{r}}(\mathcal{P}, \alpha)$, X' contains one variable π_s per state s in S with domain $[0, 1]$, D' contains the domains of these variables, and C' is composed of the following constraints for each state $s \in S$:

(16) $\neg\lambda_s \Rightarrow \pi_s = 0$ **(17)** $\lambda_s \Rightarrow \pi_s = 1$, if $\alpha = V(s)$

(18) $\lambda_s \Rightarrow \pi_s = \Sigma_{s' \in \mathrm{Succ}(s)} \pi_{s'} \theta_{s'}^s$,   if $\alpha \neq V(s)$

As a consequence, variables π_s encode the probability with which state s eventually reaches α when s is reachable from the initial state and 0 otherwise.

Let $p \in [0, 1] \subseteq \mathbb{R}$ be a probability bound. Adding the constraint $\pi_{s_0} \leq p$ (resp. $\pi_{s_0} \geq p$) to the previous $\mathbf{C}_{\exists\overline{\mathbf{r}}}$ encoding allows to determine if there exists a MC $\mathcal{M} \models_{\mathrm{pI}}^a \mathcal{P}$ such that $\mathbb{P}^{\mathcal{M}}(\Diamond\alpha) \leq p$ (resp $\geq p$). Formally, let $\sim \in \{\leq, <, \geq, >\}$ be a comparison operator, and we write $\not\sim$ for its negation (e.g., $\not\leq$ is $>$). This leads to the following theorem.

Theorem 2. *Let* $\mathcal{P} = (S, s_0, P, V, Y)$ *be a pIMC,* $\alpha \subseteq A$ *be a label,* $p \in [0, 1]$, $\sim \in \{\leq, <, \geq, >\}$ *be a comparison operator, and* (X, D, C) *be* $\mathbf{C}_{\exists\overline{\mathbf{r}}}(\mathcal{P}, \alpha)$:

- *CSP* $(X, D, C \cup (\pi_{s_0} \sim p))$ *is satisfiable iff* $\exists \mathcal{M} \models_{\mathrm{pI}}^a \mathcal{P}$ *s.t.* $\mathbb{P}^{\mathcal{M}}(\Diamond\alpha) \sim p$
- *CSP* $(X, D, C \cup (\pi_{s_0} \not\sim p))$ *is unsatisfiable iff* $\forall \mathcal{M} \models_{\mathrm{pI}}^a \mathcal{P}$: $\mathbb{P}^{\mathcal{M}}(\Diamond\alpha) \sim p$

6   Prototype Implementation

Our results have been implemented in a prototype tool[3] which generates the above CSP encodings, and CSP encodings from [11] as well. Given a pIMC in a text format inspired from [20], our tool produces the desired CSP as a SMT instance with the QF_NRA logic (Quantifier Free Non linear Real-number Arithmetic). This instance can then be fed to any solver accepting the SMT-LIB format with QF_NRA logic [3]. For our benchmarks, we chose Z3 [8] (latest version: 4.5.0).

QF_NRA does not deal with integer variables. In practice, logics mixing integers and reals are harder than those over reals only. Thus we obtained better results by encoding integer variables into real ones. In our implementations each integer variable x is declared as a real variable whose real domain bounds are its original integer domain bounds; we also add the constraint $x < 1 \Rightarrow x = 0$. Since we only perform incrementation of x this preserves the same set of solutions.

In order to evaluate our prototype, we extend the NAND model from [17][4]. The original MC NAND model has already been extended as a pMC in [9], where the authors consider a single parameter p for the probability that each of the N *nand* gates fails during the multiplexing. We extend this model to pIMC by considering intervals for the probability that the initial inputs are stimulated and we have one parameter per *nand* gate to represent the probability that it fails. pIMCs in text format are automatically generated from the PRISM model.

Table 1 summarizes the size of the considered instances of the model (in terms of states, transitions, and parameters) and of the corresponding CSP problems (in terms of number of variables and constraints). In addition, we also present the resolution time of the given CSPs using the Z3 solver. Our experiments were performed on a 2.4 GHz Intel Core i5 processor with time out set to 10 min and memory out set to 2 Gb.

Table 1. Benchmarks

Benchmark	pIMC			$C_{\exists c}$			$C_{\exists r}$			$C_{\exists F}$		
	#states	#trans	#par	#var	#cstr	time	#var	#cstr	time	#var	#cstr	time
NAND $K = 1; N = 2$	104	147	4	255	1,526	0.17s	170	1,497	0.19s	296	2,457	69.57s
NAND $K = 1; N = 3$	252	364	5	621	3,727	0.24s	406	3,557	0.30s	703	5,828	31.69s
NAND $K = 1; N = 5$	930	1,371	7	2,308	13,859	0.57s	1,378	12,305	0.51s	2,404	20,165	T.O.
NAND $K = 1; N = 10$	7,392	11,207	12	18,611	111,366	9.46s	9,978	89,705	13.44s	17,454	147,015	T.O

7   Conclusion and Future Work

In this paper, we have compared several Markov Chain abstractions in terms of succinctness and we have shown that Parametric Interval Markov Chain is a

---

[3] All resources, benchmarks, and source code are available online as a Python library at https://github.com/anicet-bart/pimc_pylib.

[4] Available online at http://www.prismmodelchecker.com.

strictly more succinct abstraction formalism than other existing formalisms such as Parametric Markov Chains and Interval Markov Chains. In addition, we have proposed constraint encodings for checking several properties over pIMC. In the context of qualitative properties such as existencial consistency or consistent reachability, the size of our encodings is significantly smaller than other existing solutions. In the quantitative setting, we have compared the two usual semantics for IMCs and pIMCs and showed that both semantics are equivalent with respect to quantitative reachability properties. As a side effect, this result ensures that all existing tools and algorithms solving reachability problems in IMCs under the once-and-for-all semantics can safely be extended to the at-every-step semantics with no changes. Based on this result, we have then proposed CSP encodings addressing quantitative reachability in the context of pIMCs regardless of the chosen semantics. Finally, we have developed a prototype tool that automatically generates our CSP encodings and that can be plugged to any constraint solver accepting the SMT-LIB format as input.

We plan to develop our tool for pIMC verification in order to manage other, more complex, properties (*e.g.*, supporting the LTL-language in the spirit of what Tulip [20] does). We also plan on investigating a practical way of computing and representing the set of *all solutions* to the parameter synthesis problem.

References

1. Alur, R., Henzinger, T.A., Vardi, M.Y.: Parametric real-time reasoning. In: STOC, pp. 592–601. ACM (1993)
2. Baier, C., Katoen, J.P.: Principles of Model Checking (Representation and Mind Series). The MIT Press, Cambridge (2008)
3. Barrett, C., Fontaine, P., Tinelli, C.: The Satisfiability Modulo Theories Library (SMT-LIB) (2016). www.SMT-LIB.org
4. Bart, A., Delahaye, B., Lime, D., Monfroy, E., Truchet, C.: Reachability in Parametric Interval Markov Chains using Constraints (2017). https://hal.archives-ouvertes.fr/hal-01529681 (long version)
5. Benedikt, M., Lenhardt, R., Worrell, J.: LTL model checking of interval Markov chains. In: Piterman, N., Smolka, S.A. (eds.) TACAS 2013. LNCS, vol. 7795, pp. 32–46. Springer, Heidelberg (2013). doi:10.1007/978-3-642-36742-7_3
6. Cantor, G.: Über unendliche, lineare punktmannigfaltigkeiten v (on infinite, linear point-manifolds). Math. Ann. **21**, 545–591 (1883)
7. Courcoubetis, C., Yannakakis, M.: The complexity of probabilistic verification. J. ACM **42**(4), 857–907 (1995)
8. Moura, L., Bjørner, N.: Z3: an efficient SMT solver. In: Ramakrishnan, C.R., Rehof, J. (eds.) TACAS 2008. LNCS, vol. 4963, pp. 337–340. Springer, Heidelberg (2008). doi:10.1007/978-3-540-78800-3_24
9. Dehnert, C., Junges, S., Jansen, N., Corzilius, F., Volk, M., Bruintjes, H., Katoen, J.-P., Ábrahám, E.: PROPhESY: A PRObabilistic ParamEter SYnthesis Tool. In: Kroening, D., Păsăreanu, C.S. (eds.) CAV 2015. LNCS, vol. 9206, pp. 214–231. Springer, Cham (2015). doi:10.1007/978-3-319-21690-4_13
10. Delahaye, B.: Consistency for parametric interval Markov chains. In: SynCoP, pp. 17–32 (2015)

11. Delahaye, B., Lime, D., Petrucci, L.: Parameter synthesis for parametric interval Markov chains. In: Jobstmann, B., Leino, K.R.M. (eds.) VMCAI 2016. LNCS, vol. 9583, pp. 372–390. Springer, Heidelberg (2016). doi:10.1007/978-3-662-49122-5_18

12. Chakraborty, S., Katoen, J.-P.: Model checking of open Interval Markov chains. In: Gribaudo, M., Manini, D., Remke, A. (eds.) ASMTA 2015. LNCS, vol. 9081, pp. 30–42. Springer, Cham (2015). doi:10.1007/978-3-319-18579-8_3

13. Hahn, E.M., Hermanns, H., Wachter, B., Zhang, L.: PARAM: a model checker for parametric Markov models. In: Touili, T., Cook, B., Jackson, P. (eds.) CAV 2010. LNCS, vol. 6174, pp. 660–664. Springer, Heidelberg (2010). doi:10.1007/978-3-642-14295-6_56

14. Husmeier, D., Dybowski, R., Roberts, S.: Probabilistic Modeling in Bioinformatics and Medical Informatics. Springer Publishing Company, Incorporated, London (2010)

15. Jonsson, B., Larsen, K.G.: Specification and refinement of probabilistic processes. In: LICS, pp. 266–277 (1991)

16. Kwiatkowska, M., Norman, G., Parker, D.: PRISM 4.0: verification of probabilistic real-time systems. In: Gopalakrishnan, G., Qadeer, S. (eds.) CAV 2011. LNCS, vol. 6806, pp. 585–591. Springer, Heidelberg (2011). doi:10.1007/978-3-642-22110-1_47

17. Norman, G., Parker, D., Kwiatkowska, M., Shukla, S.: Evaluating the reliability of NAND multiplexing with PRISM. IEEE Trans. Comput.-Aided Des. Integr. Circ. Syst. **24**(10), 1629–1637 (2005)

18. Puggelli, A., Li, W., Sangiovanni-Vincentelli, A.L., Seshia, S.A.: Polynomial-time verification of PCTL properties of MDPs with convex uncertainties. In: Sharygina, N., Veith, H. (eds.) CAV 2013. LNCS, vol. 8044, pp. 527–542. Springer, Heidelberg (2013). doi:10.1007/978-3-642-39799-8_35

19. Rossi, F., Beek, P.V., Walsh, T.: Handbook of Constraint Programming (Foundations of Artificial Intelligence). Elsevier Science Inc., Amsterdam (2006)

20. Wongpiromsarn, T., Topcu, U., Ozay, N., Xu, H., Murray, R.M.: Tulip: a software toolbox for receding horizon temporal logic planning (2011)

Mean-Payoff Optimization in Continuous-Time Markov Chains with Parametric Alarms

Christel Baier[1], Clemens Dubslaff[1], Ľuboš Korenčiak[2(✉)], Antonín Kučera[2], and Vojtěch Řehák[2]

[1] TU Dresden, Dresden, Germany
{christel.baier,clemens.dubslaff}@tu-dresden.de
[2] Masaryk University, Brno, Czech Republic
{korenciak,kucera,rehak}@fi.muni.cz

Abstract. Continuous-time Markov chains with alarms (ACTMCs) allow for alarm events that can be non-exponentially distributed. Within *parametric* ACTMCs, the parameters of alarm-event distributions are not given explicitly and can be subject of parameter synthesis. An algorithm solving the ε-optimal parameter synthesis problem for parametric ACTMCs with long-run average optimization objectives is presented. Our approach is based on reduction of the problem to finding long-run average optimal strategies in semi-Markov decision processes (semi-MDPs) and sufficient discretization of parameter (i.e., action) space. Since the set of actions in the discretized semi-MDP can be very large, a straightforward approach based on explicit action-space construction fails to solve even simple instances of the problem. The presented algorithm uses an enhanced policy iteration on symbolic representations of the action space. The soundness of the algorithm is established for parametric ACTMCs with alarm-event distributions satisfying four mild assumptions that are shown to hold for uniform, Dirac, exponential, and Weibull distributions in particular, but are satisfied for many other distributions as well. An experimental implementation shows that the symbolic technique substantially improves the efficiency of the synthesis algorithm and allows to solve instances of realistic size.

1 Introduction

Mean-payoff is widely accepted as an appropriate concept for measuring long-run average performance of systems with rewards or costs. In this paper, we study the problem of synthesizing parameters for (possibly *non-exponentially* distributed) events in a given stochastic system to achieve an ε-optimal mean-payoff. One simple example of such events are *timeouts* widely used, e.g., to prevent deadlocks or to ensure some sort of progress in distributed systems. In practice, timeout

The authors are partly supported by the Czech Science Foundation, grant No. 15-17564S, by the DFG through the Collaborative Research Center SFB 912 – HAEC, the Excellence Initiative by the German Federal and State Governments (cluster of excellence cfAED), and the DFG-projects BA-1679/11-1 and BA-1679/12-1.

© Springer International Publishing AG 2017
N. Bertrand and L. Bortolussi (Eds.): QEST 2017, LNCS 10503, pp. 190–206, 2017.
DOI: 10.1007/978-3-319-66335-7_12

durations are usually determined in an ad-hoc manner, requiring a considerable amount of expertise and experimental effort. This naturally raises the question of automating this design step, i.e., is there an algorithm synthesizing *optimal* timeouts?

The underlying stochastic model this paper relies on is provided by *continuous-time Markov chains with alarms (ACTMCs)*. Intuitively, ACTMCs extend continuous-time Markov chains by generally distributed *alarm events*, where at most one alarm is active during a system execution and non-alarm events can disable the alarm. In *parametric ACTMCs*, every alarm distribution depends on one single parameter ranging over a given interval of eligible values. For example, a timeout is a Dirac-distributed alarm event where the parameter specifies its duration. A *parameter function* assigning to every alarm a parameter value within the allowed interval yields a (non-parametric) ACTMC. We aim towards an algorithm that synthesizes a parameter function for an arbitrarily small $\varepsilon > 0$ achieving ε-optimal mean-payoff.

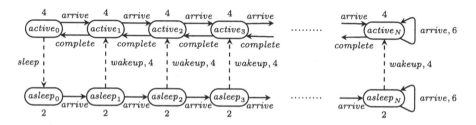

Fig. 1. Dynamic power manager of a disk drive.

Motivating Example. To get some intuition about the described task, consider a dynamic power management of a disk drive inspired by [28]. The behavior of the disk drive can be described as follows (see Fig. 1): At every moment, the drive is either *active* or *asleep*, and it maintains a queue of incoming I/O operations of capacity N. The events of *arriving* and *completing* an I/O operation have exponential distributions with rates 1.39 and 12.5, respectively. When the queue is full, all newly arriving I/O operations are rejected. The I/O operations are performed only in the *active* mode. When the drive is *active* and the queue becomes empty, an internal clock is set to d_s. If then no further I/O request is received within the next d_s time units, the *sleep* event changes the mode to *asleep*. When the drive is *asleep* and some I/O operation arrives, the internal clock is set to d_w and after d_w time the *wakeup* event changes the mode to *active*. We annotate costs in terms of energy per time unit or instantaneous energy costs for events. The power consumption is 4 and 2 per time unit in the states *active* and *asleep*, respectively. Moving from *asleep* to *active* requires 4 units of energy. Rejecting a newly arrived I/O request when the queue is full is undesirable, penalized by costs of 6. All other transitions incur with cost 1. Obviously, the designer of the disk drive controller has some freedom in choosing the delays

d_s and d_w, i.e., they are free parameters of Dirac distribution. However, d_w cannot be lower than the minimal time required to wake up the drive, which is constrained by the physical properties of the hardware used in the drive. Further, there is also a natural upper bound on d_s and d_w given by the capacity of the internal clock. Observe that if d_s is too small, then many costly transitions from *asleep* to *active* are performed; and if d_s is too large, a lot of time is wasted in the more power consuming *active* state. Similarly, if d_w is too small, a switch to the *active* mode is likely to be invoked with a few I/O operations in the queue, and more energy could have been saved by waiting somewhat longer; and if d_w is too large, the risk of rejecting newly arriving I/O operations increases. Now we may ask the following instance of an optimal parameter synthesis problem we deal with in this paper:

> *What values should a designer assign to the delays d_s and d_w such that the long-run average power consumption is minimized?*

Contribution. The main result of our paper is a *symbolic* algorithm for ε-optimal parameter synthesis that is *generic* in the sense that it is applicable to all systems where the optimized alarm events satisfy four abstractly formulated criteria. We show that these criteria are fulfilled, e.g., for timeout events modeled by Dirac distributions, uniformly distributed alarms (used in, e.g., in variants of the CSMA/CD protocol [5]), and Weibull distributions (used to model hardware failures [25]). For a given $\varepsilon > 0$, our algorithm first computes a sufficiently small discretization step such that an ε-optimal parameter function exists even when its range is restricted to the discretized parameter values. Since the discretization step is typically very small, an *explicit* construction of all discretized parameter values and their effects is computationally infeasible. Instead, our algorithm employs a symbolic variant of the standard policy iteration technique for optimizing the mean-payoff. It starts with *some* parameter function which is gradually improved until a fixed point is reached. In each improvement step, our algorithm computes a small *candidate subset* of the discretized parameter values such that a possible improvement is realizable by one of these candidate values. This is achieved by designing a suitable *ranking* function for each of the optimized events, such that an optimal parameter value is the minimal value of the ranking function in the interval of eligible parameter values. Then, the algorithm approximates the roots of the symbolic derivative of the ranking function, and constructs the candidate subset by collecting all discretized parameter values close to the approximated roots. This leads to a drastic efficiency improvement, which makes the resulting algorithm applicable to problems of realistic size.

Some proofs are omitted due to space constraints. Full details can be found in the accompanied technical report [3].

Related Work. Synthesis of optimal timeouts guaranteeing quantitative properties in timed systems was considered in [11]. There are various parametric formalisms for timed systems that deal with some sort of synthesis, such as parametric timed automata [1,18,19], parametric one-counter automata [14], parametric timed Petri nets [29], or parametric Markov models [15]. However, all works referenced above do not consider models with continuous-time distributions, thus they synthesize different parameters than we do. Contrary, the synthesis of appropriate rates in CTMCs was efficiently solved in [8,10,16,17]. A special variant of ACTMC, where only alarms with Dirac distributions are allowed, has been considered in [6,20,21]. Their algorithms synthesize ε-optimal alarm parameters towards an expected reachability objective. Using a simulation-based approach, the optimization environment of the tool TIMENET is able to approximate locally optimal distribution parameters in stochastic Petri nets, e.g., using methods as simulated annealing, hill climbing or genetic algorithms. To the best of our knowledge, we present the first algorithm that approximates globally mean-payoff optimal parameters of non-exponential distributions in continuous-time models.

The (non-parametric) ACTMCs form a subclass of Markov regenerative processes (MRP) [2,9,24]. Alternatively, ACTMCs can be also understood as a generalized semi-Markov processes (GSMPs) with at most one non-exponential event enabled in each state or as bounded stochastic Petri nets (SPNs) [13] with at most one non-exponential transition enabled in any reachable marking [9]. Note that ACTMCs are analytically tractable thanks to methods for subordinated Markov-chain (SMC) that allow for efficient computation of transient and steady-state distributions [9,22]. Recently, methods for computing steady-state distributions in larger classes of regenerative GSMPs or SPNs have been presented in [23]. We did not incorporate this method into our approach as our methods to compute sufficiently small discretization and approximation precisions to guarantee ε-optimal mean-payoffs are not directly applicable for this class of systems. To the best of our knowledge there are no efficient algorithms with a guaranteed error for computation of steady-state distribution for a general GSMP (or SPN). For some cases it is even known that the steady-state distribution does not exist [7].

2 Preliminaries

Let $\mathbb{N}$, $\mathbb{N}_0$, $\mathbb{Q}_{\geq 0}$, $\mathbb{Q}_{>0}$, $\mathbb{R}_{\geq 0}$, and $\mathbb{R}_{>0}$ denote the set of all positive integers, non-negative integers, non-negative rational numbers, positive rational numbers, non-negative real numbers, and positive real numbers, respectively. For a countable set A, we denote by $\mathcal{D}(A)$ the set of discrete probability distributions over A, i.e., functions $\mu\colon A \to \mathbb{R}_{\geq 0}$ where $\sum_{a \in A} \mu(a) = 1$. The *support* of μ is the set of all $a \in A$ with $\mu(a) > 0$. A *probability matrix* over some finite A is a function $M\colon A \times A \to \mathbb{R}_{\geq 0}$ where $M(a, \cdot) \in \mathcal{D}(A)$ for all $a \in A$.

2.1 Continuous-Time Markov Chains with Alarms

A *continuous-time Markov chain (CTMC)* is a triple $\mathcal{C} = (S, \lambda, P)$, where S is a finite set of states, $\lambda \in \mathbb{R}_{>0}$ is a common exit rate[1], and P is a probability

---

[1] We can assume without restrictions that the parameter λ is the same for all states of S since every CTMC can be effectively transformed into an equivalent CTMC satisfying this property by the standard uniformization method (see, e.g., [26]).

matrix over S. Transitions in C are exponentially distributed over the time, i.e., the probability of moving from s to s' within time τ is $P(s, s') \cdot (1 - e^{-\lambda \cdot \tau})$.

We extend CTMCs by generally distributed events called *alarms*. A *CTMC with alarms (ACTMC)* over a finite set of alarms A is a tuple

$$\mathcal{A} = (S, \lambda, P, A, \langle S_a \rangle, \langle P_a \rangle, \langle F_a \rangle),$$

where (S, λ, P) is a CTMC and $\langle S_a \rangle$, $\langle P_a \rangle$, and $\langle F_a \rangle$ are tuples defined as follows: $\langle S_a \rangle = (S_a)_{a \in A}$ where S_a is the set of states where an alarm $a \in A$ is enabled; $\langle P_a \rangle = (P_a)_{a \in A}$ where P_a is a probability matrix of some alarm $a \in A$ for which $P_a(s, s) = 1$ if $s \in S \setminus S_a$; and $\langle F_a \rangle = (F_a)_{a \in A}$ where F_a is the cumulative distribution function (CDF) according to which the ringing time of an alarm $a \in A$ is distributed. We assume that each distribution has finite mean and $F_a(0) = 0$, i.e., a positive ringing time is chosen almost surely. Furthermore, we require $S_a \cap S_{a'} = \emptyset$ for $a \neq a'$, i.e., in each state at most one alarm is enabled. The set of states where some alarm is enabled is denoted by S_{on}, and we also use S_{off} to denote the set $S \setminus S_{\text{on}}$. The pairs $(s, s') \in S \times S$ with $P(s, s') > 0$ and $P_a(s, s') > 0$ are referred to as *delay transitions* and *a-alarm transitions*, respectively.

Operational Behavior. Since in every state only one alarm is active, the semantics of an ACTMC can be seen as an infinite CTMC amended with a timer that runs backwards and is set whenever a new alarm is set or the alarm gets disabled. A *run* of the ACTMC $\mathcal{A}$ is an infinite sequence $(s_0, \eta_0), t_0, (s_1, \eta_1), t_1, \ldots$ where η_i is the current value of the timer and t_i is the time spent in s_i. If $s_0 \in S_{\text{off}}$, then $\eta_0 = \infty$. Otherwise, $s_0 \in S_a$ for some $a \in A$ and the value of η_0 is selected randomly according to F_a. In a current configuration (s_i, η_i), a random delay t is chosen according to the exponential distribution with rate λ. Then, the time t_i and the next configuration (s_{i+1}, η_{i+1}) are determined as follows:

- If $s_i \in S_a$ and $\eta_i \leq t$, then $t_i = \eta_i$ and s_{i+1} is selected randomly according to $P_a(s_i, \cdot)$. The value of η_{i+1} is either set to ∞ or selected randomly according to F_b for some $b \in A$, depending on whether the chosen s_{i+1} belongs to S_{off} or S_b, respectively (note that it may happen that $b = a$).
- If $t < \eta_i$, then $t_i = t$ and s_{i+1} is selected randomly according to $P(s_i, \cdot)$. Clearly, if $s_{i+1} \in S_{\text{off}}$, then $\eta_{i+1} = \infty$. Further, if $s_{i+1} \in S_b$ and $s_i \notin S_b$ for some $b \in A$, then η_{i+1} is selected randomly according to F_b. Otherwise, $\eta_{i+1} = \eta_i - t$ (where $\infty - t = \infty$).

Similarly as for standard CTMCs, we define a probability space over all runs initiated in a given $s_0 \in S$. We say that $\mathcal{A}$ is *strongly connected* if its underlying graph is, i.e., for all $s, s' \in S$, where $s \neq s'$, there is a finite sequence $s_0, \ldots, s_n$ of states such that $s = s_0$, $s' = s_n$, and $P(s_i, s_{i+1}) > 0$ or $P_a(s_i, s_{i+1}) > 0$ (for some $a \in A$) for all $0 \leq i < n$.

Note that the timer is set to a new value in a state s only if $s \in S_a$ for some $a \in A$, and the previous state either does not belong to S_a or the transition used to enter s was an alarm transition[2]. Formally, we say that $s \in S_a$ is an a-*setting state* if there exists $s' \in S$ such that either $P_b(s', s) > 0$ for some $b \in A$, or $s' \notin S_a$ and $P(s', s) > 0$. The set of all alarm-setting states is denoted by S_{set}. If $S_{\mathrm{set}} \cap S_a$ is a singleton for each $a \in A$, we say that the alarms in $\mathcal{A}$ are *localized*.

Cost Structures and Mean-Payoff for ACTMCs. We use the standard cost structures that assign non-negative cost values to both states and transitions (see, e.g., [27]). More precisely, we consider the following cost functions: $\mathcal{R} \colon S \to \mathbb{R}_{\geq 0}$, which assigns a cost rate $\mathcal{R}(s)$ to every state s such that the cost $\mathcal{R}(s)$ is paid for every time unit spent in s, and functions $\mathcal{I}, \mathcal{I}_A \colon S \times S \to \mathbb{R}_{\geq 0}$ that assign to each delay transition and each alarm-setting transition, respectively, an instant execution cost. For every run $\omega = (s_0, \eta_0), t_0, (s_1, \eta_1), t_1, \ldots$ of $\mathcal{N}$ we define the associated *mean-payoff* by

$$\mathrm{MP}(\omega) \quad = \quad \limsup_{n \to \infty} \frac{\sum_{i=0}^{n} \left(\mathcal{R}(s_i) \cdot t_i + \mathcal{J}(s_i, s_{i+1}) \right)}{\sum_{i=0}^{n} t_i}.$$

Here, $\mathcal{J}(s_i, s_{i+1})$ is either $\mathcal{I}(s_i, s_{i+1})$ or $\mathcal{I}_A(s_i, s_{i+1})$ depending on whether $t_i < \eta_i$ or not, respectively. We use $\mathbb{E}[\mathrm{MP}]$ to denote the expectation of MP. In general, MP may take more than one value with positive probability. However, if the graph of the underlying ACTMC is strongly connected, almost all runs yield the same mean-payoff value independent of the initial state [9].

2.2 Parametric ACTMCs

In ACTMCs, the distribution functions for the alarms are already fixed. For example, if alarm a is a timeout, it is set to some concrete value d, i.e., the associated F_a is a Dirac distribution such that $F_a(\tau) = 1$ for all $\tau \geq d$ and $F_a(\tau) = 0$ for all $0 \leq \tau < d$. Similarly, if a is a random delay selected uniformly in the interval $[0.01, d]$, then $F_a(\tau) = 0$ for all $\tau < 0.01$ and $F_a(\tau) = \min\{1, (\tau - 0.01)/(d - 0.01)\}$ for all $\tau \geq 0.01$. We also consider alarms with Weibull distributions, where $F_a(\tau) = 0$ for all $\tau \leq 0$ and $F_a(\tau) = 1 - e^{-(\tau/d)^k}$ for all $\tau > 0$, where $k \in \mathbb{N}$ is a fixed constant.[3]

In the above examples, we can interpret d as a *parameter* and ask what parameter values minimize the expected long-run average costs. For simplicity, we restrict our attention to distributions with only *one* parameter.[4] A *parametric ACTMC* is defined similarly as an ACTMC, but instead of the concrete distribution function F_a, we specify a *parameterized distribution function* $F_a[x]$ together with the interval $[\ell_a, u_a]$ of eligible parameter values for every $a \in A$. For every

---

[2] In fact, another possibility (which does not require any special attention) is that s is the initial state of a run.

[3] Note that a Weibull distribution with $k = 1$ is an exponential distribution.

[4] In our current setting, distribution functions with several parameters can be accommodated by choosing the parameter to optimize and fixing the others. In some cases we can also use simple extensions to synthesize, e.g., both d_1 and d_2 for the uniform distribution in $[d_1, d_2]$ (see Appendix B in [3]).

$d \in [\ell_a, u_a]$, we use $F_a[d]$ to denote the distribution obtained by instantiating the parameter x with d. Formally, a parametric ACTMC is a tuple

$$\mathcal{N} = (S, \lambda, P, A, \langle S_a \rangle, \langle P_a \rangle, \langle F_a[x] \rangle, \langle \ell_a \rangle, \langle u_a \rangle)$$

where all components are defined in the same way as for ACTMC except for the tuples $\langle F_a[x] \rangle$, $\langle \ell_a \rangle$, and $\langle u_a \rangle$ of all $F_a[x]$, ℓ_a, and u_a discussed above. Strong connectedness, localized alarms, and cost structures are defined as for (non-parametric) ACTMCs.

A *parameter function* for $\mathcal{N}$ is a function $\mathbf{d} \colon A \to \mathbb{R}$ such that $\mathbf{d}(a) \in [\ell_a, u_a]$ for every $a \in A$. For every parameter function $\mathbf{d}$, we use $\mathcal{N}^{\mathbf{d}}$ to denote the ACTMC obtained from $\mathcal{N}$ by replacing each $F_a[x]$ with the distribution function $F_a[\mathbf{d}(a)]$. We allow only parametric ACTMCs that for each parametric function yield ACTMC. When cost structures are defined on $\mathcal{N}$, we use $\mathbb{E}[\mathrm{MP}^{\mathbf{d}}]$ to denote the expected mean-payoff in $\mathcal{N}^{\mathbf{d}}$. For a given $\varepsilon > 0$, we say that a parameter function $\mathbf{d}$ is ε-*optimal* if

$$\mathbb{E}[\mathrm{MP}^{\mathbf{d}}] \leq \inf_{\mathbf{d}'} \mathbb{E}[\mathrm{MP}^{\mathbf{d}'}] + \varepsilon,$$

where $\mathbf{d}'$ ranges over all parameter functions for $\mathcal{N}$.

2.3   Semi-Markov Decision Processes

A *semi-Markov decision process (semi-MDP)* is a tuple $\mathcal{M} = (M, Act, Q, t, c)$, where M is a finite set of states, $Act = \biguplus_{m \in M} Act_m$ is a set of actions where $Act_m \neq \emptyset$ is a subset of actions enabled in a state m, $Q \colon Act \to \mathcal{D}(M)$ is a function assigning the probability $Q(b)(m')$ to move from $m \in M$ to $m' \in M$ executing $b \in Act_m$, and functions $t, c \colon Act \to \mathbb{R}_{\geq 0}$ provide the expected time and costs when executing an action, respectively.[5] A *run* in $\mathcal{M}$ is an infinite sequence $\omega = m_0, b_0, m_1, b_1, \ldots$ where $b_i \in Act_{m_i}$ for every $i \geq 0$. The mean-payoff of ω is

$$\mathrm{MP}(\omega) = \limsup_{n \to \infty} \left(\sum\nolimits_{i=0}^{n} c(b_i) \right) \Big/ \left(\sum\nolimits_{i=0}^{n} t(b_i) \right).$$

A (stationary and deterministic) *strategy* for $\mathcal{M}$ is a function $\sigma \colon M \to Act$ such that $\sigma(m) \in Act_m$ for all $m \in M$. Applying σ to $\mathcal{M}$ yields the standard probability measure Pr^{σ} over all runs initiated in a given initial state m_{in}. The expected mean-payoff achieved by σ is denoted by $\mathbb{E}[\mathrm{MP}^{\sigma}_{\mathcal{M}}]$. An *optimal*[6] strategy achieving the *minimal* expected mean-payoff is guaranteed to exist, and it is computable by a simple *policy iteration algorithm* (see, e.g., [27]).

κ-Approximations of Semi-MDPs. Let $\mathcal{M} = (M, Act, Q, t, c)$ be a semi-MDP, and $\kappa \in \mathbb{Q}_{>0}$. We say that $Q^{\kappa} \colon Act \to \mathcal{D}(M)$ and $t^{\kappa}, c^{\kappa} \colon Act \to \mathbb{R}_{\geq 0}$

---

[5] For our purposes, the actual distribution of the time and costs spent before executing some action is irrelevant, only their expectations matter, see Sect. 11.4 in [27].

[6] This strategy is optimal not only among stationary and deterministic strategies, but even among all randomized and history-dependent strategies.

are κ-*approximations* of Q, t, c, if for all $m, m' \in M$ and $b \in Act_m$ it holds that $Q(b)$ and $Q^\kappa(b)$ have the same support, $|Q(b)(m') - Q^\kappa(b)(m')| \leq \kappa$, $|t(b) - t^\kappa(b)| \leq \kappa$, and $|c(b) - c^\kappa(b)| \leq \kappa$. A κ-*approximation* of $\mathcal{M}$ is a semi-MDP $(M, Act, Q^\kappa, t^\kappa, c^\kappa)$ where Q^κ, t^κ, c^κ are κ-approximations of Q, t, c. We denote by $[\mathcal{M}]_\kappa$ the set of all κ-approximations of $\mathcal{M}$.

3   Synthesizing ε-optimal Parameter Functions

In the following, we fix a strongly connected parametric ACTMC $\mathcal{N} = (S, \lambda, P, A, \langle S_a \rangle, \langle P_a \rangle, \langle F_a[x] \rangle, \langle \ell_a \rangle, \langle u_a \rangle)$ with localized alarms and cost functions $\mathcal{R}$, $\mathcal{I}$, and $\mathcal{I}_A$, and aim towards an algorithm synthesizing an ε-optimal parameter function for $\mathcal{N}$. Here, ε-optimality is understood with respect to the expected mean-payoff. That is, we deal with the following computational problem:

ε-*optimal parameter synthesis for parametric ACTMCs with localized alarms.*

Input: $\varepsilon \in \mathbb{Q}_{>0}$, a strongly connected parametric ACTMC $\mathcal{N}$ with localized alarms, rational transition probabilities, rate λ, bounds $\langle \ell_a \rangle, \langle u_a \rangle$, and cost functions $\mathcal{R}, \mathcal{I}$, and $\mathcal{I}_A$.

Output: An ε-optimal parameter function $\mathbf{d}$.

3.1   The Set of Semi-Markov Decision Processes $[\mathcal{M}_{\mathcal{N}} \langle \delta \rangle]_\kappa$

Our approach to solve the above problem is based on a reduction to the problem of synthesizing expected mean-payoff optimal strategies in semi-MDPs. Let $a \in A$, and let $s \in S_a \cap S_{\text{set}}$. Recall that $\mathcal{N}$ is localized and thus, s is the uniquely defined a-setting state. Then, for every $d \in [\ell_a, u_a]$, consider runs initiated in a configuration (s, η), where η is chosen randomly according to $F_a[d]$. Almost all such runs eventually visit a *regenerative* configuration (s', η') where either $s' \in S_{\text{off}}$ or η' is chosen randomly in $s' \in S_{\text{set}}$, i.e., either all alarms are disabled or one is newly set. We use $\Pi_s(d)$ to denote the associated probability distribution over $S_{\text{set}} \cup S_{\text{off}}$, i.e., $\Pi_s(d)(s')$ is the probability of visiting a regenerative configuration of the form (s', η') from s without previously visiting another regenerative configuration. Further, we use $\mathbf{\large€}_{s(d)}$ and $\Theta_s(d)$ to denote the expected accumulated costs and the expected time elapsed until visiting a regenerative configuration, respectively. We use the same notation also for $s \in S_{\text{off}}$, where $\Pi_s(d) = P(s, \cdot)$, $\mathbf{\large€}_s(d) = \mathcal{R}(s)/\lambda + P(s, \cdot) \cdot \mathcal{I}_P$, and $\Theta_s(d) = 1/\lambda$ are independent of d. The semi-MDP $\mathcal{M}_{\mathcal{N}} = (S_{\text{set}} \cup S_{\text{off}}, Act, Q, t, c)$ is defined over actions

$$Act = \{ \langle\!\langle s, d \rangle\!\rangle : d \in [\ell_a, u_a], s \in S_{\text{set}} \cap S_a, a \in A \} \cup \{ \langle\!\langle s, 0 \rangle\!\rangle : s \in S_{\text{off}} \},$$

where for all $\langle\!\langle s, d \rangle\!\rangle \in Act$ we have $Q(\langle\!\langle s, d \rangle\!\rangle) = \Pi_s(d)$, $t(\langle\!\langle s, d \rangle\!\rangle) = \Theta_s(d)$, and $c(\langle\!\langle s, d \rangle\!\rangle) = \mathbf{\large€}_s(d)$. Note that the action space of $\mathcal{M}_{\mathcal{N}}$ is dense and that $\Pi_s(d)$, $\Theta_s(d)$, and $\mathbf{\large€}_s(d)$ might be irrational. For our algorithms, we have to ensure a finite action space and rational probability and expectation values. We thus define the δ-discretization of $\mathcal{M}_{\mathcal{N}}$ as $\mathcal{M}_{\mathcal{N}} \langle \delta \rangle = (S_{\text{set}} \cup S_{\text{off}}, Act^\delta, Q^\delta, t^\delta, c^\delta)$ for a given discretization function $\delta\colon S_{\text{set}} \to \mathbb{Q}_{>0}$. $\mathcal{M}_{\mathcal{N}}\langle\delta\rangle$ is defined as $\mathcal{M}_{\mathcal{N}}$ above, but over the action space $Act^{\delta} = \bigcup_{s \in S_{\text{set}} \cup S_{\text{off}}} Act^{\delta}_s$ with

$$Act^{\delta}_s = \big\{ \langle\!\langle s, d\rangle\!\rangle : d = \ell_a + i \cdot \delta(s) < u_a, i \in \mathbb{N}_0 \big\} \cup \big\{ \langle\!\langle s, u_a \rangle\!\rangle \big\}$$

for $s \in S_{\text{set}} \cap S_a$ and $Act^{\delta}_s = \{\langle\!\langle s, 0\rangle\!\rangle\}$ otherwise.

To ensure rational values of $\varPi_s(d)$, $\varTheta_s(d)$, and $\mathbf{\epsilon}_s(d)$, we consider the set of κ-approximations $[\mathcal{M}_{\mathcal{N}}\langle\delta\rangle]_{\kappa}$ of $\mathcal{M}_{\mathcal{N}}\langle\delta\rangle$ for any $\kappa \in \mathbb{Q}_{>0}$. Note that, as $\mathcal{N}$ is strongly connected, every $\mathcal{M} \in [\mathcal{M}_{\mathcal{N}}\langle\delta\rangle]_{\kappa}$ is also strongly connected.

3.2 An Explicit Parameter Synthesis Algorithm

Every strategy σ minimizing the expected mean-payoff in $\mathcal{M}_{\mathcal{N}}$ yields an optimal parameter function $\mathbf{d}^{\sigma}$ for $\mathcal{N}$ defined by $\mathbf{d}^{\sigma}(a) = d$ where $\sigma(s) = \langle\!\langle s, d\rangle\!\rangle$ for the unique a-setting state s. A naive approach towards an ε-optimal parameter function minimizing the expected mean-payoff in $\mathcal{N}$ is to compute a sufficiently small discretization function δ, approximation constant κ, and some $\mathcal{M} \in [\mathcal{M}_{\mathcal{N}}\langle\delta\rangle]_{\kappa}$ such that synthesizing an optimal strategy in $\mathcal{M}$ yields an ε-optimal parameter function for $\mathcal{N}$. As $\mathcal{M}$ is finite and contains only rational probability and expectation values, the synthesis of an optimal strategy for $\mathcal{M}$ can then be carried out using standard algorithms for semi-MDP (see, e.g., [27]). This approach is applicable under the following mild assumptions:

1. For every $\varepsilon \in \mathbb{Q}_{>0}$, there are computable $\delta\colon S_{\text{set}} \to \mathbb{Q}_{>0}$ and $\kappa \in \mathbb{Q}_{>0}$ such that for every $\mathcal{M} \in [\mathcal{M}_{\mathcal{N}}\langle\delta\rangle]_{\kappa}$ and every optimal strategy σ for $\mathcal{M}$, the associated parameter function $\mathbf{d}^{\sigma}$ is ε-optimal for $\mathcal{N}$.
2. For all $\kappa \in \mathbb{Q}_{>0}$ and $s \in S_{\text{set}}$, there are computable rational κ-approximations $\varPi^{\kappa}_s(d)$, $\varTheta^{\kappa}_s(d)$, $\mathbf{\epsilon}^{\kappa}_s(d)$ of $\varPi_s$, $\varTheta_s$, $\mathbf{\epsilon}_s$.

Assumption 1 usually follows from perturbation bounds on the expected mean-payoff using a straightforward error-propagation analysis. Assumption 2 can be obtained, e.g., by first computing $\kappa/2$-approximations of $\varPi_s$, $\varTheta_s$, and $\mathbf{\epsilon}_s$ for $s \in S_{\text{set}} \cap S_a$, considering a as alarm with Dirac distribution, and then integrate the obtained functions over the probability measure determined by $F_a[x]$ to get the resulting κ-approximation (see also [6,9]). Hence, Assumptions 1 and 2 rule out only those types of distributions that are rarely used in practice. In particular, the assumptions are satisfied for uniform, Dirac, and Weibull distributions. Note that Assumption 2 implies that for all $\delta\colon S_{\text{set}} \to \mathbb{Q}_{>0}$ and $\kappa \in \mathbb{Q}_{>0}$, there is a computable $\mathcal{M} \in [\mathcal{M}_{\mathcal{N}}\langle\delta\rangle]_{\kappa}$. Usually, this naive *explicit approach* to parameter synthesis is computationally infeasible due the large number of actions in $\mathcal{M}$.

3.3 A Symbolic Parameter Synthesis Algorithm

Our symbolic parameter synthesis algorithm computes the set of states of some $\mathcal{M} \in [\mathcal{M}_{\mathcal{N}}\langle\delta\rangle]_{\kappa}$ (see Assumption 1) but avoids computing the set of all actions of $\mathcal{M}$ and their effects. The algorithm is obtained by modifying the standard policy iteration [27] for semi-MDPs.

Standard Policy Iteration Algorithm. When applied to $\mathcal{M}$, standard policy iteration starts by picking an arbitrary strategy σ, which is then repeatedly improved until a fixed point is reached. In each iteration, the current strategy σ is first evaluated by computing the associated *gain* g and *bias* $\mathbf{h}$.[7] Then, for each state $s \in S_{\text{set}}$, every outgoing action $\langle\!\langle s, d \rangle\!\rangle$ is ranked by the function

$$F_s^\kappa[g, \mathbf{h}](d) \quad = \quad \mathbf{\in}_s^\kappa(d) - g \cdot \varTheta_s^\kappa(d) + \varPi_s^\kappa(d) \cdot \mathbf{h} \qquad (\times)$$

where $\mathbf{\in}_s^\kappa$, $\varTheta_s^\kappa$, and $\varPi_s^\kappa$ are the determining functions of $\mathcal{M}$. If the action chosen by σ at s does not have the best (minimal) rank, it is improved by redefining $\sigma(s)$ to some best-ranked action. The new strategy is then evaluated by computing its gain and bias and possibly improved again. The standard algorithm terminates when for all states the current strategy σ is no improvement to the previous.

Symbolic κ-approximations. In many cases, $\varPi_s(d)$, $\varTheta_s(d)$, and $\mathbf{\in}_s(d)$ for $s \in S_{\text{set}}$ are expressible as infinite sums where the summands comprise elementary functions such as polynomials or $\exp(\cdot)$. Given κ, one may effectively truncate these infinite sums into finitely many initial summands such that the obtained expressions are differentiable in the interval $[\ell_a, u_a]$ and yield the *analytical κ-approximations* $\varPi_s^\kappa(d)$, $\varTheta_s^\kappa(d)$, and $\mathbf{\in}_s^\kappa(d)$, respectively. Now we can analytically approximate $F_s^\kappa[g, \mathbf{h}](d)$ by the value $\boldsymbol{F}_s^\kappa[g, \mathbf{h}](d)$ obtained from $(\times)$ by using the analytical κ-approximations:

$$\boldsymbol{F}_s^\kappa[g, \mathbf{h}](d) \quad = \quad \boldsymbol{\in}_s^\kappa(d) - g \cdot \boldsymbol{\varTheta}_s^\kappa(d) + \boldsymbol{\varPi}_s^\kappa(d) \cdot \mathbf{h}. \qquad (\star)$$

This function is differentiable for $d \in [\ell_a, u_a]$ when g and $\mathbf{h}$ are constant. Note that the discretized parameters minimizing $F_s^\kappa[g, \mathbf{h}](d)$ are either close to ℓ_a, u_a, or roots of the derivative of $\boldsymbol{F}_s^\kappa[g, \mathbf{h}](d)$. Using the isolated roots and bounds ℓ_a and u_a, we identify a small set of candidate actions and explicitly evaluate only those instead of all actions. Note, that $\boldsymbol{\varPi}_s^\kappa(d)$, $\boldsymbol{\varTheta}_s^\kappa(d)$, $\boldsymbol{\in}_s^\kappa(d)$ may return *irrational* values for rational arguments. Hence, they cannot be evaluated precisely even for the discretized parameter values. However, when Assumption 2 is fulfilled, it is safe to use *rational* κ-approximations $\varPi_s^\kappa(d)$, $\varTheta_s^\kappa(d)$, $\mathbf{\in}_s^\kappa(d)$ for this purpose. Before we provide our symbolic algorithm, we formally state the additional assumptions required to guarantee its soundness:

3. For all $a \in A$, $s \in S_{\text{set}} \cap S_a$, $\delta\colon S_{\text{set}} \to \mathbb{Q}_{>0}$ and $\kappa \in \mathbb{Q}_{>0}$, there are analytical κ-approximations $\boldsymbol{\varPi}_s^\kappa$, $\boldsymbol{\varTheta}_s^\kappa$, $\boldsymbol{\in}_s^\kappa$ of $\varPi_s$, $\varTheta_s$, $\mathbf{\in}_s$, respectively, such that the function $\boldsymbol{F}_s^\kappa[g, \mathbf{h}](d)$, where $g \in \mathbb{Q}$ and $\mathbf{h}\colon S_{\text{set}} \cup S_{\text{off}} \to \mathbb{Q}$ are constant, is differentiable for $d \in [\ell_a, u_a]$. Further, there is an algorithm approximating the roots of the derivative of $\boldsymbol{F}_s^\kappa[g, \mathbf{h}](d)$ in the interval $[\ell_a, u_a]$ up to the absolute error $\delta(s)$.

---

[7] Here, it suffices to know that g is a scalar and $\mathbf{h}$ is a vector assigning numbers to states; for more details, see Sects. 8.2.1 and 8.6.1 in [27].

Algorithm 1. Symbolic policy iteration

---

input : A strongly connected parametric ACTMC $\mathcal{N}$ with localized alarms,
rational cost functions $\mathcal{R}$, $\mathcal{I}_P$, $\mathcal{I}_{P_a}$, and $\varepsilon \in \mathbb{Q}_{>0}$ such that
Assumptions 1–4 are fulfilled.

output: An ε-optimal parameter function $\mathbf{d}$.

1   compute the sets S_{set} and S_{off}

2   compute δ, κ, and $\Pi_s^{\min}$ of Assumptions 1 and 4

3   let $\xi = \min\{\kappa/4, \Pi_s^{\min}/3 : \text{where } s \in S_{\text{set}}\}$

4   fix the functions $\Pi_s^\xi, \Theta_s^\xi, \mathfrak{C}_s^\xi$ of Assumption 2 determining $\mathcal{M}_\xi \in [\mathcal{M}_\mathcal{N}\langle\delta\rangle]_\xi$

5   choose an arbitrary state $s' \in S_{\text{set}} \cup S_{\text{off}}$ and a strategy σ' for $\mathcal{M}_\xi$

6   **repeat**

7       $\sigma := \sigma'$

       `// policy evaluation`

8   compute the *gain*, i.e., the scalar $g := \mathbb{E}[\text{MP}^\sigma]$

9   compute the *bias*, i.e., the vector $\mathbf{h} \colon S \to \mathbb{Q}$ satisfying $\mathbf{h}(s') = 0$ and for each
       $s \in S_{\text{set}} \cup S_{\text{off}}$, $\mathbf{h}(s) = \mathfrak{C}_s^\xi(d) - g \cdot \Theta_s^\xi(d) + \Pi_s^\xi(d) \cdot \mathbf{h}$, where $\sigma(s) = \langle\!\langle s, d \rangle\!\rangle$

10   **foreach** $a \in A$ *and* $s \in S_{\text{set}} \cap S_a$ **do**

         `// policy improvement`

11   compute the set R of $\delta(s)/2$-approximations of the roots of the
          derivative of $\boldsymbol{F}_s^\xi[g, \mathbf{h}](d)$ in $[\ell_a, u_a]$ using Assumption 3

12         $C := \{\sigma(s)\} \cup \{\langle\!\langle s, d \rangle\!\rangle \in \text{Act}_s^\delta : |d - r| \leq 3 \cdot \delta(s)/2, \text{ for } r \in R \cup \{\ell_a, u_a\}\}$

13         $B := \underset{\langle\!\langle s, d\rangle\!\rangle \in C}{\operatorname{argmin}} \ F_s^\xi[g, \mathbf{h}](d)$

14   **if** $\sigma(s) \in B$ **then** $\sigma'(s) := \sigma(s)$

15   **else** $\sigma'(s) := \langle\!\langle s, d \rangle\!\rangle$ where $\langle\!\langle s, d \rangle\!\rangle \in B$

16 **until** $\sigma = \sigma'$

17 **return** $\mathbf{d}^\sigma$

---

4. For each $s \in S_{\text{set}}$ (let a be the alarm of s) there is a computable constant
$\Pi_s^{\min} \in \mathbb{Q}_{>0}$ such that for all $d \in [\ell_a, u_a]$ and $s' \in S_{\text{set}} \cup S_{\text{off}}$ we have that
$\Pi_s(d)(s') > 0$ implies $\Pi_s(d)(s') \geq \Pi_s^{\min}$.

Note that compared to Assumption 2, the κ-approximations of Assumption 3
are harder to construct: we require closed forms for $\boldsymbol{\Pi}_s^\kappa$, $\boldsymbol{\Theta}_s^\kappa$, and $\boldsymbol{\mathfrak{C}}_s^\kappa$ making
the symbolic derivative of $\boldsymbol{F}_s^\kappa[g, \mathbf{h}](d)$ computable and suitable for effective root
approximation.

Symbolic Policy Iteration Algorithm. Algorithm 1 closely mimics the standard policy iteration algorithm except for the definition of new precision ξ at
line 3 and the policy improvement part. The local extrema points of $\boldsymbol{F}_s^\xi[g, \mathbf{h}](d)$
(cf. Eq. $(\star)$) in the interval $[\ell_a, u_a]$ are identified by computing roots of its
symbolic derivative (line 11). Then, we construct a small set C of *candidate actions* that are close to these roots and the bounds ℓ_a, u_a (line 12).
Each given candidate action is then evaluated using the function $F_s^\xi[g, \mathbf{h}](d) =$

$\mathfrak{E}_s^\xi(d) - g \cdot \Theta_s^\xi(d) + \Pi_s^\xi(d) \cdot \mathbf{h}$ (cf. Eq. (×)). An improving candidate action is chosen based on the computed values (lines 14–15).

Theorem 1 (Correctness of Algorithm 1). *The symbolic policy iteration algorithm effectively solves the ε-optimal parameter synthesis problem for parametric ACTMCs and cost functions that fulfill Assumptions 1–4.*

Proof (Sketch). Since the number of actions of $\mathcal{M}_\xi$ is finite, Algorithm 1 terminates. A challenging point is that we compute only approximate minima of the function $\boldsymbol{F}_s^\xi[g, \mathbf{h}](d)$, which is *different* from the function $F_s^\xi[g, \mathbf{h}](d)$ used to evaluate the candidate actions. There may exist an action that is not in the candidate set C even if it has minimal $F_s^\xi[g, \mathbf{h}](d)$. Hence, the strategy computed by Algorithm 1 is not necessarily optimal for $\mathcal{M}_\xi$. Fortunately, due to Assumption 1, the strategy induces ε-optimal parameters for any parametric ACTMC if it is optimal for *some* $\mathcal{M}' \in [\mathcal{M}_\mathcal{N}\langle\delta\rangle]_\kappa$. Therefore, for each $s \in S_{\text{set}}$ we construct Π_s', Θ_s', and $\mathfrak{E}_s'$ determining such $\mathcal{M}'$. Omitting the details, the functions Π_s', Θ_s', $\mathfrak{E}_s'$ are constructed from Π_s^ξ, $\mathfrak{E}_s^\xi$, Θ_s^ξ and slightly (by at most 2ξ) shifted Π_s^ξ, $\mathfrak{E}_s^\xi$, Θ_s^ξ. The constant ξ was chosen sufficiently small such that the shifted Π_s^ξ, $\mathfrak{E}_s^\xi$, Θ_s^ξ are still κ-approximations of Π_s, Θ_s, $\mathfrak{E}_s$ and the shifted $\Pi_s^\xi(d)(\cdot)$ is a correct distribution for each $d \in [\ell_a, u_a]$. The technical details of the construction are provided in Appendix A of [3]. □

The following theorem implies that the explicit and symbolic algorithms are applicable to parametric ACTMCs with uniform, Dirac, exponential, or Weibull distributions. The proof is technical, see Appendix B of [3].

Theorem 2. *Assumptions 1–4 are fulfilled for parametric ACTMCs with rational cost functions where for all $a \in A$ we have that $F_a[x]$ is either a uniform, Dirac, exponential, or Weibull distribution.*

4   Experimental Evaluation

We demonstrate feasibility of the symbolic algorithm presented in Sect. 3 on the running example of Fig. 1 and on a preventive maintenance model inspired by [12]. The experiments were carried out[8] using our prototype implementation of the symbolic algorithm implemented in MAPLE [4]. MAPLE is appropriate as it supports the root isolation of univariate polynomials with arbitrary high precision due to its symbolic engine. The implementation currently supports Dirac and uniform distributions only, but could be easily extended by other distributions fulfilling Assumptions 1–4.

Disk Drive Model. In the running example of this paper (see Sect. 1 and Fig. 1) we aimed towards synthesizing delays d_s and d_w such that the long-run average power consumption of the disk drive is ε-optimal. Let us describe the impact

---

[8] All the computations were run on a machine equipped with Intel Core[TM] i7-3770 CPU processor at 3.40 GHz and 8 GiB of DDR RAM.

of choosing delay values d_s and d_w on the expected mean-payoff in more detail. In Fig. 2 (left), we illustrate the trade-off between choosing different delays d_w depending on delays $d_s \in \{0.1, 10\}$ and queue sizes $N \in \{2, 8\}$. When the queue is small, e.g., $N = 2$ (dashed curves), the expected mean-payoff is optimal for large d_s (here, $d_s = 10$). Differently, when the queue size is large, e.g., $N = 8$ (solid curves), it is better to choose small d_s (here, $d_s = 0.1$) to minimize the expected mean-payoff with d_w chosen at the minimum of the solid curve at around 3.6. This illustrates that the example is non-trivial.

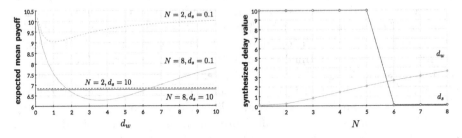

Fig. 2. Results for the disk drive example: optimal expected mean-payoff (left), and trade-off illustrated by the synthesized delay values (right)

The results of applying our synthesis algorithm for determining ε-optimal delays d_s and d_w depending on different queue sizes $N \in \{1, ..., 8\}$ with common delay bounds $\ell = 0.1$ and $u = 10$ are depicted in Fig. 2 (right). From this figure we observe that for increasing queue sizes, also the synthesized value d_w increases, whereas the optimal value for d_s is u in case $N < 6$ and ℓ otherwise.

N	ε	creating time [s]	solving time [s]	poly degree
2	0.1	0.15	0.24	46
	0.01	0.15	0.25	46
	0.001	0.16	0.28	53
	0.0005	0.16	0.33	53
4	0.1	0.14	0.25	46
	0.01	0.16	0.25	46
	0.001	0.16	0.28	53
	0.0005	0.16	0.33	53
6	0.1	0.16	0.35	46
	0.01	0.16	0.35	46
	0.001	0.17	0.40	53
	0.0005	0.18	0.43	53
8	0.1	0.19	0.35	46
	0.01	0.19	0.35	46
	0.001	0.20	0.43	53
	0.0005	0.22	0.44	53

N	ε	creating time [s]	solving time [s]	poly degree	results	
2	0.1	0.15	1.80	86	$E[MP]$	0.85524
	0.01	0.15	2.57	92	d_o	1.82752
	0.001	0.15	2.97	96	d_p	0.66167
	0.0001	0.15	3.84	101	d_q	2.05189
4	0.1	0.83	1.92	86	$E[MP]$	0.46127
	0.01	0.92	2.40	92	d_o	1.92513
	0.001	1.04	3.06	97	d_p	0.66167
	0.0001	1.04	4.24	101	d_q	2.05189
6	0.1	2.25	2.18	87	$E[MP]$	0.33060
	0.01	2.36	2.53	92	d_o	1.95764
	0.001	2.37	3.83	97	d_p	0.66167
	0.0001	2.41	4.41	101	d_q	2.05189
8	0.1	17.08	2.22	87	$E[MP]$	0.29536
	0.01	17.60	2.81	93	d_o	1.96540
	0.001	17.78	3.33	97	d_p	0.66167
	0.0001	17.87	4.48	102	d_q	2.05189

Fig. 3. Statistics of the symbolic algorithm applied to the disk drive example

Fig. 4. Results and statistics of the symbolic algorithm applied to the preventive maintenance example

The table in Fig. 3 shows the running time of creation and solving of the MAPLE models, as well as the largest polynomial degrees for selected queue sizes $N = \{2, 4, 6, 8\}$ and error bounds $\varepsilon = \{0.1, 0.01, 0.001, 0.0005\}$. In all cases, discretization step sizes of $10^{-6} \cdot 10^{-19} < \delta(\cdot) < 10^{-19}$ were required to obtain results guaranteeing ε-optimal parameter functions. These small discretization constants underpin that the ε-optimal parameter synthesis problem cannot be carried out using the explicit approach (our implementation of the explicit algorithm runs out of memory for all of the listed instances). However, the symbolic algorithm evaluating roots of polynomials with high degree is capable to solve the problem within seconds in all cases. This can be explained through the small number of candidate actions we had to consider (always at most 200).

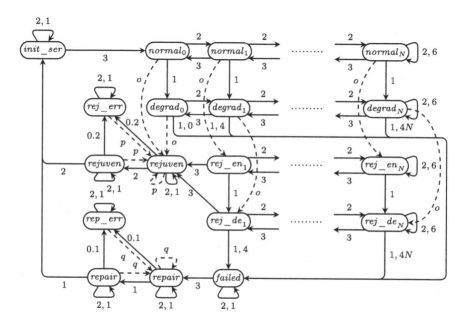

Fig. 5. Preventive maintenance of a server.

Preventive Maintenance. As depicted in Fig. 5, we consider a slightly modified model of a server that is susceptible to software faults [12]. A *rejuvenation* is the process of performing *preventive maintenance* of the server after a fixed period of time (usually during night time) to prevent performance degradation or even failure of the server. The first row of states in Fig. 5 represents the normal behavior of the server. Jobs arrive with rate 2 and are completed with rate 3. If job arrives and queue is full, it is rejected what is penalized by cost 6. Degradation of server is modeled by delay transitions of rates 1 leading to *degrad* states of the second row or eventually leading to the *failed* state. The failure causes rejection of all jobs in the queue and incurs cost 4 for each rejected job. After the failure is reported (delay event with rate 3), the repair process is initiated and completed after two exponentially distributed steps of rate 1. The repair can also fail with a certain probability (rate 0.1), thus after uniformly distributed time, the repair process is restarted. After each successful repair, the server is initialized by an exponential event with rate 3. The rejuvenation procedure is enabled after staying in *normal* or *degrad* states for time d_o. Then the rejuvenation itself is initiated after all jobs in the queue are completed. The rejuvenation procedure behaves similarly as the repair process, except that it is two times faster (all rates are multiplied by two).

First, we want to synthesize the value of the delay after which the rejuvenation is enabled, i.e., we aim towards the optimal schedule for rejuvenation. Furthermore, we synthesize the shifts d_p and d_q of the uniform distributions with length 2 associated with rejuvenation and repair, respectively, i.e., the corresponding uniform distribution function is $F_x[d_x](\tau) = \min\{1, \max\{0, \tau - d_x/2\}\}$, where $x \in \{p, q\}$. The interval of eligible values is $[0.1, 10]$ for all synthesized parameters. Similarly as for previous example we show results of experiments for queue sizes $N = \{2, 4, 6, 8\}$ and error bounds $\varepsilon = \{0.1, 0.01, 0.001, 0.0001\}$ in Fig. 4. The CPU time of model creation grows (almost quadratically) to the number of states, caused by multiplication of large matrices in MAPLE. As within the disk-drive example, we obtained the solutions very fast since we had to consider small number of candidate actions (always at most 500).

Optimizations in the Implementation. For the sake of a clean presentation in this paper, we established *global* theoretical upper bounds on δ and κ sufficient to guarantee ε-optimal solutions, see Appendix B of [3]. The theoretical bounds assume the worst underlying transition structure of a given ACTMC. In the prototype implementation, we applied some optimizations mainly computing *local* upper bounds for each state in the constructed semi-MDP. Also, to achieve better perturbation bounds on the expected mean-payoff, i.e., to compute bounds on expected time and cost to reach some state from all other states, we rely on techniques presented in [6,20]. Using these optimizations, for instance in the experiment of disk drive model, some discretization bounds δ were improved from $2.39 \cdot 10^{-239}$ to $7.03 \cdot 10^{-19}$. Note that even with these optimizations, the explicit algorithm for parameter synthesis would not be feasible as, more than 10^{18} actions would have to be considered for each state. This would clearly exceed the memory limit of state-of-the art computers.

References

1. Alur, R., Henzinger, T.A., Vardi, M.Y.: Parametric real-time reasoning. In: STOC, pp. 592–601. ACM (1993)
2. Amparore, E.G., Buchholz, P., Donatelli, S.: A structured solution approach for Markov regenerative processes. In: Norman, G., Sanders, W. (eds.) QEST 2014. LNCS, vol. 8657, pp. 9–24. Springer, Cham (2014). doi:10.1007/978-3-319-10696-0_3

3. Baier, C., Dubslaff, C., Korenčiak, Ľ., Kučera, A., Řehák, V.: Mean-payoff optimization in continuous-time Markov chains with parametric alarms. CoRR, abs/1706.06486 (2017)
4. Bernardin, L., et al.: Maple 16 Programming Guide (2012)
5. Bertsekas, D.P., Gallager, R.G.: Data Networks, 2nd edn. Prentice-Hall International, Upper Saddle River (1992)
6. Brázdil, T., Korenčiak, Ľ., Krčál, J., Novotný, P., Řehák, V.: Optimizing performance of continuous-time stochastic systems using timeout synthesis. In: Campos, J., Haverkort, B.R. (eds.) QEST 2015. LNCS, vol. 9259, pp. 141–159. Springer, Cham (2015). doi:10.1007/978-3-319-22264-6_10
7. Brázdil, T., Krčál, J., Křetínský, J., Řehák, V.: Fixed-delay events in generalized semi-Markov processes revisited. In: Katoen, J.-P., König, B. (eds.) CONCUR 2011. LNCS, vol. 6901, pp. 140–155. Springer, Heidelberg (2011). doi:10.1007/978-3-642-23217-6_10
8. Češka, M., Dannenberg, F., Kwiatkowska, M., Paoletti, N.: Precise parameter synthesis for stochastic biochemical systems. In: Mendes, P., Dada, J.O., Smallbone, K. (eds.) CMSB 2014. LNCS, vol. 8859, pp. 86–98. Springer, Cham (2014). doi:10.1007/978-3-319-12982-2_7
9. Choi, H., Kulkarni, V.G., Trivedi, K.S.: Markov regenerative stochastic Petri nets. Perform. Eval. 20(1–3), 337–357 (1994)
10. Alfaro, L.: Stochastic transition systems. In: Sangiorgi, D., Simone, R. (eds.) CONCUR 1998. LNCS, vol. 1466, pp. 423–438. Springer, Heidelberg (1998). doi:10.1007/BFb0055639
11. Diciolla, M., Kim, C.H.P., Kwiatkowska, M., Mereacre, A.: Synthesising optimal timing delays for timed I/O automata. In: EMSOFT, pp. 1–10. ACM (2014)
12. German, R.: Performance Analysis of Communication Systems with Non-Markovian Stochastic Petri Nets. Wiley, Hoboken (2000)
13. Haas, P.J.: Stochastic Petri Nets: Modelling, Stability, Simulation. Springer, New York (2010)
14. Haase, C., Kreutzer, S., Ouaknine, J., Worrell, J.: Reachability in succinct and parametric one-counter automata. In: Bravetti, M., Zavattaro, G. (eds.) CONCUR 2009. LNCS, vol. 5710, pp. 369–383. Springer, Heidelberg (2009). doi:10.1007/978-3-642-04081-8_25
15. Hahn, E.M., Hermanns, H., Zhang, L.: Probabilistic reachability for parametric Markov models. STTT 13(1), 3–19 (2011)
16. Han, T., Katoen, J.P., Mereacre, A.: Approximate parameter synthesis for probabilistic time-bounded reachability. In: RTSS, pp. 173–182. IEEE (2008)
17. Jha, S.K., Langmead, C.J.: Synthesis and infeasibility analysis for stochastic models of biochemical systems using statistical model checking and abstraction refinement. TCS 412(21), 2162–2187 (2011)
18. Jovanović, A., Kwiatkowska, M.: Parameter synthesis for probabilistic timed automata using stochastic game abstractions. In: Ouaknine, J., Potapov, I., Worrell, J. (eds.) RP 2014. LNCS, vol. 8762, pp. 176–189. Springer, Cham (2014). doi:10.1007/978-3-319-11439-2_14
19. Jovanovic, A., Lime, D., Roux, O.H.: Integer parameter synthesis for real-time systems. IEEE Trans. Softw. Eng. 41(5), 445–461 (2015)
20. Korenčiak, Ľ., Kučera, A., Řehák, V.: Efficient timeout synthesis in fixed-delay CTMC using policy iteration. In: MASCOTS, pp. 367–372. IEEE (2016)
21. Korenčiak, Ľ., Řehák, V., Farmadin, A.: Extension of PRISM by synthesis of optimal timeouts in fixed-delay CTMC. In: iFM, pp. 130–138 (2016)

22. Lindemann, C.: An improved numerical algorithm for calculating steady-state solutions of deterministic and stochastic Petri net models. Perform. Eval. **18**(1), 79–95 (1993)

23. Martina, S., Paolieri, M., Papini, T., Vicario, E.: Performance evaluation of Fischer's protocol through steady-state analysis of Markov regenerative processes. In: MASCOTS, pp. 355–360. IEEE (2016)

24. Minh, D.L.P., Minh, D.D.L., Nguyen, A.L.: Regenerative Markov chain Monte Carlo for any distribution. Commun. Stat.-Simul. C. **41**(9), 1745–1760 (2012)

25. Nelson, W.: Weibull analysis of reliability data with few or no failures. J. Qual. Technol. **17**, 140–146 (1985)

26. Norris, J.R.: Markov Chains. Cambridge University Press, Cambridge (1998)

27. Puterman, M.L.: Markov Decision Processes. Wiley, Hoboken (1994)

28. Qiu, Q., Wu, Q., Pedram, M.: Stochastic modeling of a power-managed system: construction and optimization. In: ISLPED, pp. 194–199. ACM Press (1999)

29. Traonouez, L.-M., Lime, D., Roux, O.H.: Parametric model-checking of time Petri nets with stopwatches using the state-class graph. In: Cassez, F., Jard, C. (eds.) FORMATS 2008. LNCS, vol. 5215, pp. 280–294. Springer, Heidelberg (2008). doi:10.1007/978-3-540-85778-5_20

Multi-objective Robust Strategy Synthesis for Interval Markov Decision Processes

Ernst Moritz Hahn[1,2], Vahid Hashemi[1], Holger Hermanns[1],
Morteza Lahijanian[3], and Andrea Turrini[2(✉)]

[1] Saarland University, Saarland Informatics Campus, Saarbrücken, Germany
[2] State Key Laboratory of Computer Science,
Institute of Software Chinese Academy of Sciences, Beijing, China
`turrini@ios.ac.cn`
[3] Department of Computer Science, University of Oxford, Oxford, UK

Abstract. Interval Markov decision processes (*IMDPs*) generalise
classical MDPs by having interval-valued transition probabilities. They
provide a powerful modelling tool for probabilistic systems with an additional variation or uncertainty that prevents the knowledge of the exact
transition probabilities. In this paper, we consider the problem of multi-
objective robust strategy synthesis for interval MDPs, where the aim is to
find a robust strategy that guarantees the satisfaction of multiple properties at the same time in face of the transition probability uncertainty.
We first show that this problem is **PSPACE**-hard. Then, we provide a
value iteration-based decision algorithm to approximate the Pareto set
of achievable points. We finally demonstrate the practical effectiveness
of our proposals by applying them on several real-world case studies.

1 Introduction

Interval Markov Decision Processes (*IMDPs*) extend the classical *Markov Decision Processes* (*MDPs*) by including uncertainty over the transition probabilities.
Instead of a single value for the probability of taking a transition, *IMDPs* allow
ranges of probabilities given as closed intervals. *IMDPs* are thus a powerful modelling tool for probabilistic systems with an additional variation or uncertainty
concerning the knowledge of exact transition probabilities. They are well suited
to represent realistic stochastic systems that, for instance, evolve in unknown
environments with bounded behaviour or do not preserve the Markov property.

Since their introduction (under the name of bounded-parameter *MDPs*) [15],
IMDPs have been receiving a lot of attention in the formal verification community. They are particularly viewed as the appropriate abstraction model for

This work is supported by the ERC Advanced Investigators Grant 695614
(POWVER), by the CAS/SAFEA International Partnership Program for Creative
Research Teams, by the National Natural Science Foundation of China (Grants No.
61550110506 and 61650410658), by the Chinese Academy of Sciences Fellowship for
International Young Scientists, by the CDZ project CAP (GZ 1023), and by EPSRC
Mobile Autonomy Program Grant EP/M019918/1.

© Springer International Publishing AG 2017
N. Bertrand and L. Bortolussi (Eds.): QEST 2017, LNCS 10503, pp. 207–223, 2017.
DOI: 10.1007/978-3-319-66335-7_13

uncertain systems with large state spaces, including continuous dynamical systems, for the purpose of analysis, verification, and control synthesis. Several model checking and control synthesis techniques have been developed [31,32,34] causing a boost in the applications of *IMDPs*, ranging from verification of continuous stochastic systems (e.g., [22]) to robust strategy synthesis for robotic systems (e.g., [24–26,34]).

In recent years, there has been an increasing interest in multi-objective strategy synthesis for probabilistic systems [5,10,13,14,21,27,29,30,33]. The goal is first to provide a complete trade-off analysis of several, possibly conflicting, quantitative properties and then to synthesise a strategy that guarantees the desired behaviour. Such properties, for instance, ask to "find a robot strategy that maximises p_{safe}, the probability of successfully completing a track by safely maneuvering between obstacles, while minimising t_{travel}, the total expected travel time". This example has competing objectives: maximising p_{safe}, which requires the robot to be conservative, and minimising t_{travel}, which causes the robot to be reckless. In such contexts, the interest is in the *Pareto curve* of the possible solution points: the set of all pairs of (p_{safe}, t_{travel}) for which an increase in the value of p_{safe} must induce an increase in the value of t_{travel}, and vice versa. Given a point on the curve, the computation of the corresponding strategy is asked.

Existing multi-objective synthesis frameworks are limited to *MDP* models. The algorithms use iterative methods (similar to value iteration) for the computation of the Pareto curve and rely on reductions to linear programming for strategy synthesis. As discussed above, *MDPs*, however, are constrained to single-valued transition probabilities, posing severe limitations for many real-world systems.

In this paper, we present a novel technique for multi-objective strategy synthesis for *IMDPs*. Our aim is to synthesise a robust strategy that guarantees the satisfaction of the multi-objective property, despite the additional uncertainty over the transition probabilities. Our approach views the uncertainty as making adversarial choices among the available transition probability distributions induced by the intervals, as the system evolves along state transitions. We refer to this as the *controller synthesis* semantics. We first analyse the problem complexity, proving that it is **PSPACE**-hard and then develop a value iteration-based decision algorithm to approximate the Pareto curve. We present promising results on a variety of case studies, obtained by prototypical implementations of all algorithms, to show the effectiveness of our approach.

Related Work. Related work can be grouped into two main categories: uncertain Markov model formalisms and model checking/synthesis algorithms.

Firstly, regarding the modelling frameworks, various probabilistic modelling formalisms with uncertain transitions are studied in the literature. Interval Markov Chains (*IMCs*) [19,20] or abstract Markov chains [12] extend standard discrete-time Markov Chains (*MCs*) with interval uncertainties. They do not feature the non-deterministic choices of transitions. Uncertain *MDPs* [32] allow more general sets of distributions to be associated with each transition, not only those described by intervals. They usually are restricted to *rectangular uncertainty sets* requiring that the uncertainty is linear and independent for any two transitions of any two states. Parametric *MDPs* [16], to the contrary, allow such dependencies as every probability is described as a rational function of a finite set of global parameters. *IMDPs* extend *IMCs* by inclusion of nondeterminism and are a subset of uncertain *MDPs* and parametric *MDPs*.

Secondly, regarding the algorithms, several verification methods for uncertain Markov models have been proposed. The problems of computing reachability probabilities and expected total reward for *IMCs* and *IMDPs* were first investigated in [8,35]. Then, several of their PCTL and LTL model checking algorithms were introduced in [2,6,8,22,32,34], respectively. As regards to strategy synthesis algorithms, the work in [16,28] considered synthesis for parametric *MDPs* and *MDPs* with ellipsoidal uncertainty in the verification community. In the control community, such synthesis problems were mostly studied for uncertain Markov models in [15,28,35] with the aim to maximise expected finite-horizon (un)discounted rewards. All these works, however, consider solely single objective properties, and their extension to multi-objective synthesis is not trivial.

Multi-objective model checking of probabilistic models with respect to various quantitative objectives has been recently investigated in a few works. The works in [11,13,14,21] focused on multi-objective verification of ordinary *MDPs*. In [7], these algorithms were extended to the more general models of 2-player stochastic games. These models, however, cannot capture the continuous uncertainty in the transition probabilities as *IMDPs* do. For the purposes of synthesis though, it is possible to transform an *IMDP* into a 2-player stochastic game; nevertheless, such a transformation raises an extra exponential factor to the complexity of the decision problem. This exponential blowup has been avoided in our setting.

2 Preliminaries

For a set X, denote by $\mathrm{Disc}(X)$ the sets of discrete probability distributions over X. A discrete probability distribution ρ is a function $\rho \colon X \to \mathbb{R}_{\geq 0}$ such that $\sum_{x \in X} \rho(x) = 1$; for $X' \subseteq X$, we write $\rho(X')$ for $\sum_{x \in X'} \rho(x)$. Given $\rho \in \mathrm{Disc}(X)$, we denote by $\mathrm{Supp}(\rho)$ the set $\{\, x \in X \mid \rho(x) > 0 \,\}$, and by δ_x, where $x \in X$, the *Dirac* distribution such that $\delta_x(y) = 1$ for $y = x$, 0 otherwise. For a distribution ρ, we also write $\rho = \{\, (x, p_x) \mid x \in X \,\}$ where p_x is the probability of x.

For a vector $\mathbf{x} \in \mathbb{R}^n$ we denote by x_i, its i-th component, and we call $\mathbf{x}$ a *weight vector* if $x_i \geq 0$ for all i and $\sum_{i=1}^{n} x_i = 1$. The Euclidean inner product $\mathbf{x} \cdot \mathbf{y}$ of two vectors $\mathbf{x}, \mathbf{y} \in \mathbb{R}^n$ is defined as $\sum_{i=1}^{n} x_i \cdot y_i$. For a set of vectors $S = \{\mathbf{s}_1, \ldots, \mathbf{s}_t\} \subseteq \mathbb{R}^n$, we say that $\mathbf{s} \in \mathbb{R}^n$ is a *convex combination* of elements of S, if $\mathbf{s} = \sum_{i=1}^{t} w_i \cdot \mathbf{s}_i$ for some weight vector $\mathbf{w} \in \mathbb{R}^t_{\geq 0}$. Furthermore, we denote by $S\!\downarrow$ the *downward closure* of the convex hull of $\bar{S}$ which is defined as $S\!\downarrow = \{\, \mathbf{y} \in \mathbb{R}^n \mid \mathbf{y} \leq \mathbf{z} \text{ for some convex combination } \mathbf{z} \text{ of } S \,\}$. For a given convex set X, we say that a point $\mathbf{x} \in X$ is on the boundary of X, denoted by $\mathbf{x} \in \partial X$, if for every $\varepsilon > 0$ there is a point $\mathbf{y} \notin X$ such that the Euclidean distance between $\mathbf{x}$ and $\mathbf{y}$ is at most ε. Given a downward closed set $X \in \mathbb{R}^n$, for any

$\mathbf{z} \in \mathbb{R}^n$ such that $\mathbf{z} \in \partial X$ or $\mathbf{z} \notin X$, there is a weight vector $\mathbf{w} \in \mathbb{R}^n$ such that $\mathbf{w} \cdot \mathbf{z} \geq \mathbf{w} \cdot \mathbf{x}$ for all $\mathbf{x} \in X$ [3]. We say that $\mathbf{w}$ separates $\mathbf{z}$ from $X{\downarrow}$. Given a set $Y \subseteq \mathbb{R}^k$, we call a vector $\mathbf{y} \in Y$ *Pareto optimal* in Y if there does not exist a vector $\mathbf{z} \in Y$ such that $\mathbf{y} \leq \mathbf{z}$ and $\mathbf{y} \neq \mathbf{z}$. We define the *Pareto set* or *Pareto curve* of Y to be the set of all Pareto optimal vectors in Y, i.e., Pareto set $\mathcal{Y} = \{\, \mathbf{y} \in Y \mid \mathbf{y} \text{ is Pareto optimal} \,\}$.

2.1   Interval Markov Decision Processes

We now define *Interval Markov Decision Processes* (*IMDPs*) as an extension of *MDPs*, which allows for the inclusion of transition probability uncertainties as *intervals*. *IMDPs* belong to the family of uncertain *MDPs* and allow to describe a set of *MDPs* with identical (graph) structures that differ in distributions associated with transitions. Formally,

Definition 1 (*IMDPs*). *An* Interval Markov Decision Process *(IMDP)* $\mathcal{M}$ *is a tuple* $(S, \bar{s}, \mathcal{A}, I)$, *where* S *is a finite set of* states, $\bar{s} \in S$ *is the initial state,* $\mathcal{A}$ *is a finite set of* actions, *and* $I \colon S \times \mathcal{A} \times S \to \mathbb{I} \cup \{[0,0]\}$ *is a total interval transition probability function where* $\mathbb{I} = \{\, [a, b] \mid 0 < a \leq b \leq 1 \,\}$.

Given $s \in S$ and $a \in \mathcal{A}$, we call $\mathfrak{h}_s^a \in \mathrm{Disc}(S)$ a *feasible distribution* reachable from s by a, denoted by $s \xrightarrow{a} \mathfrak{h}_s^a$, if, for each state $s' \in S$, we have $\mathfrak{h}_s^a(s') \in I(s, a, s')$. We denote the set of feasible distributions for state s and action a by $\mathcal{H}_s^a$, i.e., $\mathcal{H}_s^a = \{\, \mathfrak{h}_s^a \in \mathrm{Disc}(S) \mid s \xrightarrow{a} \mathfrak{h}_s^a \,\}$ and we denote the set of available actions at state $s \in S$ by $\mathcal{A}(s)$, i.e., $\mathcal{A}(s) = \{\, a \in \mathcal{A} \mid \mathcal{H}_s^a \neq \emptyset \,\}$. We assume that $\mathcal{A}(s) \neq \emptyset$ for all $s \in S$. We define the *size* of $\mathcal{M}$, written $|\mathcal{M}|$, as the number of non-zero entries of I, i.e., $|\mathcal{M}| = |\{\, (s, a, s', \iota) \in S \times \mathcal{A} \times S \times \mathbb{I} \mid I(s, a, s') = \iota \,\}| \in \mathcal{O}(|S|^2 \cdot |\mathcal{A}|)$.

A *path* ξ in $\mathcal{M}$ is a finite or infinite sequence of alternating states and actions $\xi = s_0 a_0 s_1 \ldots$, ending with a state if finite, such that for each $i \geq 0$, $I(s_i, a_i, s_{i+1}) \in \mathbb{I}$. The i-th state (action) along the path ξ is denoted by $\xi[i]$ ($\xi(i)$) and, if the path is finite, we denote by $last(\xi)$ its last state. The sets of all finite and infinite paths in $\mathcal{M}$ are denoted by *FPaths* and *IPaths*, respectively.

The nondeterministic choices between available actions and feasible distributions present in an *IMDP* are resolved by strategies and natures, respectively.

Definition 2 (Strategy and Nature in *IMDPs*). *Given an IMDP* $\mathcal{M}$, *a* strategy *is a function* $\sigma \colon \mathit{FPaths} \to \mathrm{Disc}(\mathcal{A})$ *such that for each* $\xi \in \mathit{FPaths}$, $\sigma(\xi) \in \mathrm{Disc}(\mathcal{A}(last(\xi)))$. *A* nature *is a function* $\pi \colon \mathit{FPaths} \times \mathcal{A} \to \mathrm{Disc}(S)$ *such that for each* $\xi \in \mathit{FPaths}$ *and* $a \in \mathcal{A}(s)$, $\pi(\xi, a) \in \mathcal{H}_s^a$ *where* $s = last(\xi)$. *The sets of all strategies and all natures are denoted by* Σ *and* Π, *respectively.*

Given a finite path ξ of an *IMDP*, a strategy σ, and a nature π, the system evolution proceeds as follows: let $s = last(\xi)$. First, an action $a \in \mathcal{A}(s)$ is chosen probabilistically by σ. Then, π resolves the uncertainties and chooses one feasible distribution $\mathfrak{h}_s^a \in \mathcal{H}_s^a$. Finally, the next state s' is chosen according to the distribution $\mathfrak{h}_s^a$, and the path ξ is extended by s'.

A strategy σ and a nature π induce a probability measure over paths as follows. The basic measurable events are the cylinder sets of finite paths, where the *cylinder set* of a finite path ξ is the set $Cyl_\xi = \{\xi' \in IPaths \mid \xi \text{ is a prefix of } \xi'\}$. The probability $\Pr_{\mathcal{M}}^{\sigma,\pi}$ of a state s' is defined to be $\Pr_{\mathcal{M}}^{\sigma,\pi}[Cyl_{s'}] = \delta_{\bar{s}}(s')$ and the probability $\Pr_{\mathcal{M}}^{\sigma,\pi}[Cyl_{\xi as'}]$ of traversing a finite path $\xi as'$ is defined to be $\Pr_{\mathcal{M}}^{\sigma,\pi}[Cyl_{\xi as'}] = \Pr_{\mathcal{M}}^{\sigma,\pi}[Cyl_\xi] \cdot \sigma(\xi)(a) \cdot \pi(\xi, a)(s')$. Then, $\Pr_{\mathcal{M}}^{\sigma,\pi}$ extends uniquely to the σ-field generated by cylinder sets.

In order to model additional quantitative measures of an *IMDP*, we associate rewards to the enabled actions. This is done by means of *reward structures*.

Definition 3 (Reward Structure). *A* reward structure *for an IMDP is a function* $\mathbf{r}\colon S \times A \to \mathbb{R}$ *that assigns to each state-action pair* (s, a)*, where* $s \in S$ *and* $a \in A(s)$*, a reward* $\mathbf{r}(s, a) \in \mathbb{R}$*. Given a path* ξ *and* $k \in \mathbb{N} \cup \{\infty\}$*, the total accumulated reward in* k *steps for* ξ *over* $\mathbf{r}$ *is* $\mathbf{r}[k](\xi) = \sum_{i=0}^{k-1} \mathbf{r}(\xi[i], \xi(i))$.

Note that we allow negative rewards in this definition, but that due to later assumptions their use is restricted.

As an example of *IMDP* with a reward structure, consider the *IMDP* $\mathcal{M}$ depicted in Fig. 1. The set of states is $S = \{s, t, u\}$ with s being the initial one. The set of actions is $\mathcal{A} = \{a, b\}$, and the non-zero transition probability intervals are $I(s, a, t) = [\frac{1}{3}, \frac{2}{3}]$, $I(s, a, u) = [\frac{1}{10}, 1]$, $I(s, b, t) = [\frac{2}{5}, \frac{3}{5}]$, $I(s, b, u) = [\frac{1}{4}, \frac{2}{3}]$, $I(t, a, t) = I(u, b, u) = [1, 1]$, and $I(t, b, t) = I(u, a, u) = [0, 0]$. The underlined numbers indicate the reward structure $\mathbf{r}$ such that $\mathbf{r}(s, a) = 3$, $\mathbf{r}(s, b) = 1$, and $\mathbf{r}(t, a) = \mathbf{r}(u, b) = 0$. Note that since $\mathcal{H}_t^b = \mathcal{H}_u^a = \emptyset$, then $\mathbf{r}(t, b)$ and $\mathbf{r}(u, a)$ are undefined.

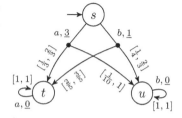

Fig. 1. An example of *IMDP*.

3   Multi-objective Robust Strategy Synthesis for IMDPs

In this paper, we consider two main classes of properties for *IMDPs*; the *probability of reaching a target* and the *expected total reward*. The reason that we focus on these properties is that their algorithms usually serve as the basis for more complex properties. For instance, they can be easily extended to answer queries with linear temporal logic properties as shown in [11]. To this aim, we lift the satisfaction definitions of these two classes of properties from *MDPs* in [13,14] to *IMDPs* by encoding the notion of robustness for strategies.

Note that all proofs are contained in the extended version of the paper [17].

Definition 4 (Reachability Predicate & its Robust Satisfaction). *A* reachability predicate $[T]_{\sim p}^{\leq k}$ *consists of a set of target states* $T \subseteq S$*, a relational operator* $\sim \in \{\leq, \geq\}$*, a rational probability bound* $p \in [0, 1] \cap \mathbb{Q}$ *and a time bound* $k \in \mathbb{N} \cup \{\infty\}$*. It indicates that the probability of reaching* T *within* k *time steps satisfies* $\sim p$.

Robust satisfaction of $[T]_{\sim p}^{\leq k}$ *by IMDP* $\mathcal{M}$ *under strategy* $\sigma \in \Sigma$ *is denoted by* $\mathcal{M}|_\sigma \models_\Pi [T]_{\sim p}^{\leq k}$ *and indicates that the probability of the set of all paths that reach* T *under* σ *satisfies the bound* $\sim p$ *for every choice of nature* $\pi \in \Pi$. *Formally,* $\mathcal{M}|_\sigma \models_\Pi [T]_{\sim p}^{\leq k}$ *iff* $\mathrm{Pr}_\mathcal{M}^\sigma(\Diamond^{\leq k} T) \sim p$ *where* $\mathrm{Pr}_\mathcal{M}^\sigma(\Diamond^{\leq k} T) = \mathrm{opt}_{\pi \in \Pi} \mathrm{Pr}_\mathcal{M}^{\sigma,\pi}\{\xi \in IPaths \mid \exists i \leq k : \xi[i] \in T\}$ *and* $\mathrm{opt} = \min$ *if* $\sim\,=\,\geq$ *and* $\mathrm{opt} = \max$ *if* $\sim\,=\,\leq$. *Furthermore,* σ *is referred to as a robust strategy.*

Definition 5 (Reward Predicate & its Robust Satisfaction). *A* reward predicate $[\mathbf{r}]_{\sim r}^{\leq k}$ *consists of a reward structure* $\mathbf{r}$, *a time bound* $k \in \mathbb{N} \cup \{\infty\}$, *a relational operator* $\sim\,\in \{\leq, \geq\}$ *and a reward bound* $r \in \mathbb{Q}$. *It indicates that the expected total accumulated reward within* k *steps satisfies* $\sim r$.

Robust satisfaction of $[\mathbf{r}]_{\sim r}^{\leq k}$ *by IMDP* $\mathcal{M}$ *under strategy* $\sigma \in \Sigma$ *is denoted by* $\mathcal{M}|_\sigma \models_\Pi [\mathbf{r}]_{\sim r}^{\leq k}$ *and indicates that the expected total reward over the set of all paths under* σ *satisfies the bound* $\sim r$ *for every choice of nature* $\pi \in \Pi$. *Formally,* $\mathcal{M}|_\sigma \models_\Pi [\mathbf{r}]_{\sim r}^{\leq k}$ *iff* $ExpTot_\mathcal{M}^{\sigma,k}[\mathbf{r}] \sim r$ *where* $ExpTot_\mathcal{M}^{\sigma,k}[\mathbf{r}] = \mathrm{opt}_{\pi \in \Pi} \int_\xi \mathbf{r}[k]\xi \, d\mathrm{Pr}_\mathcal{M}^{\sigma,\pi}$ *and* $\mathrm{opt} = \min$ *if* $\sim\,=\,\geq$ *and* $\mathrm{opt} = \max$ *if* $\sim\,=\,\leq$. *Furthermore,* σ *is referred to as the robust strategy.*

For the purpose of algorithm design, we also consider weighted sum of rewards.

Definition 6 (Weighted Reward Sum). *Given a weight vector* $\mathbf{w} \in \mathbb{R}^n$, *vector of time bounds* $\mathbf{k} = (k_1, \ldots, k_n) \in (\mathbb{N} \cup \{\infty\})^n$ *and reward structures* $\mathbf{r} = (\mathbf{r}_1, \ldots, \mathbf{r}_n)$ *for IMDP* $\mathcal{M}$, *the* weighted reward sum $\mathbf{w} \cdot \mathbf{r}[\mathbf{k}]$ *over a path* ξ *is defined as* $\mathbf{w} \cdot \mathbf{r}[\mathbf{k}](\xi) = \sum_{i=1}^n w_i \cdot \mathbf{r}_i[k](\xi)$. *The* expected total weighted sum *is defined as* $ExpTot_\mathcal{M}^{\sigma,\mathbf{k}}[\mathbf{w} \cdot \mathbf{r}] = \max_{\pi \in \Pi} \int_\xi \mathbf{w} \cdot \mathbf{r}[\mathbf{k}](\xi) \, d\mathrm{Pr}_\mathcal{M}^{\sigma,\pi}$ *for bounds* $\leq$ *and accordingly minimises over natures for* $\geq$; *for a given strategy* σ, *we have:* $ExpTot_\mathcal{M}^{\sigma,\mathbf{k}}[\mathbf{w} \cdot \mathbf{r}] = \sum_{i=1}^n w_i \cdot ExpTot_\mathcal{M}^{\sigma,k_i}[\mathbf{r}_i]$.

3.1   Multi-objective Queries

Multi-objective properties for *IMDPs* essentially require multiple predicates to be satisfied at the same time under the same strategy for every choice of the nature. We now explain how to formalise multi-objective queries for *IMDPs*.

Definition 7 (Multi-objective Predicate). *A* multi-objective predicate *is a vector* $\varphi = (\varphi_1, \ldots, \varphi_n)$ *of reachability or reward predicates. We say that* φ *is satisfied by IMDP* $\mathcal{M}$ *under strategy* σ *for every choice of nature* $\pi \in \Pi$, *denoted by* $\mathcal{M}|_\sigma \models_\Pi \varphi$ *if, for each* $1 \leq i \leq n$, *it is* $\mathcal{M}|_\sigma \models_\Pi \varphi_i$. *We refer to* σ *as a robust strategy. Furthermore, we call* φ *a basic multi-objective predicate if it is of the form* $([\mathbf{r}_1]_{\geq r_1}^{\leq k_1}, \ldots, [\mathbf{r}_n]_{\geq r_n}^{\leq k_n})$, *i.e., it includes only lower-bounded reward predicates.*

We formulate multi-objective queries for *IMDPs* in three ways, namely *synthesis queries*, *quantitative queries* and *Pareto queries*. Due to lack of space, we only focus on the synthesis queries and discuss the other types of queries in [17, Appendix C].

Definition 8 (Synthesis Query). *Given an IMDP $\mathcal{M}$ and a multi-objective predicate φ, the synthesis query asks if there exists a robust strategy $\sigma \in \Sigma$ such that $\mathcal{M}|_\sigma \models_\Pi \varphi$.*

Note that the synthesis queries check for the existence of a robust strategy that satisfies a multi-objective predicate φ for every resolution of nature. In order to avoid unusual behaviours in strategy synthesis such as infinite total expected reward, we restrict the usage of rewards by assuming reward-finiteness for the strategies that satisfy the reachability predicates in φ.

Assumption 1 (Reward-finiteness). *Suppose that an IMDP $\mathcal{M}$ and a synthesis query φ are given. Let $\varphi = ([T_1]^{\leq k_1}_{\sim p_1}, \ldots, [T_n]^{\leq k_n}_{\sim p_n}, [\mathbf{r}_{n+1}]^{\leq k_{n+1}}_{\sim r_{n+1}}, \ldots, [\mathbf{r}_m]^{\leq k_m}_{\sim r_m})$. We say that φ is reward-finite if for each $n + 1 \leq i \leq m$ such that $k_i = \infty$, $\sup\{ \mathit{ExpTot}^{\sigma,k_i}_{\mathcal{M}}[\mathbf{r}_i] \mid \mathcal{M}|_\sigma \models_\Pi ([T_1]^{\leq k_1}_{\sim p_1}, \ldots, [T_n]^{\leq k_n}_{\sim p_n})\} < \infty$.*

Due to lack of space, we provide in [17, Appendix B] a method to check for this assumption, a preprocessing procedure that removes actions with non-zero rewards from the end components of the *IMDP*, and a proof for its correctness. Therefore, in the rest of the paper, we assume that all queries are reward-finite, and the *IMDP* does not include actions with non-zero rewards in its end components. Furthermore, for the soundness of our analysis we also require that for any *IMDP* $\mathcal{M}$ and φ given as in Assumption 1: *(i)* each reward structure $\mathbf{r}_i$ assigns only non-negative values; *(ii)* φ is reward-finite; and *(iii)* for indices $n + 1 \leq i \leq m$ such that $k_i = \infty$, either all $\sim_i$s are $\leq$ or all are $\geq$.

3.2   Robust Strategy Synthesis

We first study the computational complexity of multi-objective robust strategy synthesis problem for *IMDP*s. Formally,

Theorem 9. *Given an IMDP $\mathcal{M}$ and a multi-objective predicate φ, the problem of synthesising a strategy $\sigma \in \Sigma$ such that $\mathcal{M}|_\sigma \models_\Pi \varphi$ is **PSPACE**-hard.*

As the first step towards derivation of a solution approach for the robust strategy synthesis problem, we need to convert all reachability predicates to reward predicates and therefore, to transform an arbitrarily given query to a query over a basic predicate on a modified *IMDP*. This can be simply done by adding, once for all, a reward of one at the time of reaching the target set and also negating the objective of predicates with upper-bounded relational operators. We correct and extend the procedure in [14] to reduce a general multi-objective predicate on an *IMDP* model to a basic form.

Proposition 10. *Given an IMDP $\mathcal{M} = (S, \bar{s}, \mathcal{A}, I)$ and a multi-objective predicate $\varphi = ([T_1]^{\leq k_1}_{\sim_1 p_1}, \ldots, [T_n]^{\leq k_n}_{\sim_n p_n}, [\mathbf{r}_{n+1}]^{\leq k_{n+1}}_{\sim_{n+1} r_{n+1}}, \ldots, [\mathbf{r}_m]^{\leq k_m}_{\sim_m r_m})$, let $\mathcal{M}' = (S', \bar{s}', \mathcal{A}', I')$ be the IMDP whose components are defined as follows: $S' =$*

$S \times 2^{\{1,\dots,n\}}$; $\bar{s}' = (\bar{s}, \emptyset)$; $\mathcal{A}' = \mathcal{A} \times 2^{\{1,\dots,n\}}$; and for all $s, s' \in S$, $a \in \mathcal{A}$, and $v, v', v'' \subseteq \{1, \dots, n\}$,

$$I'((s,v),(a,v'),(s',v'')) = \begin{cases} I(s,a,s') & if\ v' = \{ i \mid s \in T_i \} \setminus v\ and\ v'' = v \cup v', \\ [0,0] & otherwise. \end{cases}$$

Now, let $\varphi' = ([\mathbf{r}_{T_1}]^{\leq k_1+1}_{\geq p_1'}, \dots, [\mathbf{r}_{T_n}]^{\leq k_n+1}_{\geq p_n'}, [\bar{\mathbf{r}}_{n+1}]^{\leq k_{n+1}}_{\geq r_{n+1}'}, \dots, [\bar{\mathbf{r}}_m]^{\leq k_m}_{\geq r_m'})$ where, for each $i \in \{1, \dots, n\}$,

$$p_i' = \begin{cases} p_i & if \sim_i\ = \geq, \\ -p_i & if \sim_i\ = \leq; \end{cases} \quad and \quad \mathbf{r}_{T_i}((s,v),(a,v')) = \begin{cases} 1 & if\ i \in v'\ and \sim_i\ = \geq, \\ -1 & if\ i \in v'\ and \sim_i\ = \leq, \\ 0 & otherwise; \end{cases}$$

and, for each $j \in \{n+1, \dots, m\}$,

$$r_j' = \begin{cases} r_j & if \sim_j\ = \geq, \\ -r_j & if \sim_j\ = \leq; \end{cases} \quad and \quad \bar{\mathbf{r}}_j((s,v),(a,v')) = \begin{cases} r_j(s,a) & if \sim_j\ = \geq, \\ -r_j(s,a) & if \sim_j\ = \leq. \end{cases}$$

Then φ is satisfiable in $\mathcal{M}$ if and only if φ' is satisfiable in $\mathcal{M}'$.

We therefore need to only consider the basic multi-objective predicates of the form $([\mathbf{r}_1]^{\leq k_1}_{\geq r_1}, \dots, [\mathbf{r}_n]^{\leq k_n}_{\geq r_n})$ for the purpose of robust strategy synthesis. For a basic multi-objective predicate, we define its Pareto curve as follows.

Definition 11 (Pareto Curve of a Multi-objective Predicate). *Given an IMDP $\mathcal{M}$ and a basic multi-objective predicate $\varphi = ([\mathbf{r}_1]^{\leq k_1}_{\geq r_1}, \dots, [\mathbf{r}_n]^{\leq k_n}_{\geq r_n})$, we define the set of achievable values with respect to φ as $A_{\mathcal{M},\varphi} = \{ (r_1, \dots, r_n) \in \mathbb{R}^n \mid ([\mathbf{r}_1]^{\leq k_1}_{\geq r_1}, \dots, [\mathbf{r}_n]^{\leq k_n}_{\geq r_n})$ is satisfiable $\}$. We define the Pareto curve of φ to be the Pareto curve of $A_{\mathcal{M},\varphi}$ and denote it by $\mathcal{P}_{\mathcal{M},\varphi}$.*

To illustrate the transformation presented in Proposition 10, consider again the IMDP depicted in Fig. 1. Assume that the target set is $T = \{t\}$ and consider the property $\varphi = ([T]^{\leq 1}_{\geq \frac{1}{3}}, [\mathbf{r}]^{\leq 1}_{\geq \frac{1}{4}})$. The reduction converts φ to the property $\varphi' = ([\mathbf{r}_T]^{\leq 2}_{\geq \frac{1}{3}}, [\mathbf{r}]^{\leq 1}_{\geq \frac{1}{4}})$ on the modified $\mathcal{M}'$ depicted in Fig. 2a. We show two different reward structures $\bar{\mathbf{r}}$ and $\mathbf{r}_T$ besides each action, respectively. In Fig. 2b we show the Pareto curve for this property. As we see, until required probability $\frac{1}{3}$ to reach T, the maximal reward value is 3. Afterwards, the reward obtainable linearly decreases, until at required probability $\frac{2}{5}$ it is just 1. For higher required probabilities, the problem becomes infeasible. The reason for this behaviour is that, up to minimal probability $\frac{1}{3}$, action a can be chosen in state s, because the lower interval bound to reach t is $\frac{1}{3}$, which in turn leads to a reward of 3 being obtained. For higher reachability probabilities required, choosing action b with a certain probability is required, which however provides a lower reward. There is no strategy with which t is reached with a probability larger than $\frac{2}{5}$.

It is not difficult to see that the Pareto curve is in general an infinite set, and therefore, it is usually not possible to derive an exact representation of it

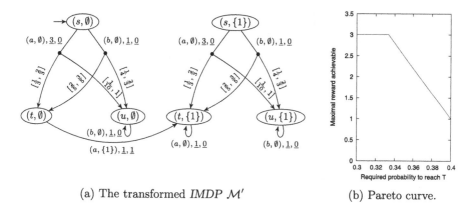

(a) The transformed IMDP $\mathcal{M}'$ (b) Pareto curve.

Fig. 2. Example of *IMDP* transformation. (a) The *IMDP* $\mathcal{M}'$ generated from $\mathcal{M}$ shown in Fig. 1. (b) Pareto curve for the property $([\mathbf{r}_T]^{\leq 2}_{\max}, [\mathbf{r}]^{\leq 1}_{\max})$.

in polynomial time. However, it can be shown that an ε-approximation of it can be computed efficiently [11]. In the rest of this section, we describe an algorithm to solve the synthesis query. We follow the well-known *normalisation* approach in order to solve the multi-objective predicate which is essentially based on normalising multiple objectives into one single objective. It is known that the optimal solution of the normalised (single-objective) predicate, if it exists, is the Pareto optimal solution of the multi-objective predicate [9].

The robust synthesis procedure is detailed in Algorithm 1. It basically aims to construct a sequential approximation to the Pareto curve $\mathcal{P}_{\mathcal{M},\varphi}$ while the quality of approximations gets better and more precise along the iterations. In other words, along the course of Algorithm 1 a sequence of weight vectors **w** are generated and corresponding to each of them, a **w**-weighted sum of n objectives is optimised through lines 6–7. The optimal strategy σ is then used to generate a point **g** on the Pareto curve $\mathcal{P}_{\mathcal{M},\varphi}$. We collect all these points in the set X. The multi-objective predicate φ is satisfiable once we realise that **r** belongs to $X\!\downarrow$.

The optimal strategies for the multi-objective robust synthesis queries are constructed following the approach of [14] and as a result of termination of Algorithm 1. In particular, when Algorithm 1 terminates, a sequence of points $\mathbf{g}^1,\ldots,\mathbf{g}^t$ on the Pareto curve $\mathcal{P}_{\mathcal{M},\varphi}$ are generated each of which corresponds to a deterministic strategy $\sigma_{\mathbf{g}^j}$ for the current point $\mathbf{g}^j$. The resulting optimal strategy σ_{opt} is subsequently constructed from these using a randomised weight vector $\alpha \in \mathbb{R}^t$ satisfying $r_i \leq \sum_{j=1}^{t} \alpha_i \cdot g_i^j$ [17, Appendix E].

Remark 12. It is worthwhile to mention that the synthesis query for *IMDP*s cannot be solved on the *MDP*s generated from *IMDP*s by computing all feasible extreme transition probabilities and then applying the algorithm in [14]. The latter is a valid approach provided the cooperative semantics is applied for resolving the two sources of nondeterminism in *IMDP*s. With respect to the competitive

---

Algorithm 1. Algorithm for solving robust synthesis queries

Input: An *IMDP* $\mathcal{M}$, multi-objective predicate $\varphi = ([\mathbf{r}_1]_{\geq r_1}^{\leq k_1}, \ldots, [\mathbf{r}_n]_{\geq r_n}^{\leq k_n})$

Output: true if there exists a strategy $\sigma \in \Sigma$ such that $\mathcal{M}|_\sigma \models_\Pi \varphi$, false if not.

1 **begin**
2   $X := \emptyset$; $\mathbf{r} := (\mathbf{r}_1, \ldots, \mathbf{r}_n)$;
3   $\mathbf{k} := (k_1, \ldots, k_n)$; $\mathbf{r} := (r_1, \ldots, r_n)$;
4   **while** $\mathbf{r} \notin X{\downarrow}$ **do**
5   Find $\mathbf{w}$ separating $\mathbf{r}$ from $X{\downarrow}$;
6   Find strategy σ maximising $ExpTot_{\mathcal{M}}^{\sigma,\mathbf{k}}[\mathbf{w} \cdot \mathbf{r}]$;
7    $\mathbf{g} := (ExpTot_{\mathcal{M}}^{\sigma,k_i}[\mathbf{r}_i])_{1 \leq i \leq n}$;
8   **if** $\mathbf{w} \cdot \mathbf{g} < \mathbf{w} \cdot \mathbf{r}$ **then**
9   **return** false;
10    $X := X \cup \{\mathbf{g}\}$;
11   **return** true;

---

semantics needed here, one can instead transform *IMDP*s to $2\frac{1}{2}$-player games [1] and then along the lines of the previous approach apply the algorithm in [7]. Unfortunately, the transformation to (*MDP*s or) $2\frac{1}{2}$-player games induces an exponential blowup, adding an exponential factor to the worst case time complexity of the decision problem. Our algorithm avoids this by solving the robust synthesis problem directly on the *IMDP* so that the core part, i.e., lines 6–7 of Algorithm 1 can be solved with time complexity polynomial in $|\mathcal{M}|$.

Algorithm 2 represents a value iteration-based algorithm which extends the value iteration-based algorithm in [14] and adjusts it for *IMDP* models by encoding the notion of robustness. The core difference is indicated in lines 6 and 16 where the optimal strategy is computed so as to be robust against any choice of nature.

Theorem 13. *Algorithm 1 is sound, complete and has runtime exponential in* $|\mathcal{M}|$, $\mathbf{k}$, *and* n.

Remark 14. It is worthwhile to mention that our robust strategy synthesis approach can also be applied to *MDP*s with richer formalisms for uncertainties such as likelihood or ellipsoidal uncertainties while preserving the computational complexity. In particular, in every inner optimisation problem in Algorithm 1, the optimality of a Markovian deterministic strategy and nature is guaranteed as long as the uncertainty set is convex, the set of actions is finite and the inner optimisation problem which minimises/maximises the objective function over the choices of nature achieves its optimum (cf. [31, Proposition 4.1]). Furthermore, due to the convexity of the generated optimisation problems, the computational complexity of our approach remains intact.

Algorithm 2. Value iteration algorithm to solve lines 6–7 of Algorithm 1

Input: An *IMDP* $\mathcal{M}$, weight vector w, reward structures $\mathbf{r} = (\mathbf{r}_1, \ldots, \mathbf{r}_n)$, time-bound vector $\mathbf{k} \in (\mathbb{N} \cup \{\infty\})^n$, threshold ε

Output: strategy σ maximising $ExpTot_{\mathcal{M}}^{\sigma,\mathbf{k}}[\mathbf{w} \cdot \mathbf{r}]$, $\mathbf{g} := (ExpTot_{\mathcal{M}}^{\sigma,k_i}[\mathbf{r}_i])_{1 \leq i \leq n}$

1 **begin**

2    $\mathbf{x} := 0$; $\mathbf{x}^1 := 0$; $\ldots$; $\mathbf{x}^n := 0$; $\mathbf{y} := 0$; $\mathbf{y}^1 := 0$; $\ldots$; $\mathbf{y}^n := 0$;

3    $\sigma^\infty(s) := \perp$ for all $s \in S$

4   **while** $\delta > \varepsilon$ **do**

5   **foreach** $s \in S$ **do**

6        $y_s := \max\limits_{a \in \mathcal{A}(s)} (\sum_{\{i | k_i = \infty\}} w_i \cdot \mathbf{r}_i(s,a) + \min\limits_{\mathfrak{h}_s^a \in \mathcal{H}_s^a} \sum_{s' \in S} \mathfrak{h}_s^a(s') \cdot x_{s'})$;

7        $\sigma^\infty(s) := \arg \max\limits_{a \in \mathcal{A}(s)} (\sum_{\{i | k_i = \infty\}} w_i \cdot \mathbf{r}_i(s,a) + \min\limits_{\mathfrak{h}_s^a \in \mathcal{H}_s^a} \sum_{s' \in S} \mathfrak{h}_s^a(s') \cdot x_{s'})$

8        $\bar{\mathfrak{h}}_s^{\sigma^\infty(s)}(s') := \arg \min_{\mathfrak{h}_s^a \in \mathcal{H}_s^a} \sum_{s' \in S} \mathfrak{h}_s^a(s') \cdot x_{s'}$

9      $\delta := \max_{s \in S}(y_s - x_s)$; $\mathbf{x} := \mathbf{y}$;

10   **while** $\delta > \varepsilon$ **do**

11   **foreach** $s \in S$ *and* $i \in \{1, \ldots, n\}$ **where** $k_i = \infty$ **do**

12        $y_s^i := \mathbf{r}_i(s, \sigma^\infty(s)) + \sum_{s' \in S} \bar{\mathfrak{h}}_s^{\sigma^\infty(s)}(s') \cdot x_{s'}^i$;

13      $\delta := \max_{i=1}^n \max_{s \in S}(y_s^i - x_s^i)$; $\mathbf{x}^1 := \mathbf{y}^1$; $\ldots$; $\mathbf{x}^n := \mathbf{y}^n$;

14   **for** $j = \max\{ k_b < \infty \mid b \in \{1, \ldots, n\}\}$ **down to 1 do**

15   **foreach** $s \in S$ **do**

16        $y_s := \max_{a \in \mathcal{A}(s)}(\sum_{\{i | k_i \geq j\}} w_i \cdot \mathbf{r}_i(s,a) + \min_{\mathfrak{h}_s^a \in \mathcal{H}_s^a} \sum_{s' \in S} \mathfrak{h}_s^a(s') \cdot x_{s'})$;

17        $\sigma^j(s) := \arg \max\limits_{a \in \mathcal{A}(s)} (\sum_{\{i | k_i \geq j\}} w_i \cdot \mathbf{r}_i(s,a) + \min\limits_{\mathfrak{h}_s^a \in \mathcal{H}_s^a} \sum_{s' \in S} \mathfrak{h}_s^a(s') \cdot x_{s'})$;

18        $\bar{\mathfrak{h}}_s^{\sigma^j(s)}(s') := \arg \min_{\mathfrak{h}_s^a \in \mathcal{H}_s^a} \sum_{s' \in S} \mathfrak{h}_s^a(s') \cdot x_{s'}$;

19   **foreach** $i \in \{1, \ldots, n\}$ **where** $k_i \geq j$ **do**

20          $y_s^i := \mathbf{r}_i(s, \sigma^j(s)) + \sum_{s' \in S} \bar{\mathfrak{h}}_s^{\sigma^j(s)}(s') \cdot x_{s'}^i$;

21      $\mathbf{x} := \mathbf{y}$; $\mathbf{x}^1 := \mathbf{y}^1$; $\ldots$; $\mathbf{x}^n := \mathbf{y}^n$;

22   **for** $i = 1$ **to** n **do**

23      $g_i := y_{\bar{s}}^i$;

24   σ acts as σ^j in j^{th} step when $j < \max_{i \in \{1, \ldots, n\}} k_i$ and as σ^∞ afterwards;

25   **return** $\sigma, \mathbf{g}$

---

4 Case Studies

We implemented the proposed multi-objective robust strategy synthesis algorithm and applied them to two case studies: (1) motion planning for a robot with noisy continuous dynamics and (2) autonomous nondeterministic tour guides drawn from [4,18]. All experiments completed in few seconds on a standard laptop PC.

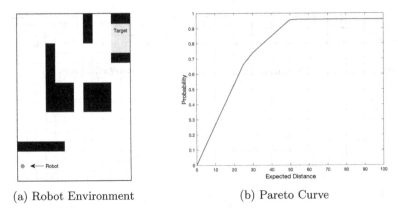

(a) Robot Environment   (b) Pareto Curve

Fig. 3. Robotic Scenario. (a) Environment map, with black obstacles and gray target area. (b) Pareto curve for the property $([\mathbf{r}_p]_{\max}^{\leq\infty}, [\mathbf{r}_d]_{\min}^{\leq\infty})$.

4.1   Robot Motion Planning Under Uncertainty

In robot motion planning, designers often seek a plan that simultaneously satisfies multiple objectives [23], e.g., *maximising the chances of reaching the target while minimising the energy consumption.* These objectives are usually in conflict with each other; hence, presenting the Pareto curve, i.e., the set of achievable points with optimal trade-off between the objectives, is helpful to the designers. They can then choose a point on the curve according to their desired guarantees and obtain the corresponding plan (strategy) for the robot. In this case study, we considered such a motion planning problem for a noisy robot with continuous dynamics in an environment with obstacles and a target region, as depicted in Fig. 3a. The robot's motion model was a single integrator with additive Gaussian noise. The initial state of the robot was on the bottom-left of the environment. The objectives were to reach the target safely while reducing the energy consumption, which is proportional to the travelled distance.

We approached this problem by first abstracting the motion of the noisy robot in the environment as an *IMDP* $\mathcal{M}$ and then computing strategies on $\mathcal{M}$ as in [24–26]. The abstraction was achieved by partitioning the environment into a grid and computing local (continuous) controllers to allow transitions from every cell to each of its neighbours. The cells and the local controllers were then associated to the states and actions of the *IMDP*, respectively, resulting in 204 states (cells) and 4 actions per state. The boundaries of the environment were also associated with a state. Note that the transition probabilities between cells were raised by the noise in the dynamics and their ranges were due to variation of the possible initial robot (continuous) state within each cell.

The *IMDP* states corresponding to obstacles (including boundaries) were given deterministic self-transitions, modelling robot termination as the result of a collision. To allow for the computation of the probability of reaching target, we included an extra state in the *IMDP* with a deterministic self-transition and then added incoming deterministic transitions to this state from the target states. A reward structure $\mathbf{r}_p$, which assigns a reward of 1 to these transitions and 0 to all the others, in fact, computes the probability of reaching the target. To capture the travelled distance, we defined a reward structure $\mathbf{r}_d$ assigning a reward of 0 to the state-action pairs with self-transitions and 1 to the other pairs.

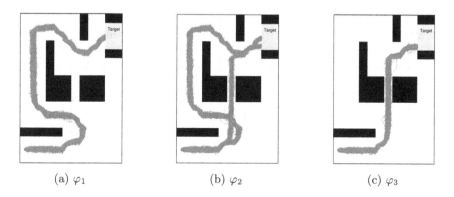

(a) φ_1 (b) φ_2 (c) φ_3

Fig. 4. Robot sample paths under strategies for φ_1, φ_2, and φ_3

The two robot objectives then can be expressed as: $([\mathbf{r}_p]_{\max}^{\leq \infty}, [\mathbf{r}_d]_{\min}^{\leq \infty})$ – see [17, Appendix C] for Pareto queries. We first computed the Pareto curve for the property, which is shown in Fig. 3b, to find the set of all achievable values (optimal trade-offs) for the reachability probability and expected travelled distance. The Pareto curve shows that there is clearly a trade-off between the two objectives. To achieve high probability of reaching target safely, the robot needs to travel a longer distance, i.e., spend more energy, and vice versa. We chose three points on the curve and computed the corresponding robust strategies for

$$\varphi_1 = ([\mathbf{r}_p]_{\geq 0.95}^{\leq \infty}, [\mathbf{r}_d]_{\leq 50}^{\leq \infty}), \quad \varphi_2 = ([\mathbf{r}_p]_{\geq 0.90}^{\leq \infty}, [\mathbf{r}_d]_{\leq 45}^{\leq \infty}), \quad \varphi_3 = ([\mathbf{r}_p]_{\geq 0.66}^{\leq \infty}, [\mathbf{r}_d]_{\leq 25}^{\leq \infty}).$$

We then simulated the robot under each strategy 500 times. The statistical results of these simulations are consistent with the bounds in φ_1, φ_2, and φ_3. The collision-free robot trajectories are shown in Fig. 4. These trajectories illustrate that the robot is conservative under φ_1 and takes a longer route with open spaces around it to go to target in order to be safe (Fig. 4a), while it becomes reckless under φ_3 and tries to go through a narrow passage with the knowledge that its motion is noisy and could collide with the obstacles (Fig. 4c). This risky behaviour, however, is required in order to meet the bound on the expected travelled distance in φ_3. The sample trajectories for φ_2 (Fig. 4b) demonstrate the stochastic nature of the strategy. That is, the robot probabilistically chooses between being safe and reckless in order to satisfy the bounds in φ_2.

4.2   The Model of Autonomous Nondeterministic Tour Guides

Our second case study is inspired by "Autonomous Nondeterministic Tour Guides" (ANTG) in [4,18], which models a complex museum with a variety of collections. We note that the model introduced in [4] is an *MDP*. In this case study, we use an *IMDP* model by inserting uncertainties into the *MDP*. Due to the popularity of the museum, there are many visitors at the same time. Different visitors may have different preferences of arts. We assume the museum divides all collections into different categories so that visitors can choose what they would like to visit and pay tickets according to their preferences. In order to obtain the best experience, a visitor can first assign certain weights to all categories denoting their preferences to the museum, and then design the best strategy for a target. However, the preference of a sort of arts to a visitor may depend on many factors including price or length of queue at that moment etc., hence it is hard to assign fixed values to these preferences. In our model we allow uncertainties of preferences such that their values may lie in an interval.

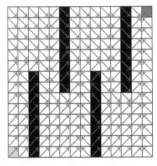

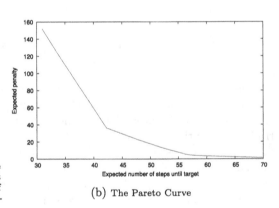

(a) The ANTG model for $n = 14$. The yellow, black and green cells represent the entrance, closed and exit parts of the museum, respectively. The red arrows indicate an example strategy.

(b) The Pareto Curve

Fig. 5. The ANTG case study: model and analysis (Color figure online)

For simplicity we assume all collections are organised in an $n \times n$ square with $n \geq 10$, with $(0,0)$ being the south-west corner of the museum and $(n - 1, n - 1)$ the north-east one. Let $c = \frac{n-1}{2}$; note that (c, c) is at the centre of the museum. We assume all collections at (x, y) are assigned with a weight interval $[3, 4]$ if $\max\{|x - c|, |y - c|\} \leq \frac{n}{10}$, with a weight 2 if $\frac{n}{10} < \max\{|x - c|, |y - c|\} \leq \frac{n}{5}$, and a weight 1 if $\max\{|x - c|, |y - c|\} > \frac{n}{5}$. In other words, we expect collections in the centre to be more popular and subject to more uncertainties than others. Furthermore, we assume that people at each location (x, y) have four nondeterministic choices of moving to (x', y') in the north east, south east, north west, and south west of (x, y) (limited to the boundaries of the museum). The outcome of these choices, however, is not deterministic. That is, deciding to go to (x', y') takes the visitor to either (x, y') or (x', y) depending on the weight intervals of (x, y') and (x', y). Thus, the actual outcome of the move is probabilistic to north, south, east or west. To obtain an *IMDP*, weights are normalised. For instance, if the visitor chooses to go to the north east and on $(x, y + 1)$ there is a weight interval of $[3, 4]$ and on $(x + 1, y)$ there is a weight interval of $[2, 2]$, it will go to $(x, y+1)$ with probability interval $[\frac{3}{3+2}, \frac{4}{4+2}]$ and to $(x + 1, y)$ with probability interval $[\frac{2}{2+4}, \frac{2}{2+3}]$. Therefore a model with parameter n has n^2 states in total and roughly $4n^2$ transitions, a few of which are associated with uncertain transition probabilities. An instance of the museum model for $n = 14$ is depicted in Fig. 5a. In this instantiation, we assume that the visitor starts in the lower left corner (marked yellow) and wants to move to the upper right corner (marked green) with as few steps as possible. On the other hand, it wants to avoid moving to the black cells, because they correspond to exhibitions which are closed. For closed exhibitions located at $x = 2$, the visitor receive a penalty of 2, for those at $x = 5$ it receives a penalty of 4, for $x = 8$ one of 16 and for $x = 11$ one of 64. Therefore, there is a tradeoff between leaving the museum as fast as possible and minimising the penalty received. With $\mathbf{r}_s$ being the reward structure for the number of steps and $\mathbf{r}_p$ denoting the penalty accumulated, $([\mathbf{r}_s]_{\leq 40}^{\leq \infty}, [\mathbf{r}_p]_{\leq 70}^{\leq \infty})$ requires that we leave the museum within 40 steps but with a penalty of no more than 70. The red arrows indicate a strategy which has been used when computing the Pareto curve by our tool. Here, the tourist mostly ignores closed exhibitions at $x = 2$ but avoids them later. In [17, Appendix D], we provide a few more strategies occurring during the computation. We provide the Pareto curve for this situation in Fig. 5b. With an increasing step bound considered acceptable, the optimal accumulated penalty decreases. This is expected, since with a larger step bound, the visitor has more time to walk around more of the closed exhibitions, thus facing a lower penalty.

5 Concluding Remarks

In this paper, we have analysed *IMDPs* under controller synthesis semantics in a dynamic setting; we discussed the multi-objective robust strategy synthesis problem for *IMDPs*, aiming for strategies that satisfy a given multi-objective predicate under all resolutions of the uncertainty in the transition probabilities. We showed that this problem is **PSPACE**-hard and introduced a value iteration-based decision algorithm to approximate the Pareto set. We finally presented the effectiveness of the proposed algorithms on several real-world case studies.

Even though we focused here on *IMDPs* with multi-objective reachability and reward properties, the proposed robust synthesis algorithm can also handle *MDPs* with convex uncertain sets and any ω-regular properties such as LTL. For future work, we aim to explore the upper bound of the time complexity of the multi-objective robust strategy synthesis which is left open in this paper.

References

1. Basset, N., Kwiatkowska, M., Wiltsche, C.: Compositional controller synthesis for stochastic games. In: Baldan, P., Gorla, D. (eds.) CONCUR 2014. LNCS, vol. 8704, pp. 173–187. Springer, Heidelberg (2014). doi:10.1007/978-3-662-44584-6_13

2. Benedikt, M., Lenhardt, R., Worrell, J.: LTL model checking of interval Markov chains. In: Piterman, N., Smolka, S.A. (eds.) TACAS 2013. LNCS, vol. 7795, pp. 32–46. Springer, Heidelberg (2013). doi:10.1007/978-3-642-36742-7_3

3. Boyd, S., Vandenberghe, L.: Convex Optimization. Cambridge University Press, Cambridge (2004)

4. Cantino, A.S., Roberts, D.L., Isbell, C.L.: Autonomous nondeterministic tour guides: improving quality of experience with TTD-MDPs. In: AAMAS, p. 22 (2007)

5. Chatterjee, K., Majumdar, R., Henzinger, T.A.: Markov decision processes with multiple objectives. In: Durand, B., Thomas, W. (eds.) STACS 2006. LNCS, vol. 3884, pp. 325–336. Springer, Heidelberg (2006). doi:10.1007/11672142_26

6. Chatterjee, K., Sen, K., Henzinger, T.A.: Model-checking ω-regular properties of interval Markov chains. In: Amadio, R. (ed.) FoSSaCS 2008. LNCS, vol. 4962, pp. 302–317. Springer, Heidelberg (2008). doi:10.1007/978-3-540-78499-9_22

7. Chen, T., Forejt, V., Kwiatkowska, M., Simaitis, A., Wiltsche, C.: On stochastic games with multiple objectives. In: Chatterjee, K., Sgall, J. (eds.) MFCS 2013. LNCS, vol. 8087, pp. 266–277. Springer, Heidelberg (2013). doi:10.1007/978-3-642-40313-2_25

8. Chen, T., Han, T., Kwiatkowska, M.: On the complexity of model checking interval-valued discrete time Markov chains. Inf. Proc. Lett. **113**(7), 210–216 (2013)

9. Ehrgott, M.: Multicriteria Optimization. Springer Science & Business Media, Heidelberg (2006)

10. Esteve, M.-A., Katoen, J.-P., Nguyen, V.Y., Postma, B., Yushtein, Y.: Formal correctness, safety, dependability and performance analysis of a satellite. In: ICSE, pp. 1022–1031 (2012)

11. Etessami, K., Kwiatkowska, M., Vardi, M.Y., Yannakakis, M.: Multi-objective model checking of Markov decision processes. In: Grumberg, O., Huth, M. (eds.) TACAS 2007. LNCS, vol. 4424, pp. 50–65. Springer, Heidelberg (2007). doi:10.1007/978-3-540-71209-1_6

12. Fecher, H., Leucker, M., Wolf, V.: *Don't Know* in probabilistic systems. In: Valmari, A. (ed.) SPIN 2006. LNCS, vol. 3925, pp. 71–88. Springer, Heidelberg (2006). doi:10.1007/11691617_5

13. Forejt, V., Kwiatkowska, M., Norman, G., Parker, D., Qu, H.: Quantitative multi-objective verification for probabilistic systems. In: Abdulla, P.A., Leino, K.R.M. (eds.) TACAS 2011. LNCS, vol. 6605, pp. 112–127. Springer, Heidelberg (2011). doi:10.1007/978-3-642-19835-9_11

14. Forejt, V., Kwiatkowska, M., Parker, D.: Pareto curves for probabilistic model checking. In: Chakraborty, S., Mukund, M. (eds.) ATVA 2012. LNCS, pp. 317–332. Springer, Heidelberg (2012). doi:10.1007/978-3-642-33386-6_25

15. Givan, R., Leach, S.M., Dean, T.L.: Bounded-parameter Markov decision processes. AI **122**(1–2), 71–109 (2000)

16. Hahn, E.M., Han, T., Zhang, L.: Synthesis for PCTL in parametric Markov decision processes. In: Bobaru, M., Havelund, K., Holzmann, G.J., Joshi, R. (eds.) NFM 2011. LNCS, vol. 6617, pp. 146–161. Springer, Heidelberg (2011). doi:10.1007/978-3-642-20398-5_12

17. Hahn, E.M., Hashemi, V., Hermanns, H., Lahijanian, M., Turrini, A.: Multi-objective robust strategy synthesis for interval Markov decision processes (2017). http://arxiv.org/abs/1706.06875
18. Hashemi, V., Hermanns, H., Song, L.: Reward-bounded reachability probability for uncertain weighted MDPs. In: Jobstmann, B., Leino, K.R.M. (eds.) VMCAI 2016. LNCS, vol. 9583, pp. 351–371. Springer, Heidelberg (2016). doi:10.1007/978-3-662-49122-5_17
19. Jonsson, B., Larsen, K.G.: Specification and refinement of probabilistic processes. In: LICS, pp. 266–277. IEEE Computer Society (1991)
20. Kozine, I., Utkin, L.V.: Interval-valued finite Markov chains. Reliable Comput. 8(2), 97–113 (2002)
21. Kwiatkowska, M., Norman, G., Parker, D., Qu, H.: Compositional probabilistic verification through multi-objective model checking. I&C 232, 38–65 (2013)
22. Lahijanian, M., Andersson, S.B., Belta, C.: Formal verification and synthesis for discrete-time stochastic systems. IEEE Tr. Autom. Contr. 60(8), 2031–2045 (2015)
23. Lahijanian, M., Kwiatkowska, M.: Specification revision for Markov decision processes with optimal trade-off. In: CDC, pp. 7411–7418 (2016)
24. Luna, R., Lahijanian, M., Moll, M., Kavraki, L.E.: Asymptotically optimal stochastic motion planning with temporal goals. In: Akin, H.L., Amato, N.M., Isler, V., Stappen, A.F. (eds.) WAFR 2014. STAR, vol. 107, pp. 335–352. Springer, Cham (2015). doi:10.1007/978-3-319-16595-0_20
25. Luna, R., Lahijanian, M., Moll, M., Kavraki, L.E.: Fast stochastic motion planning with optimality guarantees using local policy reconfiguration. In: ICRA, pp. 3013–3019 (2014)
26. Luna, R., Lahijanian, M., Moll, M., Kavraki, L.E.: Optimal and efficient stochastic motion planning in partially-known environments. In: AAAI, pp. 2549–2555 (2014)
27. Mouaddib, A.: Multi-objective decision-theoretic plan problem. In: ICRA, pp. 2814–2819 (2004)
28. Nilim, A., El Ghaoui, L.: Robust control of Markov decision processes with uncertain transition matrices. Oper. Res. 53(5), 780–798 (2005)
29. Ogryczak, W., Perny, P., Weng, P.: A compromise programming approach to multiobjective Markov decision processes. IJITDM 12(5), 1021–1054 (2013)
30. Perny, P., Weng, P., Goldsmith, J., Hanna, J.P.: Approximation of Lorenz-optimal solutions in multiobjective Markov decision processes. In: AAAI, pp. 92–94 (2013)
31. Puggelli, A.: Formal techniques for the verification and optimal control of probabilistic systems in the presence of modeling uncertainties. Ph.D. thesis, UC Berkeley (2014)
32. Puggelli, A., Li, W., Sangiovanni-Vincentelli, A.L., Seshia, S.A.: Polynomial-time verification of PCTL properties of MDPs with convex uncertainties. In: Sharygina, N., Veith, H. (eds.) CAV 2013. LNCS, vol. 8044, pp. 527–542. Springer, Heidelberg (2013). doi:10.1007/978-3-642-39799-8_35
33. Randour, M., Raskin, J.-F., Sankur, O.: Percentile queries in multi-dimensional Markov decision processes. In: Kroening, D., Păsăreanu, C.S. (eds.) CAV 2015. LNCS, vol. 9206, pp. 123–139. Springer, Cham (2015). doi:10.1007/978-3-319-21690-4_8
34. Wolff, E.M., Topcu, U., Murray, R.M.: Robust control of uncertain Markov decision processes with temporal logic specifications. In: CDC, pp. 3372–3379 (2012)
35. Wu, D., Koutsoukos, X.D.: Reachability analysis of uncertain systems using bounded parameter Markov decision processes. AI 172(9), 945–954 (2008)

Investigating Parametric Influence on Discrete Synchronisation Protocols Using Quantitative Model Checking

Paul Gainer[✉], Sven Linker, Clare Dixon, Ullrich Hustadt, and Michael Fisher

University of Liverpool, Liverpool, UK
{p.gainer,s.linker,cldixon,u.hustadt,mfisher}@liverpool.ac.uk

Abstract. Synchronisation is an emergent phenomenon observable in nature. Natural synchronising systems have inspired the development of protocols for achieving coordination in a diverse range of distributed dynamic systems. Spontaneously synchronising systems can be mathematically modelled as coupled oscillators. In this paper we present a novel approach using model checking to reason about achieving synchrony for different models of synchronisation. We describe a general, formal population model where oscillators interact at discrete moments in time, and whose cycles are sequences of discrete states. Using the probabilistic model checker PRISM, we investigate the influence of various parameters of the model on the likelihood of, and time required for, achieving synchronisation.

1 Introduction

Synchronisation is an emergent phenomenon observable throughout nature; pacemaker cells in the sinoatrial node of the heart synchronise to set the rate and rhythm of a heartbeat, and populations of fireflies synchronise their flashing to attract mates [6]. These decentralised natural systems have inspired the development of protocols for achieving synchrony in a diverse range of artificial decentralised systems; in particular swarm robotic systems and wireless sensor networks (WSNs). Applications include detecting faults in members of a robotic swarm [8], synchronising the duty cycles of sensor nodes in a network [19], auto-tuning mobile networks to save energy [4], and coordinating data dissemination for a WSN [5].

The cyclic behaviour of systems where synchrony spontaneously occurs can be modelled as networks of coupled oscillators with similar frequencies. Oscillators are *coupled* when some process results in the transferral of energy between them. In some systems oscillators are coupled such that their oscillations have a continuous influence on each other. The strength of the coupling between oscillators is determined by some global constant. When the mutual agitation

This work was supported by both the Sir Joseph Rotblat Alumni Scholarship at Liverpool and the EPSRC Research Programme EP/N007565/1 *Science of Sensor System Software*.

© Springer International Publishing AG 2017
N. Bertrand and L. Bortolussi (Eds.): QEST 2017, LNCS 10503, pp. 224–239, 2017.
DOI: 10.1007/978-3-319-66335-7_14

of oscillators takes place only at discrete instances in time the oscillators are *pulse-coupled* [15,17]. At some distinguished point in the oscillation cycle a pulse-coupled oscillator *fires* and influences other nearby oscillators. An oscillator that is perturbed by another oscillator shifts or resets the phase of its own oscillation cycle to more closely match that of its neighbour. Over time this can lead to all oscillators matching phase, and synchronisation is achieved if all oscillators fire synchronously. In nature, the oscillation cycle of oscillators often includes a *refractory period*. The refractory period is an interval in the oscillation cycle during which its phase cannot be perturbed by other firing oscillators. This refractory period can prevent spurious mutual stimulation of the oscillators, which could lead to perpetual asynchrony. The introduction of a refractory period to oscillators in artificial systems not only helps to achieve synchrony, but can also be thought of as a period during which robots in a swarm, or nodes in a WSN, can turn off their wireless antennas and save energy.

The emergence of synchronisation in robot swarms, WSNs, and other distributed and dynamic systems is often investigated by designing and analysing simulations. Simulations can give detailed insight into how global behaviours of these systems emerge over time. Formal approaches can complement simulations, where desirable properties for the system can be unambiguously formulated, and rigorously checked against some formal model of the system, often finding corner cases that may not be covered even by a large number of tests of a simulation. Model checking has been successfully applied to qualitatively, and quantitatively, analyse control algorithms and protocols for both swarm systems [2,11,14] and WSNs [7,13]. In particular, model checking has been used to formally investigate the emergence of synchronisation in networks of oscillators. A general model of synchronisation for oscillators was introduced in [3], where oscillators were modelled as timed automata [1], and a model checking algorithm was used to determine the reachability of a synchronised state for distinguished runs of the model. More recently, Heidarian et al. [13] also used timed automata to exhaustively analyse a specific clock synchronisation protocol.

Our contribution in this paper is the development of a formal and general model for oscillator synchronisation, which is parameterised by a synchronisation model and a configuration for both oscillators and the network. In contrast to previous applications of model checking to detect synchronisation, our model is discrete. That is, the oscillators interact at discrete moments in time, and their oscillation cycles are defined as sequences of discrete states. Given an instantiation of our general model we automatically generate a discrete time Markov chain (DTMC). We discuss the results of model checking two instantiations of our model with regards to energy consumption.

In Sect. 2 we present the dynamics of individual oscillators with discrete oscillation cycles. We formally define our general, parameterised population model for a network of oscillators in Sect. 3, and describe how we construct the corresponding DTMC in Sect. 4. We discuss the results of checking synchronisation properties for two concrete instantiations of our formal model in Sect. 5. Concluding remarks and suggestions for further work are given in Sect. 6.

2   Discrete Oscillator Model

We consider a fully-connected network of N pulse-coupled oscillators with identical dynamics over discrete time. We denote the set of these oscillators by $\mathcal{O} = \{1, \ldots, N\}$, where each $i \in \mathcal{O}$ corresponds to a single pulse-coupled oscillator. The *value* or *phase* of an oscillator i in $\mathcal{O}$ at time t is denoted by $\phi_i(t)$. The phase of an oscillator progresses through a sequence of discrete integer values bounded by some $T \geq 1$.

Definition 1. *The* evolution function *is a strictly increasing function* evol $:$ $\{1, \ldots, T\} \rightarrow \mathbb{N}$ *with* $\mathrm{evol}(\Phi) \geq \Phi$ *for all* $\Phi \in \{1, \ldots, T\}$, *that maps the current phase of an oscillator to its phase in the next discrete time step.*

We now introduce the *update function* and *firing predicate*, which respectively denote the updated phase of an oscillator i at time t after one time step, and the firing of oscillator i at time t,

$$update_i(t) = \mathrm{evol}(\phi_i(t)), \quad fire_i(t) = update_i(t) > T.$$

The precise evolution of phase over time for an oscillator i is then given by

$$\phi_i(t+1) = \begin{cases} 1 & \text{if } fire_i(t) \\ update_i(t) & \text{otherwise,} \end{cases}$$

where phase increases over time until $\mathrm{evol}(\phi_i(t)) > T$, at which point oscillator i *fires*, that is, $\phi_i(t+1)$ becomes 1 and the oscillator attempts to broadcast a firing signal to all other oscillators *coupled* to it. The phase progression of an uncoupled oscillator is cyclic, and we refer to one cycle as an *oscillation cycle*.

An oscillator's firing signal perturbs the phase of all coupled oscillators; we use $\alpha_i(t)$ to denote the number of all other oscillators in $\mathcal{O}$ that are coupled to i and will broadcast their firing signal at time t. Furthermore, we define $\mu \in [0, 1]$ to be the probability that a *broadcast failure* occurs when an oscillator fires, that is, the attempt to broadcast its firing signal fails (the oscillator still resets its phase to 1). Note that μ is a global parameter, hence the chance of broadcast failure is identical for all oscillators. Observe that $\alpha_i(t)$ is defined globally even though the model is not deterministic, however we defer the reader to the detailed discussion of probabilities in the following section.

Definition 2. *The* perturbation function *is an increasing function* pert $:$ $\{1, \ldots, T\} \times \mathbb{N} \times \mathbb{R}^+ \rightarrow \mathbb{N}$ *that maps the phase of an oscillator i, the number of oscillators that have fired and perturbed i, and a real value defining the strength of the coupling between oscillators, to an integer value corresponding to the induced perturbation to phase.*

We refine the update function to include the perturbation to phase induced by the firing of other oscillators that are *coupled* to oscillator i at time t, giving $update_i(t) = \mathrm{evol}(\phi_i(t)) + \mathrm{pert}(\phi_i(t), \alpha_i(t), \epsilon)$. We can introduce a refractory period into the oscillation cycle of each oscillator. A refractory period is an interval of discrete values $[1, R] \subseteq [1, T]$ where $0 \leq R \leq T$ is the size of the refractory period, such that if $\phi_i(t)$ is inside the interval, for some oscillator i at time t, then i cannot be perturbed by other oscillators to which it is coupled. If $R = 0$ then we set $[1, R] = \emptyset$, and there is no refractory period at all. To be consistent with the literature we only consider refractory periods that occur at the start of the oscillation cycle.

Definition 3. *The* refractory function ref $: \{1, \ldots, T\} \times \mathbb{N} \to \mathbb{N}$ *is defined as* $\text{ref}(\Phi, \Delta) = 0$ *if* $\Phi \in [1, R]$, *or* $\text{ref}(\Phi, \Delta) = \Delta$ *otherwise.*

Given Δ, a degree of perturbance to the phase of an oscillator, and Φ, the phase of that oscillator, $\text{ref}(\Phi, \Delta)$ returns 0 if Φ is in the refractory period defined by R, or Δ otherwise. We again amend the update function to include the refractory function, giving $update_i(t) = \text{evol}(\phi_i(t)) + \text{ref}(\phi_i(t), \text{pert}(\phi_i(t), \alpha_i(t), \epsilon))$.

3   Population Model

Let $\mathcal{O} = \{1, \ldots, N\}$ be a fully connected network of N identical oscillators with phases in the range $1, \ldots, T$, whose dynamics are determined by the functions evol and pert, with a refractory period defined by R, coupled with strength $\epsilon \in [0, 1]$, and where the probability of broadcast failure is $\mu \in [0, 1]$. The *population model* of the network $\mathcal{O}$ is defined by $\mathcal{S} = (N, T, R, \epsilon, \text{evol}, \text{pert}, \mu)$.

Since all oscillators in our model are behaviourally identical we do not need to distinguish between oscillators sharing the same phase, and can reason about groups of oscillators, instead of individuals. The global state of the model is therefore a tuple, where each element n_Φ of the tuple $\langle n_1, \ldots, n_T \rangle$ corresponds to the number of oscillators sharing a phase value of Φ. The population model does not account for the introduction of additional oscillators to a network, or the loss of existing coupled oscillators. That is, the population N remains constant.

Definition 4. *A global state of a population model* $\mathcal{S} = (N, T, R, \epsilon, \text{evol}, \text{pert}, \mu)$ *is a T-tuple* $\pi \in \{0, \ldots, N\}^T$, *where* $\pi = \langle n_1, \ldots, n_T \rangle$ *and* $\sum_{\Phi=1}^{T} n_\Phi = N$. *The set of all global states of* $\mathcal{S}$ *is* $\Gamma(\mathcal{S})$, *or simply* Γ *when* $\mathcal{S}$ *is clear from the context.*

Example 1. Figure 1 shows three global states for an instantiated population model, $\mathcal{S} = (5, 6, 2, 0.15, \text{evol}, \text{pert}, 0.1)$, where the synchronisation model described in [8] is instantiated by defining the evolution function as $\text{evol}(\Phi) = \Phi + 1$, and the perturbation function as $\text{pert}(\Phi, \alpha, \epsilon) = [\Phi \cdot \alpha \cdot \epsilon]$, where $[x]$ denotes x rounded to the nearest integer. The label for each node n_Φ is the number of oscillators with phase Φ. We omit the label if $n_\Phi = 0$. Oscillators at node n_6 are about to fire, and oscillators at nodes n_1 and n_2 are in their refractory period, and cannot be perturbed by the firing of other oscillators. The global states can be denoted by $\pi_0 = \langle 0, 1, 0, 2, 2, 0 \rangle$, $\pi_1 = \langle 0, 0, 1, 0, 2, 2 \rangle$, and $\pi_2 = \langle 5, 0, 0, 0, 0, 0 \rangle$. Later we will explain how transitions between these global states are made. Note that directional arrows indicate cyclic direction, and do not represent transitions.

With every global state π we associate a non-empty set of *failure vectors*, where each failure vector is a tuple of broadcast failures that could occur in π.

Definition 5. *A failure vector is a T-tuple $B \in (\{0,\ldots,N\} \cup \{\star\})^T$. We denote the set of all possible failure vectors by $\mathcal{B}$.*

Given a failure vector $B = \langle b_1, \ldots, b_T \rangle$, $b_\Phi \in \{0, \ldots, N\}$ indicates the number of broadcast failures that occur for all oscillators with a phase of Φ. If $b_\Phi = \star$ then no oscillators with a phase of Φ fire, for all $1 \leq \Phi \leq T$. Semantically, $b_\Phi = 0$ and $b_\Phi = \star$ differ in that the former indicates that all (if any) oscillators with phase Φ fire and no broadcast failures occur, while the latter indicates that all (if any) oscillators with a phase of Φ do not fire. If no oscillators fire at all in a global state then we have only one possible failure vector, namely $\{\star\}^T$.

3.1 Transitions

Later in this section we will describe how we can calculate the set of all possible failure vectors for a global state, and thereby identify all of its successor states. However we must first show how we can calculate the single successor state of a global state π, given some failure vector B.

Absorptions. For real deployments of synchronisation protocols it is often the case that the duration of a single oscillation cycle will be at least several seconds [8,18]. The perturbation induced by the firing of a group of oscillators may lead to groups of other oscillators to which they are coupled firing in turn. The firing of these other oscillators may then cause further oscillators to fire, and so forth, leading to a "chain reaction", where each group of oscillators triggered to fire is *absorbed* by the initial group of firing oscillators. Since the whole chain reaction of absorptions may occur within just a few milliseconds, and in our model the oscillation cycle is a sequence of discrete states, when a chain reaction occurs the phases of all perturbed oscillators are updated at one single time step.

Since we are considering a fully connected network of oscillators, two oscillators sharing the same phase will have their phase updated to the same value in the next time step. They will always perceive the same number of other oscillators firing. Therefore, for each phase Φ we define the function $\alpha^\Phi \colon \Gamma \times \mathcal{B} \to \{1, \ldots, N\}$, where $\alpha^\Phi(\pi, B)$ is the number of oscillators with a

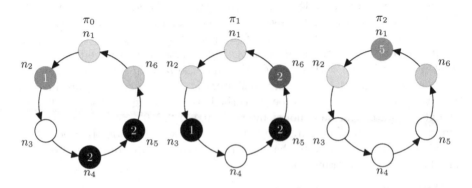

Fig. 1. Evolution of the global state over three discrete time steps.

phase greater than Φ perceived to be firing by oscillators with phase Φ, in some global state, incorporating the broadcast failures defined in the failure vector B. This allows us to encode the aforementioned chain reactions of firing oscillators. Note that our encoding of chain reactions results in a global semantics that differs from typical parallelisation operations, for example, the construction of the crossproduct of the individual oscillators.

Given a global state $\pi = \langle n, \ldots, n_T \rangle$ and a failure vector $B = \langle b_1, \ldots, b_T \rangle$, the following mutually recursive definitions show how we calculate the values $\alpha^1(\pi, B), \ldots, \alpha^T(\pi, B)$, and how functions introduced in the previous section are modified to indicate the update in phase, and firing, of *all* oscillators sharing the same phase Φ. Observe that to calculate any $\alpha^\Phi(B, \pi)$ we only refer to definitions for phases greater than Φ and the base case is $\Phi = T$, that is, values are computed from T down to 1.

$$update^\Phi(\pi, B) = evol(\Phi) + ref(\Phi, pert(\Phi, \alpha^\Phi(\pi, B), \epsilon))$$

$$fire^\Phi(\pi, B) = update^\Phi(\pi, B) > T$$

$$\alpha^\Phi(\pi, B) = \begin{cases} 0 & \text{if } \Phi = T \\ \alpha^{\Phi+1}(\pi, B) + n_{\Phi+1} - b_{\Phi+1} & \text{if } \Phi < T, b_{\Phi+1} \neq \star \text{ and } fire^{\Phi+1}(\pi, B) \\ \alpha^{\Phi+1}(\pi, B) & \text{otherwise} \end{cases}$$

Transition Function. We now define the transition function that maps phase values to their updated values in the next time step. Note that since we no longer differentiate between different oscillators with the same phase we only need to calculate a single value for their evolution and perturbation.

Definition 6. *The phase transition function $\tau : \Gamma \times \{1, \ldots, T\} \times \mathcal{B} \to \mathbb{N}$ maps a global state, a phase Φ, and some possible failure vector B for π, to the updated phase in the next discrete time step, with respect to the broadcast failures defined in B, and is defined as*

$$\tau(\pi, \Phi, B) = \begin{cases} 1 & \text{if } fire^\Phi(\pi, B) \\ update^\Phi(\pi, B) & \text{otherwise.} \end{cases}$$

Lemma 1. *The range of the function τ is bound by T. That is, for any π, for any possible failure vector B for π, and for all $\Phi \in \{1, \ldots, T\}$, we have that $1 \leq \tau(\pi, \Phi, B) \leq T$.*

Proof. By construction.

Let $\mathcal{U}_\Phi(\pi, B)$ be the set of phase values Ψ where all oscillators with phase Ψ in π will have their phase updated to Φ in the next time step, with respect to the broadcast failures defined in B. Formally,

$$\mathcal{U}_\Phi(\pi, B) = \{\Psi \mid \Psi \in \{1, \ldots, T\} \wedge \tau(\pi, \Psi, B) = \Phi\}.$$

We can now calculate the successor state of a global state π and define how the model evolves over time.

Definition 7. *The successor function* succ : $\Gamma \times \mathcal{B} \to \Gamma$ *maps a global state* π *and a failure vector* B *to a state* π', *and is defined as* succ$((n_1, \ldots, n_T), B) = (n'_1, \ldots, n'_T)$, *where* $n'_\Phi = \sum_{\Psi \in \mathcal{U}_\Phi(\pi, B)} n_\Psi$ *for* $1 \leq \Phi \leq T$.

Example 2. Consider the global state π_0 of Fig. 1 where no oscillators will fire since $n_6 = 0$. We therefore have one possible failure vector for π_0, namely $B = \{\star\}^6$. Since no oscillators fire the dynamics of the oscillators are determined solely by evol, and all oscillators simply increase their phase by 1 in the next time step. Now consider the global state π_1 and $B = (\star, \star, 0, 0, 0, 1)$, a possible failure vector for π_1, indicating that oscillators with phases of 3 to 6 will fire and one broadcast failure will occur for one of the two oscillators that will fire with phase 6. Despite the broadcast failure occurring, a chain reaction will occur as the firing of the single oscillator with phase 6 will perturb the two oscillators with phase 5 to fire also. The combined perturbation induced by the firing of all three oscillators will cause the final oscillator with phase 3 to fire. All oscillators are therefore absorbed into the initial group of firing oscillators. Since $fire^6(\pi_1, B)$ holds we have that $\alpha^5(\pi_1, B) = \alpha^6(\pi_1, B) + n_6 - b_6 = 0 + 2 - 1 = 1$. Similarly since $fire^5(\pi_1)$ holds we have that $\alpha^4(\pi_1, B) = \alpha^5(\pi_1, B) + n_5 = 1 + 2 = 3$. We then continue calculating $\alpha^\Phi(\pi_1, B)$ for $3 \geq \Phi \geq 1$, and conclude that $\mathcal{U}_1(\pi_1, B) = \{6, 5, 4, 3\}$ and $\mathcal{U}_6(\pi_1, B) = \mathcal{U}_5(\pi_1, B) = \mathcal{U}_4(\pi_1, B) = \emptyset$. Since $R = 2$ we have that $\mathcal{U}_3(\pi_1, B) = \{2\}$ and $\mathcal{U}_2(\pi_1, B) = \{1\}$. We calculate the successor of π_1 as $\pi_2 = $ succ$((0, 0, 1, 0, 2, 2), B) = (n_6 + n_5 + n_4 + n_3, n_1, n_2, 0, 0, 0) = (5, 0, 0, 0, 0, 0)$.

Lemma 2. *The number of oscillators is invariant during transitions, i.e., the successor function only creates tuples that are states of the given model. Formally, let* $\pi = (n_1, \ldots, n_T)$ *and* $\pi' = (n'_1, \ldots, n'_T)$ *be two states of a model* S *such that* $\pi' = $ succ(π, B), *where* B *is some possible failure vector for* π. *Then* $\sum_{\Phi=1}^{T} n_\Phi = \sum_{\Phi=1}^{T} n'_\Phi = N$.

Proof. By construction.

3.2   Failure Vector Calculation

We construct all possible failure vectors for a global state by considering every group of oscillators in decreasing order of phase. At each stage we determine if the oscillators would fire. If they fire then we consider each outcome where any, all, or none of the firings result in a broadcast failure. We then add a corresponding value to a partially calculated failure vector and consider the next group of oscillators with a lower phase. If the oscillators do not fire then there is nothing left to do, since by Definition 1 we know that evol is strictly increasing, and by Definition 2 we know that pert is increasing, therefore all oscillators with a lower phase will also not fire. We can then pad the partial failure vector with $\star$ appropriately to indicate that no failure could happen since no oscillators fired.

Table 1 illustrates how a possible failure vector for global state π_1 in Fig. 1 is iteratively constructed. The first three columns respectively indicate the current iteration i, the global state π_1 with the currently considered oscillators indicated,

Table 1. Construction of a possible failure vector for a global state $\pi_1 = \langle 0, 0, 1, 0, 2, 2 \rangle$.

Iteration (i)	π_1	Failure vector B	Fired	Branches
0	$\langle 0, 0, 1, 0, 2, 2 \rangle$	$\langle \rangle$	–	*false*
1	$\langle 0, 0, 1, 0, 2, \underline{2} \rangle$	$\langle 1 \rangle$	*true*	*true*
2	$\langle 0, 0, 1, 0, \underline{2}, 2 \rangle$	$\langle 0, 1 \rangle$	*true*	*true*
3	$\langle 0, 0, 1, \underline{0}, 2, 2 \rangle$	$\langle 0, 0, 1 \rangle$	*true*	*false*
4	$\langle 0, 0, \underline{1}, 0, 2, 2 \rangle$	$\langle 0, 0, 0, 1 \rangle$	*true*	*true*
5	$\langle 0, \underline{0}, 1, 0, 2, 2 \rangle$	$\langle \star, \star, 0, 0, 0, 1 \rangle$	*false*	–

and the elements of the failure vector B computed so far. The fourth column is *true* if the oscillators with phase $T + 1 - i$ would fire given the broadcast failures in the partial failure vector. We must consider all outcomes of any or all firings resulting in broadcast failure. The final column therefore indicates whether the value added to the partial failure vector in the current iteration is the only possible value (*false*), or a choice from one of several possible values (*true*).

Initially we have an empty partial failure vector. At the first iteration there are 2 oscillators with a phase of 6. These oscillators will fire so we must consider each case where $0, 1$ or 2 broadcast failures occur. Here we choose 1 broadcast failure, which is then added to the partial failure vector. At iteration 2, oscillators with a phase of 5 fire, and again we must consider each case with $0, 1$ or 2 broadcast failures occur; here we choose 0. At iteration 3 oscillators with a phase of 4 would have fired, but since there are no oscillators with a phase of 4 we only have one possible value to add to the partial failure vector, namely 0. At iteration 4 a single oscillator with a phase of 3 fires, and we choose the case where the firing did not result in a broadcast failure. In the final iteration oscillators with a phase of 2 do not fire, hence we can conclude that oscillators with a phase of 1 also do not fire, and can pad the partial failure vector appropriately with $\star$.

Formally, we define a family of functions f indexed by Φ, where each f_Φ takes as parameters some global state π, and V, a vector of length $T - \Phi$. V represents all broadcast failures for all oscillators with a phase greater than Φ. The function f_Φ then computes the set of all possible failure vectors for π with suffix V. Here we use the notation $v^\frown v'$ to indicate vector concatenation.

Definition 8. *We define* $f_\Phi : \Gamma \times \{0, \ldots, N\}^{T-\Phi} \to \mathbb{P}((\{0, \ldots, N\} \cup \{\star\})^T)$, *for* $1 \leq \Phi \leq T$, *as the family of functions indexed by* Φ, *where* $\pi = \langle n_1, \ldots, n_T \rangle$ *and*

$$
f_\Phi(\pi, V) = \begin{cases} \bigcup_{k=0}^{n_\Phi} f_{\Phi-1}(\pi, \langle k \rangle^\frown V) & \text{if } 1 < \Phi \leq T \text{ and } \textit{fire}^\Phi(\pi, \{\star\}^\Phi {}^\frown V) \\ \bigcup_{k=0}^{n_1} \{\langle k \rangle^\frown V\} & \text{if } \Phi = 1 \text{ and } \textit{fire}^1(\pi, \langle \star \rangle^\frown V) \\ \{\{\star\}^\Phi {}^\frown V\} & \textit{otherwise} \end{cases}
$$

Definition 9. *Given a global state $\pi \in \Gamma$, we define $\mathcal{B}_\pi$, the set of all possible failure vectors for that state, as $\mathcal{B}_\pi = f_T(\pi, \langle\rangle)$, and define $next(\pi)$, the set of all successor states of π, as $next(\pi) = \{\operatorname{succ}(\pi, B) \mid B \in \mathcal{B}_\pi\}$.*

Note that for some global states $|next(\pi)| < |\mathcal{B}_\pi|$, since we may have that $\operatorname{succ}(\pi, B) = \operatorname{succ}(\pi, B')$ for some $B, B' \in \mathcal{B}_\pi$ with $B \neq B'$.

Given a global state π and a failure vector $B \in \mathcal{B}_\pi$, we will now compute the probability of a transition being made to state $\operatorname{succ}(\pi, B)$ in the next time step. Recall that μ is the probability with which a broadcast failure occurs. Firstly we define the probability mass function $\mathrm{P}_{fail} : \{1, \ldots, N\} \times \{1, \ldots, N\} \to [0, 1]$, where $\mathrm{P}_{fail}(n, b)$ gives the probability of b broadcast failures occurring given that n oscillators fire, as $\mathrm{P}_{fail}(n, b) = \mu^b (1 - \mu)^{n-b} \binom{n}{b}$. We then denote by $\mathrm{P}_\tau(\pi) : \mathcal{B}_\pi \to [0, 1]$ the function mapping a possible broadcast failure vector B for π, to the probability of the failures in B occurring. That is,

$$\mathrm{P}_\tau(\langle n_1, \ldots, n_T \rangle)(\langle b_1, \ldots, b_T \rangle) = \prod_{\Phi=1}^{T} \begin{cases} \mathrm{P}_{fail}(n_\Phi, b_\Phi) & \text{if } b_\Phi \neq \star \\ 1 & \text{otherwise} \end{cases}$$

Lemma 3. *For any global state π, $\mathrm{P}_\tau(\pi)$ is a discrete probability distribution over $\mathcal{B}_\pi$. Formally, $\sum_{B \in \mathcal{B}_\pi} \mathrm{P}_\tau(\pi)(B) = 1$.*

Proof. By induction over a tree where internal nodes are partially constructed failure vectors and leaf nodes are failure vectors.

Example 3. We consider again the global states $\pi_1 = \langle 0, 0, 1, 0, 2, 2 \rangle$ and $\pi_2 = \langle 5, 0, 0, 0, 0, 0 \rangle$, given in Fig. 1, of the population model instantiated in Example 1, and the failure vector $B = \langle \star, \star, 0, 0, 0, 1 \rangle$ given in Example 2, noting that $B \in \mathcal{B}_{\pi_1}$, $\operatorname{succ}(\pi_1, B) = \pi_2$, and $\mu = 0.1$. We calculate the probability of a transition being made from π_1 to π_2 as $\mathrm{P}_\tau(\langle 0, 0, 1, 0, 2, 2 \rangle)(\langle \star, \star, 0, 0, 0, 1 \rangle) = 1 \cdot 1 \cdot \mathrm{P}_{fail}(1, 0) \cdot \mathrm{P}_{fail}(0, 0) \cdot \mathrm{P}_{fail}(2, 0) \cdot \mathrm{P}_{fail}(2, 1) = 1 \cdot 1 \cdot 0.9 \cdot 1 \cdot 0.81 \cdot 0.18 = 0.13122$.

We now have everything we need to fully describe the evolution of the global state of a population model over time. A *run* of a population model $\mathcal{S}$ is an infinite sequence of global states $\Pi = \pi_0 \to \pi_1 \to \pi_2 \to \pi_3 \cdots$, where π_0 is called the *initial state*, and for all $k \geq 0$, $\pi_k \to \pi_{k+1}$ if, and only if, $\pi_{k+1} \in next(\pi)$. We denote the set of all possible runs of $\mathcal{S}$ by $\Pi(\mathcal{S})$.

3.3   Synchronisation

When all oscillators in a population model have the same phase in a global state we say that the state is *synchronised*. Formally, a global state $\pi = \langle n_1, \ldots, n_T \rangle$ is *synchronised* if, and only if, there is some $\Phi \in \{1, \ldots, T\}$ such that $n_\Phi = N$. Hence, for all $\Phi' \neq \Phi$, we have that $n_{\Phi'} = 0$. We use the notation $synch(\pi)$ to indicate that some global state π is synchronised. We will often want to reason about whether some particular run Π of a model leads to a global state that is synchronised. We say that a run $\Pi = \pi_0 \to \pi_1 \to \ldots$ *synchronises* if, and only if, there exists some $k \geq 0$ with $synch(\pi_k)$. We use the notation $synch(\Pi)$

to indicate that some run Π synchronises. Once a synchronised global state is reached any successor states will remain synchronised. Finally we can say that a model *synchronises* if, and only if, $synch(\Pi)$ for all $\Pi \in \Pi(\mathcal{S})$. In Fig. 1 global state π_2 is synchronised, since $n_1 = N$.

4   Model Generation

We choose to use the probabilistic model checker PRISM [16] to formally verify properties of our model. Given a probabilistic model of a system, PRISM can be used to reason about both temporal and probabilistic properties of the input model, by checking some requirement expressed in a suitable formalism against all possible runs of the model. We define our input models as *Discrete Time Markov Chains* (DTMCs). A DTMC is a tuple $(Q, init, \mathrm{P})$ where Q is a set of states, $init \in Q$ is the initial state, and $\mathrm{P} : Q \times Q \rightarrow [0, 1]$ is the function mapping ordered pairs of states (q, q') to the probability with which a transition from q to q' occurs, where $\sum_{q' \in Q} \mathrm{P}(q, q') = 1$ for all $q \in Q$.

Given a population model $\mathcal{S} = (N, T, R, \epsilon, \mathrm{evol}, \mathrm{pert}, \mu)$, we construct a DTMC $(Q, init, \mathrm{P})$. We define the set of states Q to be $\Gamma(\mathcal{S}) \cup \{init\}$. In the initial state all oscillators are considered to be *unconfigured*. That is, oscillators have not yet been assigned a value for their phase. For each $q \in Q$, where $q \in \Gamma(\mathcal{S})$ and $q = \langle n_1, \ldots, n_t \rangle$, we define

$$\mathrm{P}(init, q) = \frac{1}{T^N} \cdot \prod_{i=1}^{T} \binom{N - (\sum_{j=1}^{i-1} n_j)}{n_i}$$

to be the probability of moving from $init$ to a state where the oscillators are *configured* with the phase values defined in q, since there are N choose n_1 ways to select n_1 oscillators to have a phase of 1, then $N - n_1$ choose n_2 ways to select n_2 oscillators to have a phase of 2, and so forth. For every $q \in Q \setminus init$ we consider each $q' \in Q \setminus init$ where $q' = \mathrm{succ}(q, B)$ for some $B \in \mathcal{B}_q$, and set $\mathrm{P}(q, q') = \mathrm{P}_\tau(q)(B)$. For all other $q \in Q \setminus init$ and $q' \in Q$, where $q \neq q'$ and $q' \notin next(q)$, we set $\mathrm{P}(q, q') = 0$.

A state in PRISM is a valuation for a set of variables over the domain consisting of finitely bound booleans and integers. Global states in Γ are encoded using T finitely bound integer variables ranging over N discrete values. To facilitate the analysis of many different oscillator population models we provide a Python script[1] that allows the user to define ranges for N, T, R, ϵ and μ, for some fixed definitions for evol and pert. Then, given a list of properties, for each combination of parameters the script generates a model, checks all properties using PRISM, and writes user specified output (e.g. result, model checking time, etc.) to a comma separated value file which can be used by statistical analysis tools. Even though models in PRISM can be parametric, the parameters may

---

[1] The model generation script and the results presented in this paper can be found at https://github.com/PaulGainer/mc-bio-synch.

only be used to describe probabilities. Therefore, our generated models can only be parameterised by μ, since changing the value of μ does not result in the loss or addition of any transitions to the model.

5   Evaluation

Within this section we will discuss the properties of two instantiations of the formal model defined in Sect. 3. To that end, we created concrete models for PRISM for different parameters of the models, e.g. number of oscillators and different coupling constants. Each of these models was subsequently checked by PRISM with respect to different properties. Other case studies could also be considered for alternative models of synchronisation where the dynamics of oscillators, and their interactions, can be described by some evolution and perturbation function.

Properties to be checked are specified using *Probabilistic Computation Tree Logic* (PCTL) [12]. PCTL consists of classical logical operators, temporal operators including $\Diamond\varphi$, "at some future point φ holds", and the probabilistic operator $\mathbf{P}_{\bowtie\gamma}[\varphi]$, where $\bowtie$ is a relational operator and $\gamma \in [0, 1]$ is a probability threshold. We can therefore specify properties such as $\mathbf{P}_{\geq 0.1}[\Diamond\varphi]$, "$\varphi$ holds at some future point with a probability of at least 0.1". In addition to assertions, PRISM allows the specification of properties that evaluate to a numerical value, using the syntax $\mathbf{P}_{=?}[\varphi]$, "what is the probability that φ holds?". Furthermore, rewards can be associated with states, and the reachability reward operator $\mathbf{R}$ can be used to calculate expected rewards. For example $\mathbf{R}_{=?}[\Diamond\varphi]$ expresses "what is the expected reward for reaching a state where φ holds?".

We are interested in the probability of eventual synchronisation and in the average time needed to achieve synchronisation. We formalise these properties in PCTL as $\varphi_1 = \mathbf{P}_{=?}[\Diamond \; synchronised]$, and $\varphi_2 = \mathbf{R}_{\{time\_to\_synch\}=?}[\Diamond \; synchronised]$. In these formulas *synchronised* is a name for the formula $\bigvee_{i=1}^{T} n_i = N$ used within the PRISM model, while *time_to_synch* is a reward structure associating a value of $\frac{1}{T}$ with each state where oscillators are configured (i.e., not *init*) and where the system is not synchronised, that records the number of cycles taken to achieve synchrony. As a consequence, the result of model checking with respect to φ_2 is the expected value of the reward *time_to_synch* accumulated along a path until synchrony occurs. Observe that PRISM gives a result of *Infinity* for accumulating *time_to_synch* along a run that does not synchronise. Since PRISM computes the expected value over all paths, this implies that a system with non-synchronising paths will also result in *Infinity* for φ_2.

In the following, we present the model checking results for two instantiations of our model. We will discuss these results and the resulting trade-offs for parameter choices. For a network of sensor nodes, several attributes can be weighted against each other: (i) probability of synchronisation (ii) time for achieving synchrony (iii) battery life of a single oscillator. While we get direct results for the first two properties from PRISM, battery life is dependent on the energy consumption, which can only be estimated from the parameters of the model. In WSNs, communication is costly with respect to energy consumption. Communication is either active when sending a message, i.e., when a node fires, or passive, when receiving messages from other nodes. Hence, during periods where a sensor does neither, the antenna can be shut down to save energy. In our models, this interval of inactivity corresponds to the refractory period. That is, the longer the refractory period is, the less energy will be consumed.

Mirollo and Strogatz Synchronisation Model. Here, we present the results of model checking population models where the perturbation function is a discretisation of the Mirollo and Strogatz (M&S) model of synchronisation used by Perez et al. [18], namely $\text{pert}(\Phi, \alpha, \epsilon) = [\Phi \cdot \alpha \cdot \epsilon]$. Note that, here, the perturbation induced by the firing of another oscillator increases linearly with the phase of the perturbed oscillator. The evolution function is simply the successor function, $\text{evol}(\Phi) = \Phi + 1$. With these functions fixed, we created models for different numbers of oscillators $3 \leq N \leq 7$, cycle lengths $4 \leq T \leq 10$, coupling constants $\epsilon \in \{0, 0.1, \ldots, 1.0\}$, refractory periods $0 \leq R \leq T$, and probabilities of message loss $\mu \in \{0, 0.1, \ldots, 1.0\}$. We used PRISM to analyse models with respect to φ_1 and φ_2.

Figure 2a plots the probability of synchronisation for different rates of broadcast failure against the refractory period for $N = 7$, $T = 10$, and $\epsilon = 0.1$. We can extrapolate a trade-off between a high refractory period and high synchronisation probability. As long as the refractory period is less than half the oscillation cycle, synchronisation will be achieved in almost all cases. Higher values for R result in a rapid drop in synchronisation probability. The exceptions are the edge cases $\mu = 0$ and $\mu = 1$. Unsurprisingly, if all firings result in broadcast failures ($\mu = 1$), the synchronisation probability is almost zero. In fact, the only runs that synchronise in this case are runs whose initial states are synchronised. The comparably bad synchronisation probabilities for $\mu = 0$ may seem surprising. If $\mu = 0$, a model is deterministic. This can lead to unwanted cyclic behaviour, an artefact of the discreteness of the phase values, where very minor perturbations to phase are ignored due to rounding, leading to groups of oscillators staying unsynchronised forever. Similar phenomena have also been observed in other approaches used to model emergent synchronisation [10]. When some level of uncertainty is introduced to the model perpetually asynchronous cycles no longer occur.

Figure 2b shows us that a higher refractory period results in shorter synchronisation times when the probability for broadcast failure is low. In general, a longer refractory period up to half the cycle length improves the rate of convergence to synchrony, which is consistent with the findings of [9]. Furthermore, for high values of μ the differences in synchronisation times for different refractory period lengths are negligible. Hence, a refractory period of slightly less than half the cycle, with a low coupling constant ϵ, is optimal for this model of synchronisation. As ϵ is increased the results remain similar, but with a decrease in synchronisation times.

Mean Phase Synchronisation Model. We now instantiate the evolution and perturbation functions for a model of synchronisation similar to the work of Breza [5]. To that end, we set the evolution function to be the successor function,

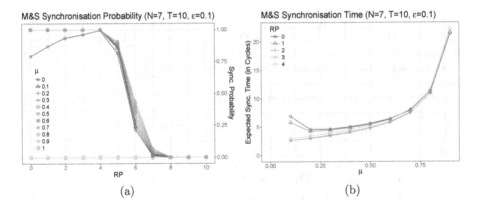

(a)   (b)

Fig. 2. Mirollo & Strogatz synchronisation: synchronisation probabilities for different refractory periods, and synchronisation times for different rates of broadcast failure.

as in the previous section; however the perturbation function is more involved. In Breza's model, an oscillator perturbed by another firing oscillator updates its phase to be the average of its current phase and the phase of the firing oscillator (fixed as T in our model). For this model of synchronisation there is no notion of coupling strength between oscillators, that is, ϵ is ignored. However our general oscillator model can still be instantiated to formalise such a protocol. We derive the following perturbation function: $\text{pert}(\Phi, \alpha, \epsilon) = \left[\frac{1}{2^\alpha}(\Phi + T(2^\alpha - 1))\right] - \Phi$. Informally, the function calculates the result of iteratively taking the mean of the phase and T, for α iterations, and returns the difference between this and the original phase. Note that the perturbation induced by the firing of another oscillator is inversely proportional to 2^Φ.

We generate models for the parameter values examined for the M&S model of synchronisation, and again analyse the models with respect to φ_1 and φ_2. Figure 3a shows the synchronisation probability for different rates of broadcast failure and lengths of refractory period. It has similar characteristics to Fig. 2a.

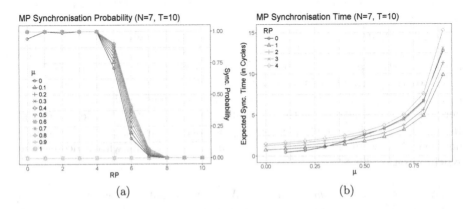

(a)   (b)

Fig. 3. Mean phase synchronisation: synchronisation probabilities for different refractory periods, and synchronisation times for different rates of broadcast failure.

That is, for almost all cases of μ, the oscillators will always synchronise when the refractory period is less than half the cycle. Again as expected, $\mu = 1$ results in almost no synchronising runs, and $\mu = 0$ creates cyclic behaviour that leads to perpetual asynchrony. We can see that the Mean Phase (MP) synchronisation model is slightly more robust in this case, than a loosely coupled oscillator with the M&S synchronisation model. If we increase the coupling strength of the latter, however, it performs even better.

We now consider the time required to achieve synchronisation. Figure 3b shows that, in most cases, a short (but non-zero) refractory period results in shorter synchronisation times. In general, it therefore seems optimal to choose a short, non-zero length refractory period. For low broadcast failure probabilities, however, there are negligible differences for refractory periods of different lengths. If we expect robust communication for a deployed network then we should choose a longer refractory period and so conserve energy.

Network Synchronisation Scalability. Figure 4a and b plot the synchronisation time against the population size for different rates of broadcast failure, for the M&S and Mean Phase synchronisation models respectively. For the M&S model we see that when $\mu > 0.3$, increasing the population size results in shorter synchronisation times, while a higher rate of broadcast failure yields longer synchronisation times. We conjecture that the surprising peaks for $\mu \leq 0.3$ are again an artefact of the rounding, resulting in cyclic behaviour, similar to that observed in deterministic models, as discussed for the M&S model. For the Mean Phase synchronisation model, we can again observe that a higher rate of broadcast failure yields longer synchronisation times. Similarly, increasing the population size results in shorter synchronisation times. However, in this case the rate at which synchronisation time decreases, given an increase in the size of the population, is more pronounced. Unlike the M&S model there are no peaks in the graphs indicating undesirable asynchronous cyclic behaviour. For the M&S model we

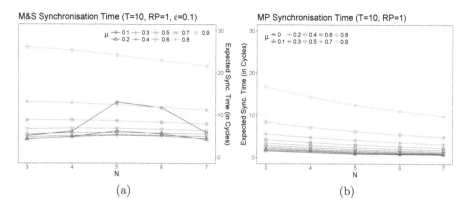

(a)   (b)

Fig. 4. Synchronisation times for different number of oscillators for Mirollo & Strogatz synchronisation (left), and Mean Phase synchronisation (right).

observed that low coupling strength resulted in minor perturbations to phase being ignored due to rounding. In the Mean Phase model this does not occur, since the fractional part of the calculated mean phase is always ≥ 0.5.

Model Checking Scalability. Using formal population models to analyse networks of indistinguishable oscillators is a promising approach. We checked networks with up to 7 oscillators, while to the best of our knowledge, other formal analyses using model checking turned out to be infeasible for more than four nodes for fully connected networks [13]. Memory and time used for model checking and construction of a single model, resp., are shown in Table 2. The increase in memory usage is as expected, and differences

Table 2. Memory for model checking (in MB) and time for model construction (in seconds), for $T = 10$, $RP = 1$, $\epsilon = 0.1$, $\mu = 0.1$.

N	M&S Mem.	M&S Time	MP Mem.	MP Time
3	124.63	0.09	131.30	0.09
4	161.33	0.37	162.42	0.43
5	262.62	1.65	261.39	1.61
6	592.94	5.28	610.20	5.42
7	1604.76	17.13	1495.59	16.88

between the two models are relatively small. The properties can be checked in under a second. While our approach allows us to postpone the state space explosion problem, we cannot escape it completely. The major bottleneck is not the model checking time itself, but rather the model construction time. For individual models this was relatively short, but greatly accumulated for the parameter combinations we investigated, where thousands of models were constructed.

6   Conclusion

In this paper we presented a formal general model for networks of pulse-coupled oscillators, whose oscillation cycles are defined as a sequence of discrete states. We instantiated the general model for two different models of synchronisation used for the coordination of wireless sensor networks and swarm robotic systems. For each instantiation, and for a range of different values for model parameters, we automatically generated input for the probabilistic model checker PRISM, encoded as a discrete time Markov chain. Finally, we used the results of model checking to analyse parametric influence on both the rate at which synchronisation occurs, and the time taken for it to occur; in particular, we discussed the trade-offs for parameter choices to minimise energy consumption in a network.

For future work, we intend to extend our current binary notion of synchronisation by introducing a metric, in the form of a reward structure, allowing us to reason about different degrees of synchronisation for global states. We also intend to formally encode energy consumption reward structures that will allow us to obtain quantitative results for those we reasoned about informally in this paper. A population model is appropriate when nodes are indistinguishable and the network is fully coupled. To analyse other network topologies, for instance a network of fully connected subcomponents, we could encode each such subcomponent as a single population model, and take the cross product of the models for all subcomponents. To accomplish this it is likely that we would need to further refine our model, as this would greatly increase the state space.

References

1. Alur, R., Dill, D.L.: A theory of timed automata. TCS **126**(2), 183–235 (1994)
2. Amin, S., Elahi, A., Saghar, K., Mehmood, F.: Formal modelling and verification approach for improving probabilistic behaviour of robot swarms. In: Proceedings of IBCAST 2017, pp. 392–400. IEEE (2017)
3. Bartocci, E., Corradini, F., Merelli, E., Tesei, L.: Detecting synchronisation of biological oscillators by model checking. TCS **411**(20), 1999–2018 (2010)
4. Bojic, I., Podobnik, V., Ljubi, I., Jezic, G., Kusek, M.: A self-optimizing mobile network: auto-tuning the network with firefly-synchronized agents. Inf. Sci. **182**(1), 77–92 (2012)
5. Breza, M.: Bio-inspired tools for a distributed wireless sensor network operating system. Ph.D. thesis, Imperial College London, March 2013
6. Buck, J.: Synchronous rhythmic flashing of fireflies. II. Q. Rev. Biol. **63**(3), 265–289 (1988)
7. Bucur, D., Kwiatkowska, M.: On software verification for sensor nodes. J. Syst. Softw. **84**(10), 1693–1707 (2011)
8. Christensen, A.L., Grady, R.O., Dorigo, M.: From fireflies to fault-tolerant swarms of robots. IEEE Trans. Evol. Comput. **13**(4), 754–766 (2009)
9. Degesys, J., Basu, P., Redi, J.: Synchronization of strongly pulse-coupled oscillators with refractory periods and random medium access. In: Proceedings of SAC 2008, pp. 1976–1980. ACM (2008)
10. Fatès, N.: Remarks on the cellular automaton global synchronisation problem. In: Kari, J. (ed.) AUTOMATA 2015. LNCS, vol. 9099, pp. 113–126. Springer, Heidelberg (2015). doi:10.1007/978-3-662-47221-7_9
11. Gainer, P., Dixon, C., Hustadt, U.: Probabilistic model checking of ant-based positionless swarming. In: Alboul, L., Damian, D., Aitken, J.M.M. (eds.) TAROS 2016. LNCS (LNAI), vol. 9716, pp. 127–138. Springer, Cham (2016). doi:10.1007/978-3-319-40379-3_13
12. Hansson, H., Jonsson, B.: A logic for reasoning about time and reliability. FAC **6**(5), 512–535 (1994)
13. Heidarian, F., Schmaltz, J., Vaandrager, F.: Analysis of a clock synchronization protocol for wireless sensor networks. TCS **413**(1), 87–105 (2012)
14. Konur, S., Dixon, C., Fisher, M.: Analysing robot swarm behaviour via probabilistic model checking. Robot. Auton. Syst. **60**(2), 199–213 (2012)
15. Kuramoto, Y.: Collective synchronization of pulse-coupled oscillators and excitable units. Physica D: Nonlin. Phenom. **50**(1), 15–30 (1991)
16. Kwiatkowska, M., Norman, G., Parker, D.: PRISM 4.0: verification of probabilistic real-time systems. In: Gopalakrishnan, G., Qadeer, S. (eds.) CAV 2011. LNCS, vol. 6806, pp. 585–591. Springer, Heidelberg (2011). doi:10.1007/978-3-642-22110-1_47
17. Mirollo, R.E., Strogatz, S.H.: Synchronization of pulse-coupled biological oscillators. SIAM J. App. Math. **50**(6), 1645–1662 (1990)
18. Perez-Diaz, F., Trenkwalder, S., Zillmer, R., Gross, R.: Emergence and inhibition of synchronization in robot swarms. In: DARS 2016. Springer Tracts in Advanced Robotics. Springer, Cham (in press)
19. Werner-Allen, G., Tewari, G., Patel, A., Welsh, M., Nagpal, R.: Firefly-inspired sensor network synchronicity with realistic radio effects. In: Proceedings of SenSys 2005, pp. 142–153. ACM (2005)

Machine Learning and Formal Methods
(Special Session)

Statistical Abstraction for Multi-scale Spatio-Temporal Systems

Michalis Michaelides[1(✉)], Jane Hillston[1], and Guido Sanguinetti[1,2]

[1] School of Informatics, University of Edinburgh, Edinburgh, UK
mic.michaelides@ed.ac.uk
[2] SynthSys, Centre for Synthetic and Systems Biology,
University of Edinburgh, Edinburgh, UK

Abstract. Spatio-temporal systems exhibiting multi-scale behaviour are common in applications ranging from cyber-physical systems to systems biology, yet they present formidable challenges for computational modelling and analysis. Here we consider a prototypic scenario where spatially distributed agents decide their movement based on external inputs and a fast-equilibrating internal computation. We propose a generally applicable strategy based on statistically abstracting the internal system using Gaussian Processes, a powerful class of non-parametric regression techniques from Bayesian Machine Learning. We show on a running example of bacterial chemotaxis that this approach leads to accurate and much faster simulations in a variety of scenarios.

1 Introduction

Modelling spatially extended dynamical systems is a task of central importance in science and engineering. Examples range from cyber-physical systems, to collective adaptive systems of human behaviour, to cellular systems. Despite their importance, computational modelling and analysis of such systems remains challenging due to a number of factors: the large number of degrees of freedom, the intrinsically hybrid nature of discrete systems existing in continuous space, and, frequently, the existence of multiple temporal scales in the system. As a result of these features, computational simulation of such systems is generally onerous, particularly in a stochastic setting [4,6].

In this paper, we consider the scenario where the system consists of multiple, spatially distributed, identical agents. The agents can sense an external, deterministic field and use this information to perform a stochastic, internal computation which determines the agent's subsequent move. The internal computation is often a system which will quickly reach a steady-state equilibrium when left unperturbed, e.g. a chemical reaction network. While this scenario is a special case as the agents do not interact with each other, it is sufficiently generic

M. Michaelides and G. Sanguinetti are supported by the European Research Council under grant MLCS 306999. J. Hillston is supported by the EU project, QUANTICOL 600708.

© Springer International Publishing AG 2017
N. Bertrand and L. Bortolussi (Eds.): QEST 2017, LNCS 10503, pp. 243–258, 2017.
DOI: 10.1007/978-3-319-66335-7_15

to cover many application scenarios, such as autonomous drones performing a task in space, or bacteria exploring a nutrient field. Such systems are cumbersome to handle computationally as the simulation of the internal computation needs to be repeated at every spatial step, so that simulating a single trajectory of the overall system may involve hundreds of simulations of the internal model.

Here we propose a novel approach to alleviate this computational burden based on emulating the statistics of the internal system. The central idea is to replace the expensive computation of the internal system with a lookup table which maps external stimulus to the output behaviour of the internal system. Crucially, we do not aim to model the detail of the internal state, but only an abstracted version capturing its qualitative behaviour (formalised as a logical property satisfied by the states). We achieve this by learning a parameters-to-behaviours regression map using Gaussian Processes (GPs), a powerful class of non-parametric Bayesian regression models. Our work is motivated by earlier work on using GPs to learn effective characterisations of system behaviour [1, 2, 11].

The rest of the paper is organised as follows: background on spatio-temporal systems and *E. coli* chemotaxis which serves as a running example (Sect. 2); the general framework for our statistical abstraction methodology, and its application to the chemotaxis system (Sect. 3); results assessing the quality and efficiency of the abstraction (Sect. 4); closing remarks about prospective expansion of the work (Sect. 5).

2   Background

2.1   Spatio-Temporal Agent Models

We start by defining the class of spatio-temporal agent models we will consider in this paper. Let $\mathcal{D}$ be a spatial domain (usually a compact subset of $\mathbb{R}^n$ with $n = 2, 3$), and let $[0, T]$ be the temporal interval of interest. We define the *spatio-temporal field* $f \colon \mathcal{D} \times [0, T] \to \mathbb{R}$ to be a real-valued function defined on the spatial and temporal domains of interest. A spatio-temporal agent model is a triple $(\mathcal{D}, f, \mathcal{A})$ where $\mathcal{A}$ is a collection of point *agents* whose location follows a stochastic process which depends on the spatio-temporal field. We note that this is not the most general case, as agents may be spatially extended, interact with each other or even influence the evolution of the spatio-temporal field. Nevertheless, such a level of abstraction is frequently adopted and justifiable in many practical applications.

Running Example: Chemotaxis in the Escherichia coli *Bacterium.* Foraging is a central problem for microbial populations. The bacterium *Escherichia coli* will normally perform a random walk within a spatial domain where nutrient concentration is constant (e.g. a Petri dish). When presented with a spatially varying nutrient field, a phenomenon known as *chemotaxis* arises. As the bacterium performs a random walk in the nutrient field encountering changing nutrient levels, its sensory pathway effectively evaluates a temporal gradient of the nutrients (or ligands) it experiences; the walk is biased so that the bacterium experiences a positive temporal gradient more often than not [18,20]. Since the bacterium is moving in the field, the temporal gradient is implicitly translated into a spatial one, so the bacterium drifts toward advantageous concentrations. Implicitly translating a temporal gradient to a spatial one through motion is necessary for the bacterium cell, because its body size is too small to allow for effective calculation of the spatial gradient of a chemical field at its location. As a result, we can safely regard the bacteria as point-like agents.

2.2 Multi-scale Models

In many practical situations, one is interested in modelling not only the movement of the agents, but also the mechanism through which sensing and decision making is carried out within each agent. This naturally leads to structured models with distinct layers of organisation, with behaviour in each layer informing the simulation that takes place at the layer above or below. We will assume that the internal workings of the agent are also stochastic, and we will model them as a *population Continuous Time Markov Chain* (pCTMC) [1]. Formally, a pCTMC is defined as follows.

Definition 1. *A population CTMC is a continuous-time Markov chain [12] with a discrete state-space $\mathcal{X}$, and an associated transition rate matrix Q. Each state in $\mathcal{X}$ counts the number of entities of each type or "species" in a population, $\mathbf{X} \in \{\mathbb{N}^0\}^d$ for d species. Transitions in this space occur according to the rates given by Q.*

The transitions can be regarded as occurrences of chemical reactions, written as

$$\sum_{i=1}^{d} r_i X_i \xrightarrow{\tau(\mathbf{X})} \sum_{i=1}^{d} s_i X_i, \tag{1}$$

where for every species X_i, r_i particles of X_i are consumed and s_i particles are created. The transition rate $\tau(\mathbf{X})$ depends upon the current state of the system, and is the rate parameter of an exponential distribution governing the waiting times for this transition. The above transition rates of allowed reactions reconstruct the rate matrix Q.

Motor Control in E. coli. An *E. coli* cell achieves motility by operating *multiple* flagellum/motor pairs (F/M), which can either drive it straight (subject to small Brownian perturbation), or rotate it in place. Thus, the cell can either be 'tumbling' (re-orienting itself while stationary) or 'running' (propelling itself forward while maintaining direction) at any time (Fig. 1: left, centre). The motility state, RUN/TUMBLE, of the cell is determined by the number of flagella found in particular conformations. The model in [15] suggests three possible conformations for a flagellum: *curly* (C), *semicoiled* (S) and *normal* (N). The associated

---

[1] The pCTMC is the internal model for a *single* agent here, not for multiple agents.

motor is modelled as a stochastic bistable system, which rotates either clockwise (CW) or counter-clockwise (CCW). Changes in motor rotation induce conformational changes on the associated flagellum. Transition rates between motor states are given by rate parameters k_+ and k_- for transitions CW → CCW and CCW → CW, respectively. The possible transitions between flagellum/motor states are summarised in the schematic diagram in Fig. 1: right. *E. coli* normally has of the order of ten flagella and associated motors; the dynamics of the pair flagellum/motor population therefore lends itself to be easily described as a pCTMC. The $k_\pm$ transition rates depend on the temporal gradient evaluated by the chemotaxis pathway, and represent the functional interface of the bacterium with its external environment.

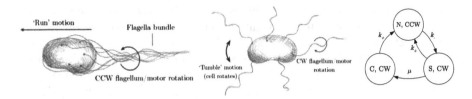

Fig. 1. The two motility modes of an *E. coli* cell. Left: the F/M are in CCW conformations, forming a helical bundle and propelling the cell. Centre: the F/M are in CW conformations, breaking the bundle apart and causing the cell to re-orient in place. Right: CTMC for a single F/M, with three conformation states and transition rates $k_\pm(m, L)$ and fixed $\mu = 5\,\mathrm{s}^{-1}$.

The classical mathematical model for the sensory response of the cell to external ligand concentration changes is provided by the Monod-Wyman-Changeux (MWC) model [9,15,17]. The model considers sensor clusters which signal information about ligand concentration changes to the motors, by triggering a biochemical response in the cell (phosphorylation of the CheY protein which binds to the motors) affecting the switching rates of rotation direction, $k_\pm$.

The full MWC model is still highly complex; in practice, we follow [15] and adopt a simplified model of sensory response to describe the dependency of motor rates $k_\pm$ on ligand concentrations. This consists in abstracting the CheY signalling pathway in an effective variable m, which represents the methylation state of the ligand receptors and whose stochastic evolution is dependent on the ligand concentration L. Since m depends on past L concentrations the cell has been in, one may think of it as a *chemical memory* of sorts which encodes the value of L at previous times. The time comparison window is determined by how fast methylation happens—faster methylation leads to a shorter memory.

Sneddon et al. [15] then resolve the entire dependency chain of the chemotaxis pathway to Eqs. 2 and 3. The motor switching rates $k_\pm(m, L)$ are given by the deterministic equation

$$k_\pm = \omega \cdot \exp\left\{ \pm \left[\frac{g_0}{4} - \frac{g_1}{2}\left(\frac{Y_p(m, L)}{Y_p(m, L) + K_D} \right) \right] \right\}, \tag{2}$$

where

$$Y_p(m, L) = \alpha \cdot \left[1 + e^{\epsilon_0 + \epsilon_1 m} \cdot \left(\frac{1 + L/K_{\text{TAR}}^{\text{off}}}{1 + L/K_{\text{TAR}}^{\text{on}}} \right)^{n_{\text{TAR}}} \cdot \left(\frac{1 + L/K_{\text{TSR}}^{\text{off}}}{1 + L/K_{\text{TAR}}^{\text{on}}} \right)^{n_{\text{TSR}}} \right]^{-1} \cdot$$

The methylation process can be naturally modelled as a birth/death process with rates depending on ligand concentration; again following [15] we take a fluid approximation of this, yielding the Ornstein-Uhlenbeck (OU) process:

$$\frac{dm}{dt} = -\frac{1}{\tau}(m - m_0(L)) + \eta_m(t). \tag{3}$$

In the above stochastic differential equation (SDE), $\eta_m = \sigma_m \sqrt{2/\tau} \Gamma(t)$, $\Gamma(t)$ is the normally distributed random process with 0 mean and unit variance, σ_m is the standard deviation of fluctuations in the methylation level, and $m_0(L)$ is an empirically derived function whose output is the methylation level required for full adaptation at the current external ligand concentration L. The adaptation rate τ, determines how fast methylation occurs and so, how long the 'chemical memory' of previous L values is in the system. The constants τ, along with mb_0 and α involved in the $m_0(L)$ function (see [15]), fully parametrise the methylation evolution. See [19] for reported values of constants used in Eq. 2 and [5,15] for a detailed derivation of the results. Equations 2 and 3 couple the transition rates of the pCTMC in Fig. 1: Right, with the external ligand concentrations, and therefore fully describe the internal model of the *E. coli* chemotactic response.

2.3   Simulating Multi-scale Systems

Multi-scale spatio-temporal systems are in general amenable to analytical techniques only in the simplest of cases. For the vast majority of real-world models, simulation-based analysis is the only option to gain behavioural insights.

Simulation of spatio-temporal systems typically employs nested algorithms: having chosen a time-discretisation for the spatial motion (which is assumed to have the slower time-scale), a spatial step is taken. Then, the value of the external field is updated, and the internal model is run for the duration of a given time-step with the new rates (corresponding to the updated value of the external field). A sample from the resulting state distribution then determines the velocity of the agent for the next time-step.

Clearly, this iterative procedure, while asymptotically exact (in the limit of small time discretisation), is computationally very demanding. This has motivated several lines of research in recent years [1,8,10,13].

Simulating Chemotaxis in E. coli. Simulations of the *E. coli* model outlined previously proceed along the general lines discussed above. Given a value of the ligand field and a characteristic time-step Δt, we draw samples of the SDE (3) using the Euler-Maruyama method, a standard method for simulating SDEs.

In the F/M pCTMC system in the reaction equation style, each species represents a different F/M conformation for a total of three species. The following transitions occur:

$$(S\_CW) \xrightarrow{\mu} (C\_CW), \quad (C\_CW) \xrightarrow{k_+} (N\_CCW),$$
$$(S\_CW) \xrightarrow{k_+} (N\_CCW), \quad (N\_CCW) \xrightarrow{k_-} (S\_CW). \tag{4}$$

Note that in the above rate transitions there are dependencies on both external (L) and internal (m) states: $k_\pm(m, L)$, where L is an external input to the system (the external chemoattractant concentration at the time) and m is the current methylation level (sampled from the OU process in Eq. 3 every Δt). Instead, the rate transition for $(S\_CW) \rightarrow (C\_CW)$ is fixed, $\mu = 5\,\text{s}^{-1}$.

Using the exact Gillespie algorithm [7], we then simulate the internal pCTMC for a length of time Δt to draw a sample configuration of the flagella/motor system. Formally, trajectories of length Δt are checked against a property specifying the motility state for the cell (RUN/TUMBLE),

$$\phi_{\text{RUN}}(\mathbf{s}) = (N \geq 2) \wedge (S = 0), \tag{5}$$

where $\mathbf{s} = (S, C, N)$ is the last state of the flagella/motor pairs in the CTMC trajectory.

The spatial location of the bacterium is then updated according to a simple rule: if the sampled internal state corresponds to RUN, the agent moves rectilinearly and updates its position $\mathbf{r} \leftarrow \mathbf{r} + \mathbf{v} \cdot \Delta t$, where $v = 20\mu\text{m/s}$, the speed of the bacterium. Otherwise, if the internal state corresponds to TUMBLE, the agent remains still and its velocity is updated $\mathbf{v} \leftarrow R(\theta) \cdot \mathbf{v}$, where $R(\theta)$ is the standard 2D unitary rotation matrix through an angle θ, and θ is a tumbling angle sampled from a Gamma distribution as reported in Sneddon et al. [15].

The above simulation scheme, outlined in Algorithm 1, produces a chemotactic response to a ligand gradient. It takes ~270 s to simulate a single cell trajectory of $t_{\text{end}} = 500\,\text{s}$ with a time-step $\Delta t = 0.05$.

3 Methodology for Statistical Abstraction

In a multi-scale system, output from a set of processes in one layer in the system is passed as input to another layer; these processes are often computationally expensive. We present a methodology to abstract away such a set of processes and replace them with a more efficient stochastic map from the input to the output, governed by an underlying probability function. We approximate this probability function using Gaussian processes after observing many input-output pairs from the processes to be abstracted. The output consists of truth evaluations of properties expressed in logical formulae, which capture some behaviour of the system that is to be preserved by the abstraction.

Algorithm 1. Simulation scheme for the *E. coli* model, based on full simulation of the pCTMC describing F/M conformation changes. Below, Δt is the fixed simulation time-step. Smaller functions called here (RUN, TUMBLE, OU-EULER-MARUYAMA) can be found in the full version of the algorithm, arXiv:1706.07005.

1: **procedure** SIMULATEFINEECOLICELL(t_{end})
2:    $t \leftarrow 0$
3:    **while** $t < t_{\mathrm{end}}$ **do**
4:       $L \leftarrow L(\boldsymbol{r}, t)$ ▷ The ligand field L value, at the cell's location $\boldsymbol{r}$.
5:       $\mathbf{s} \leftarrow \mathrm{PCTMC}(\mathbf{s}, m, L, \Delta t)$ ▷ Drawing F/M pCTMC trajectory of length Δt, with parameters $k_{\pm}(m, L)$ and initial state the last pCTMC state of the cell.
6:       $\psi \leftarrow \phi_{\mathrm{RUN}}(\mathbf{s})$ ▷ Evaluating the ϕ_{RUN} on (the final state of) the pCTMC trajectory.
7:       **if** ψ **then**
8:          $\boldsymbol{r} \leftarrow \mathrm{RUN}(\boldsymbol{r}, \boldsymbol{v}, \Delta t)$
9:       **else**
10:          $\boldsymbol{v} \leftarrow \mathrm{TUMBLE}(\boldsymbol{v}, \Delta t)$
11:       **end if**
12:       $m \leftarrow \mathrm{OU\text{-}EULER\text{-}MARUYAMA}(m, L, \Delta t)$ ▷ Evolving methylation.
13:       $t \leftarrow t + \Delta t$
14:    **end while**
15: **end procedure**

3.1   Statistical Abstraction Framework

Consider a CTMC S, which given an initial state $\mathbf{s}_0$, running time t, and input $\mathbf{q}$ which completely determines transition rates, generates a trajectory $\mathbf{s}_{[0,t]}$. The trajectory is then checked for satisfaction of a property resulting in output $y = f(\mathbf{s}_{[0,t]}), y \in \{\top, \bot\}$. This layer of the multi-scale system can therefore be described as a set of operations:

$$S(\mathbf{s}_0, t, \mathbf{q}) = \mathbf{s}_{[0,t]}; \tag{6}$$
$$f(\mathbf{s}_{[0,t]}) = y. \tag{7}$$

Note that we consider a single property here for simplicity so a single binary value, but one could generalise to multiple properties, and hence, multi-valued output. This output then becomes input to a higher layer in the multi-scale system.

Our goal is to construct a system $\tilde{S}$ that is cheaper to simulate, whose output will be consistent with the original system S. Since the system is stochastic, in this context *consistent* refers to having the same probability distribution for the output random variable y. This abstracted system should generate output y' given the last output of the system y and the same input k as before:

$$\tilde{S}(y, \mathbf{q}) = y'. \tag{8}$$

Replacing the initial state s_0 input with the previous output y allows us to substitute the whole layer of fine operations (6, 7) with the cheaper abstracted system $\tilde{S}$ (8), in the multi-scale system. We regard this abstracted system to be a stochastic map from the internal state of the system and some given input to an output; we then use Gaussian processes to estimate the underlying probability function $\Psi(y, \mathbf{q})$ which governs the output of this stochastic map over the input domain.

Abstracting the E. coli *Chemotaxis Pathway.* Returning to our model of the *E. coli* chemotaxis pathway, we associate the original system S with the pCTMC system of F/M conformations (Eq. 4), along with the OU methylation process in Eq. 3. The input starting state s_0 is the last F/M state of the pCTMC, and the last methylation level m. The simulation time T is the variable Δt from Sect. 2.1, also used for the integration step-size of the OU in the Euler-Maruyama scheme. The transition rates $k_\pm$ are calculated using the variables m and L, the last methylation level and external ligand concentration at the position of the cell, respectively. The output of this system, s_t, is then a sampled pCTMC trajectory and new methylation level. Finally, the run property (5) is evaluated on (the last state of) the drawn pCTMC trajectory and the output determines whether the cell 'runs' or 'tumbles'.

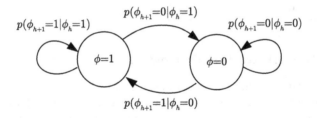

Fig. 2. DTMC with two states, $\phi_{\mathrm{RUN}} \in \{\top, \bot\}$. The transition probabilities depend on internal methylation level m and external ligand concentration L.

In observing the truth value of property ϕ_{RUN} for the state of the pCTMC at regular intervals of Δt, we cast the original pCTMC model (S) into a DTMC (Fig. 2). This DTMC has only two states, $\phi_{\mathrm{RUN}} \in \{\top, \bot\}$, and transition probabilities depending on the transition rates $k_\pm$, μ, of the original pCTMC.

Since this is only a two-state DTMC, the state at the next time-step conditioned on the current one can be modelled as a Bernoulli random variable:

$$\phi' \mid \phi \sim \mathrm{Bernoulli}(p = p_{\phi'=1|\phi}(m, L)), \qquad (9)$$

where ϕ, ϕ' are the ϕ_{RUN} DTMC states at time-steps h, $h+1$ respectively. Also, the boolean $\{\bot, \top\}$ truth values of the properties have been mapped to the standard corresponding integers $\{0, 1\}$ for mathematical ease.

We recognise that a *single step* transition of this DTMC ($\phi' \mid \phi, m, L$) is the output $y' \mid y, \mathbf{q}$ produced by the abstracted layer $\tilde{S}(y, \mathbf{q})$. Identifying the

---

Algorithm 2. Simulation scheme for the abstracted *E. coli* model, based on GP approximation for the RUN/TUMBLE probability. Steps 5, 6 here replace the expensive Steps 5, 6 in Algorithm 1.

1: **procedure** SIMULATEABSTRACTEDECOLICELL(t_{end})
2:     $t \leftarrow 0$
3:     **while** $t < t_{end}$ **do**
4:         $L \leftarrow L(\boldsymbol{r}, t)$
5:         $p \leftarrow GP_{\psi}(m, L)$
6:         $\psi \sim \text{Bernoulli}(p)$
7:         **if** ψ **then**
8:             $\boldsymbol{r} \leftarrow \text{RUN}(\boldsymbol{r}, \boldsymbol{v}, \Delta t)$
9:         **else**
10:            $\boldsymbol{v} \leftarrow \text{TUMBLE}(\boldsymbol{v}, \Delta t)$
11:        **end if**
12:        $m \leftarrow \text{OU-EULER-MARUYAMA}(m, L, \Delta t)$
13:        $t \leftarrow t + \Delta t$
14:    **end while**
15: **end procedure**

---

corresponding probability function $p_{\phi'=1|\phi}(m, L)$ as the underlying governing function $\Psi(y, \mathbf{q})$ completes the setting of *E. coli* chemotaxis model abstraction to the methodology framework given above (Sect. 3.1). Note that the OU process for methylation is retained in the abstracted model as a parallel running process in the same layer of the multi-scale system. The OU process output m, together with the ligand concentration L (output of a different layer in the multi-scale system), constitute the input $\mathbf{q}$. The altered simulation scheme for this abstracted model is outlined in Algorithm 2. Notice how Steps 5, 6 there replace the more expensive Steps 5, 6 in Algorithm 1.

3.2   Approximating the Underlying Probability Function Ψ

We use Gaussian process (GP) regression in order to infer the underlying probability function $\Psi(y, \mathbf{q})$ governing the stochastic $\tilde{S}$ mapping from internal state y and input k to output y' in Sect. 3.1. A GP models a normally distributed stochastic variable over a continuous domain. It can be thought of as a multivariate normal distribution over functions. This multivariate normal distribution can be conditioned on a finite number of (potentially noisy) observations of the function to be inferred, learning new mean and covariance parameters. These are computable at any point in the domain and correspond to the expected value of the function and associated variance at that point, respectively.

GPs are universal function approximators. The choice of covariance kernel determines the prior over the function and thus how many observations are required to get a good estimate of the underlying function. However, given enough observations, a GP with any valid kernel will approximate any smooth function arbitrarily well. We refer to [14] for a more comprehensive account of GPs.

Since training observations are binary samples of a Bernoulli distribution but GPs regress over a continuous unbounded variable, some adjustments must be made for correct evaluation of the underlying probability function Ψ. These are explained in the Gaussian process classification (GPC) method outlined in [14], and amount to identifying that class probability function with Ψ. We use Minka's Expectation-Propagation (EP) technique to approximate the posterior because it is more accurate than the Laplace approximation. Further, we use fully independent training conditional (FITC) approximation [16] to allow a large number of observations to be considered for learning the underlying function, while maintaining a low cost of predicting at any point of the domain. Note that the Bernoulli distribution likelihood, used here for GPC, is a special case result because of both the binary $y = \phi$ output and the single observation of transitions at a particular (m, L) parametrisation.[2] Lifting these restrictions would result in the more general multinomial distribution.

Constructing Ψ in E. coli *Chemotaxis.* As we mentioned, a single DTMC transition $(\phi' \mid \phi, m, L)$ corresponds to the output $y' \mid y, \mathbf{q}$ produced by the stochastic mapping $\tilde{S}(y, \mathbf{q})$. Therefore, $\tilde{S}(\phi, (m, L))$ consists of sampling from a Bernoulli distribution Bernoulli$(p = p_{\phi'=1|\phi}(m, L))$ where $p_{\phi'=1|\phi}(m, L)$ is the underlying probability function $\Psi(y = \phi, \mathbf{q} = (m, L))$ in the general formalism. We approximate $\Psi(y, \mathbf{q}) = p_{\phi'=1|\phi}(m, L)$, using GPs trained on observations from *micro-trajectories*, i.e. trajectories of the fine F/M pCTMC system which are then mapped onto the property space, $\phi \in \{0, 1\}$, to serve as training data.

Therefore, at a given (m, L) the pCTMC with transition rates $k_{\pm}(m, L)$ is at a state $\mathbf{s}_0$ which maps onto $\phi(\mathbf{s}_0)$. After a time Δt, the same CTMC is found at a state $\mathbf{s}_{\Delta t}$, which maps onto $\phi(\mathbf{s}_{\Delta t})$. An observation $\phi(\mathbf{s}_{\Delta t}) \mid \phi(\mathbf{s}_0), m, L$ is in this way recorded for every parametrisation (m, L) the bacterium has visited in the micro-trajectories.

Since the output of $\tilde{S}$ is binary $(y = \phi \in \{0, 1\})$ we construct two probability functions $\Psi_\phi(m, L) = p_{\phi'=1|\phi}(m, L)$. Each is approximated with a separate GPC function, where $\Psi_0(m, L)$ is trained on observations of transitions originating from the 'TUMBLE' state $(p_{\phi'=1|\phi=0}(m, L))$ and $\Psi_1(m, L)$ using transitions from the 'RUN' state $(p_{\phi'=1|\phi=1}(m, L))$. Notice that we need not estimate separate functions for $\phi' = \{0, 1\}$, since $p_{\phi'=1|\phi}(m, L) = 1 - p_{\phi'=0|\phi}(m, L)$. Having access to these underlying probability functions we are now able to sample the DTMC at any parametrisation (m, L) the bacterium finds itself in, by using the function estimate for $p_\phi(m, L)$ despite not having observations at that m, L.

The function $p_{\phi'=1|\phi}(m, L)$ is particularly challenging for GPs. This is due to a sharp boundary in the m, L domain, where there is a transition from $p_{\phi'=1|\phi}(m, L) \approx 0$ to $p_{\phi'=1|\phi}(m, L) \approx 1$. The bacterium has a steady state very close to this boundary, determined by the motor bias mb_0, and that is where they are most often found. Therefore, accurate estimation of this boundary is crucial for this problem. Furthermore, the low probability of finding bacteria away from

---

[2] It is highly unlikely to have more than a single transition since (m, L) are continuous values that constantly change for the bacterium.

the boundary (in a relatively smooth ligand field) gives a very narrow window of where the function is observed. To get a better overall estimate, we sporadically perturb the position of bacteria in the micro-trajectory phase of collecting observations, such that the bacterium finds itself producing observations away from the boundary for a while, before the system returns close to steady state again. Despite these difficulties, we produce a good reconstruction of the underlying functions $p_{\phi'=1|\phi=0}$ and $p_{\phi'=1|\phi=1}$ over the m, L domain (see Fig. 3).

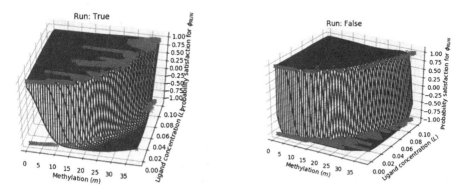

Fig. 3. The probability functions $p_{\phi'=1|\phi}(m, L)$ (left: $\phi = \top$, right: $\phi = \bot$) produced by the GP with hyperparameters $\ln(\ell) = (3.5, -2.5)$ and $\ln(\sigma) = 5$, 100 inducing points (FITC approximation), and 10000 observations (red crosses). The steep boundary is accurately captured producing a sharp switch-like transition from the run domain to the tumble domain. (Color figure online)

4 Results

When assessing performance of our method for statistical abstraction, there are two things of interest: accuracy and computational savings. Accuracy refers to how similar behaviour of the abstracted system is to the behaviour of the original system. In our case of chemotaxis in *E. coli*, this is seen by comparing population distributions in a ligand field, resulting from simulations using the original fine system and the abstracted one. We also compare run and tumble duration distributions as another metric of how closely we approximate the output and behaviour of the original model.

Learning the transition probability functions for the dual-state DTMC enabled us to simulate bacteria using our abstracted model on a host of different ligand field profiles. Beyond comparing bacteria population distributions under the original Gaussian ligand field used for learning (see L_1 below), we did the same for a linear and dynamic field (L_2, L_3 below), using the same learned functions $p_{\phi'=1|\phi}(m, L)$, $\phi \in \{0, 1\}$.

The ligand fields tested were:

$$L_1(\boldsymbol{r}) = 0.1 \cdot \exp\left[-0.5(\boldsymbol{r}^\top \boldsymbol{\Sigma}^{-1} \boldsymbol{r})\right], \qquad\qquad \boldsymbol{\Sigma} = 3 \cdot \mathbf{I}_2; \qquad\qquad (10)$$

$$L_2(\boldsymbol{r}) = \max\left(10^{-5},\ 0.1 - 0.05\sqrt{(\mathbf{A}\boldsymbol{r})^\top \mathbf{A}\boldsymbol{r}}\right), \quad \mathbf{A} = \begin{pmatrix} 1/5 & 0 \\ 0 & 1/2 \end{pmatrix}; \qquad (11)$$

$$L_3(\boldsymbol{r}, t) = 0.1 \cdot \exp\left[-0.5(\boldsymbol{r}^\top \boldsymbol{\Sigma}(t)^{-1} \boldsymbol{r})\right], \qquad\qquad \boldsymbol{\Sigma}(t) = 3(t/50 + 1) \cdot \mathbf{I}_2. \quad (12)$$

In the fields above, the maximum value is 0.1 (units are mM) and this peak concentration is at $\boldsymbol{r} = (0,0)$. The field L_2 is a static, non-isotropic, linear field, whereas L_3 is a dynamic field: a Gaussian spreading out over time, similar to what one might expect to be produced by a diffusing drop of nutrients. As expected, as long as the stimulus concentrations and their spatial gradients are within the region observed in training, the population distributions show consistency with those produced when simulating using the original full model.

Computational Cost Savings. Computational savings are given empirically here by comparing running times of simulations for both systems. A hundred (100) cells are simulated in each of the ligand fields, for a time $t_{end} = 500\,\mathrm{s}$ and a time-step of $\Delta t = 0.05$. Therefore, one million (1000000) iterations of the main while loop in Algorithms 1 and 2 are compared in the reported speed-up factor (Table 1). We observe a speed-up factor of ~ 8, reducing running times from $\sim 460\,\mathrm{m}$ to $\sim 60\,\mathrm{m}$. Table 1 reports speed-up factors for each ligand field experiment.

The reported factor values do not include the costs paid for training the GP and producing the training data. It takes $\sim 4\,\mathrm{m}$ to train GPs for both $\boldsymbol{\Psi}_\phi$ functions, and $\sim 10\,\mathrm{m}$ for producing 20000 observations of pCTMC transitions from the original fine system (10000 training points for each $\boldsymbol{\Psi}_\phi$ function). The relatively low times compared to simulation times, combined with the fact that one only pays this once, upfront, make these costs negligible.

Accuracy Evaluation. To evaluate how closely results from the abstracted model are compared to the original one, we applied the Kolmogorov-Smirnov (KS) two-sample test [3] to the population distributions of the two models at several time-points in the simulation, as well as to the distributions of running and tumbling duration. We have 100 samples from each population distribution since we simulated 100 cells. However, in the case of 'Run' and 'Tumble' duration distributions we have ~ 60000 observations from each, because we aggregate observations from the entire trajectory; we choose a random 1000 sample of these to perform the KS test.[3] In light of these difficulties, a different test which quantifies the

---

[3] We sub-sample because the KS test p-value depends heavily on sample size. Even if two distributions generating samples might be very close, in the limit of an infinite sample size one approaches the true distributions. In such a case, the KS test will reject that the two samples were produced by *the same* distribution, returning lower p-values as sample size increases (for the same KS distance). We do not expect to produce the same distributions here since we are making approximations, so comparing p-values for very large sample sizes is not of interest.

distance between the two distributions (e.g. Jensen-Shannon divergence) might be more useful here, but that requires analytic forms of the distributions.

Inspecting Table 1 we find no KS distance higher than 0.2 indicating very similar distributions, as supported by the associated high p-values. The latter do not allow rejecting the null hypothesis with the current sample, which is that the samples originate from the same distribution. An exception is the 'Tumble' duration distributions in the L_1 ligand field, where the somewhat higher KS distance of the large sample sizes gives an exaggerated p-value (see footnote 1).

We note how even in the case of the dynamic L_3 field, the resulting population behaviour of the abstracted model is preserved without any additional training necessary. The fact that the original training occurred in a static field does not affect the ability of the abstract model to cope with a dynamic one.

Table 1. KS two-sample test statistics, where the first (top) value reports KS distance and the second (in brackets, bottom) the associated p-value. One sample came from 100 trajectories of fine *E. coli* system simulations, and the other from 100 abstracted system simulations. The first four columns show KS test results of original and abstracted bacterial population distances from peak concentration at various times t (shown in Fig. 4). 'Run' and 'Tumble' columns compare the distributions of run and tumble durations respectively for 1000 samples from each system. The last column reports the observed speed-up factor based on running times and normalising for core utilisation.

Field	$t = 125\,s$	$t = 250\,s$	$t = 375\,s$	$t = 500\,s$	Run	Tumble	Speed-up
Gaussian: $L_1(r)$	0.110 (0.556)	0.160 (0.140)	0.170 (0.099)	0.160 (0.140)	0.039 (0.425)	0.101 ($7 \cdot 10^{-5}$)	7.8
Linear: $L_2(r)$	0.010 (0.677)	0.150 (0.193)	0.170 (0.100)	0.130 (0.344)	0.022 (0.967)	0.014 (0.100)	9.4
Dynamic Gaussian: $L_3(r, t)$	0.140 (0.261)	0.070 (0.961)	0.140 (0.261)	0.080 (0.894)	0.047 (0.214)	0.039 (0.425)	8.9

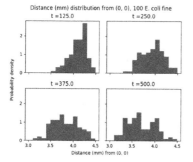
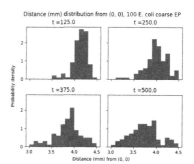

Fig. 4. Empirical distributions for the distance of bacteria populations (100 *E. coli*) at different times t of the simulation. Left: original full system simulations. Right: abstracted system simulations. Gaussian L_1 ligand field.

5   Discussion

In many domains, ranging from cyber-physical systems to biological and medical processes, consideration of spatio-temporal aspects of behaviour is essential. However, this comes at great computational expense. We have presented a methodology that allows layers of a computationally intensive multi-scale model to be replaced by more efficient abstract representations. This is a stochastic map, constructed based on some exploratory simulations of the full model and GP regression. Our results show that we are able to achieve significant speed-up without sacrificing accuracy. This establishes a framework for such statistical abstraction on which we plan to elaborate in future work.

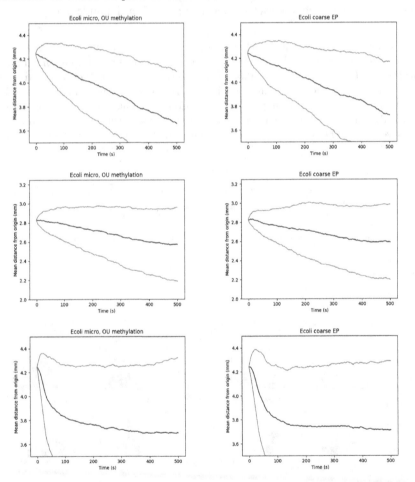

Fig. 5. Average (blue) and standard deviation (red) of distance from peak ligand concentration for a population of 100 *E. coli* over a time of 500 s. Left: original full system simulations. Right: abstracted system simulations. Rows (top to bottom): L_1, L_2, L_3 ligand fields respectively. (Color figure online)

It should be noted that the specifics of the abstraction are not automatically determined by this framework, but are left to the researcher. Having to manually specify the abstraction introduces an element of flexibility, since different abstractions may be tested and so one can see which are suitable and produce accurate approximations, indicating that pertinent elements of the original model are preserved in the coarsening. Additionally, there may be various valid ways to coarsen a model, depending on what the focus of the inquiry is. On the other hand, it shifts some of the burden of abstracting the model to the researcher, who has to find a suitable set of properties which capture the output behaviour of the layer to be abstracted.

Future work avenues include, for example, allowing more properties to be expressed and using them to guide the abstraction will capture more complex behaviours. Additionally, we could infer abstracted model parameters or underlying functions from real data, instead of synthetic ones. Finally, one would like to be able to deal with correlated agents which result in emergent behaviour at the whole population level. This may be readily achieved in this framework if the interaction between agents happens by altering their modelled external environment (e.g. by manipulating the nutrient field, or by exuding different chemical trails which can be modelled by an additional external field). However, the path is not so clear if the agents are coupled in some other way, where the internal state of one directly affects that of another.

References

1. Bortolussi, L., Milios, D., Sanguinetti, G.: Efficient stochastic simulation of systems with multiple time scales via statistical abstraction. In: Roux, O., Bourdon, J. (eds.) CMSB 2015. LNCS, vol. 9308, pp. 40–51. Springer, Cham (2015). doi:10. 1007/978-3-319-23401-4_5
2. Bortolussi, L., Milios, D., Sanguinetti, G.: Smoothed model checking for uncertain continuous-time Markov chains. Inf. Comput. **247**, 235–253 (2016)
3. Chakravarty, I.M., Laha, R.G., Roy, J.D.: Handbook of Methods of Applied Statistics. McGraw-Hill, New York (1967)
4. Dada, J.O., Mendes, P.: Multi-scale modelling and simulation in systems biology. Integr. Biol. **3**(2), 86 (2011)
5. Frankel, N.W., Pontius, W., Dufour, Y.S., Long, J., Hernandez-Nunez, L., Emonet, T.: Adaptability of non-genetic diversity in bacterial chemotaxis. eLife **3**, e03526 (2014)
6. Gilbert, D., Heiner, M., Takahashi, K., Uhrmacher, A.M.: Multiscale Spatial Computational Systems Biology (Dagstuhl Seminar 14481) (2015)
7. Gillespie, D.T.: Exact stochastic simulation of coupled chemical reactions. J. Phys. Chem. **81**(25), 2340–2361 (1977)
8. Goutsias, J.: Quasiequilibrium approximation of fast reaction kinetics in stochastic biochemical systems. J. Chem. Phys. **122**(18), 184102 (2005)
9. Hansen, C.H., Endres, R., Wingreen, N.: Chemotaxis in Escherichia coli: a molecular model for robust precise adaptation. PLoS Comput. Biol. **4**(1), e1 (2008)
10. Haseltine, E.L., Rawlings, J.B.: Approximate simulation of coupled fast and slow reactions for stochastic chemical kinetics. J. Chem. Phys. **117**(15), 6959–6969 (2002)

11. Michaelides, M., Milios, D., Hillston, J., Sanguinetti, G.: Property-driven state-space coarsening for continuous time Markov chains. In: Agha, G., Houdt, B. (eds.) QEST 2016. LNCS, vol. 9826, pp. 3–18. Springer, Cham (2016). doi:10.1007/978-3-319-43425-4_1

12. Norris, J.R.: Markov Chains. Cambridge University Press, Cambridge (1998)

13. Rao, C.V., Arkin, A.P.: Stochastic chemical kinetics and the quasi-steady-state assumption: application to the Gillespie algorithm. J. Chem. Phys. **118**(11), 4999–5010 (2003)

14. Rasmussen, C.E., Williams, C.K.I.: Gaussian Processes for Machine Learning. Adaptive Computation and Machine Learning. MIT Press, Cambridge (2006)

15. Sneddon, M.W., Pontius, W., Emonet, T.: Stochastic coordination of multiple actuators reduces latency and improves chemotactic response in bacteria. PNAS **109**(3), 805–810 (2012)

16. Snelson, E., Ghahramani, Z.: Sparse Gaussian processes using pseudo-inputs. In: Weiss, Y., Schlkopf, P.B., Platt, J.C. (eds.) Advances in Neural Information Processing Systems 18, pp. 1257–1264. MIT Press, Cambridge (2006)

17. Sourjik, V., Berg, H.C.: Functional interactions between receptors in bacterial chemotaxis. Nature **428**(6981), 437–441 (2004)

18. Sourjik, V., Wingreen, N.S.: Responding to chemical gradients: bacterial chemotaxis. Curr. Opin. Cell Biol. **24**(2), 262–268 (2012)

19. Vladimirov, N., Lebiedz, D., Sourjik, V.: Predicted auxiliary navigation mechanism of peritrichously flagellated chemotactic bacteria. PLoS Comput. Biol. **6**(3), e1000717 (2010)

20. Vladimirov, N., Lvdok, L., Lebiedz, D., Sourjik, V.: Dependence of bacterial chemotaxis on gradient shape and adaptation rate. PLoS Comput. Biol. **4**(12), e1000242 (2008)

Automated Experiment Design
for Data-Efficient Verification
of Parametric Markov Decision Processes

Elizabeth Polgreen[1(✉)], Viraj B. Wijesuriya[1], Sofie Haesaert[2],
and Alessandro Abate[1]

[1] Department of Computer Science, University of Oxford, Oxford, UK
`elizabeth.polgreen@cs.ox.ac.uk`
[2] Department of Electrical Engineering, Eindhoven University of Technology,
Eindhoven, The Netherlands

Abstract. We present a new method for statistical verification of quantitative properties over a partially unknown system with actions, utilising a parameterised model (in this work, a parametric Markov decision process) and data collected from experiments performed on the underlying system. We obtain the confidence that the underlying system satisfies a given property, and show that the method uses data efficiently and thus is robust to the amount of data available. These characteristics are achieved by firstly exploiting parameter synthesis to establish a feasible set of parameters for which the underlying system will satisfy the property; secondly, by actively synthesising experiments to increase amount of information in the collected data that is relevant to the property; and finally propagating this information over the model parameters, obtaining a confidence that reflects our belief whether or not the system parameters lie in the feasible set, thereby solving the verification problem.

1 Introduction

Formal verification relies on full access to accurate models describing the behaviour of systems in order to guarantee their correctness. Such models are often hard to obtain for systems encompassing partially understood behaviours and uncertain events. For a partially unknown system, the unknown model characteristics can be represented via non-determinism in the form of parameters. The resulting parameterised model captures all available knowledge on the underlying system of interest.

We target the verification of a fragment of Probabilistic Computation Tree Logic (PCTL) on partially unknown systems with actions. We develop a new approach that incorporates the available information captured by a parameterised model with the active collection of a limited amount of data from the underlying system. The verification problem is tackled in three phases. In the first phase we use the available parameterised model to synthesise the set of parameters for which the property of interest is satisfied (called the *feasible set*).

© Springer International Publishing AG 2017
N. Bertrand and L. Bortolussi (Eds.): QEST 2017, LNCS 10503, pp. 259–274, 2017.
DOI: 10.1007/978-3-319-66335-7_16

In the second phase a series of experiments are designed and executed on the system to update the knowledge available about the parameters of the parameterised model. More precisely, a procedure executes the designed experiments, obtains data from the system and, by means of Bayesian statistics, updates distributions over the likely parameter values of the parameterised model. This updated knowledge is returned to the experiment design module, and the process is repeated until a preset limit on the total amount of collectible data is reached. The design of such experiments is important to attain a reasonable level of confidence in the acceptance or rejection of a property with a limited amount of data.

In the final phase, we combine the output from the parameter synthesis with the updated distributions over the model parameters to quantify the confidence that the system satisfies (or does not satisfy) the property.

This work extends the contributions in [19] by focussing on systems with actions: the presence of (action) non-determinism provides the potential for experiment design, whereby we select actions to improve the accuracy of our confidence value. More precisely we design experiments that maximise the usefulness of the data collected. Intuitively, this means that we want to design experiments to prioritise the collection of data that leads to proving or disproving the satisfaction of the property. In this work, we present the complete approach, and evaluate the contribution of our experiment design procedure. We argue that *automated experiment design* allows us to draw sensible conclusions robustly with a limited amount of data.

Structure of the Paper. Section 2 provides the necessary background information for the rest of paper to build upon. Section 3 presents an overview of our algorithm. Subsequent sections detail the different phases of the algorithm: in Sect. 4 we show how we collect data; Sect. 5 provides details on the confidence computation; and the key contribution of this work is Sect. 6, which outlines our experiment design approach.

1.1   Problem Statement

Consider a partially unknown system **S**, with external non-determinism in the form of actions, and suppose we can gather a limited amount of sample trajectories from this system. Assume the partial knowledge about the system is encompassed within a parameterised model class describing the behaviour of **S**.

We investigate two sub-problems:

- Can we efficiently use this limited amount of data from a system **S** to quantify a confidence that the system **S** verifies a given PCTL property?
- How should we design an experiment on the system such that the gathered data allows us to verify the property with the greatest degree of accuracy? Let the choice of actions of system **S** be something we can control during the experiment, and let there only be a limited amount of available experiment

time; can we optimise the sequence of actions to increase the accuracy of the confidence quantification?

1.2   Related Work

We compare our work to two branches of research: Statistical Model Checking (SMC) and research concerned with learning models from system data. We contrast our experiment design method with existing strategy synthesis techniques for *fully known* Markov decision processes (MDPs).

We emphasise that we tackle a different problem than SMC: we target partially unknown systems and gather data from the underlying system; SMC [16] targets fully known models that are too big for conventional verification, and generates large amounts of data from the models themselves. When applied to model-free scenarios [23,24], SMC generates this data from the underlying system. By using partial model knowledge, we substantially reduce this data requirement. In addition, SMC for systems with non-determinism [4,12] considers only bounded-time properties, and depends on the ability to generate traces from the model of length greater than the bound. By incorporating parameter synthesis tools, we are able to consider unbounded-time properties and to draw conclusions from much shorter traces.

Research on learning models from system data is broad. [18,22] use a Bayesian approach to learn full Markov models of completely unknown systems. Our work uses a similar Bayesian method but differs because we include information from the partial model, which allows us to consider known relationships between parameters and thus reduce the amount of data needed for inference. [1,3] use active learning to discover full MDP models from data, prioritising actions by *variance minimisation* or *KL divergence*. The inclusion of a partial model in our method allows us to instead prioritise gathering data that contributes to the acceptance or rejection of a given property over the system. Although [3] learns the model with the goal of system verification, the authors provide no means of quantifying a confidence that the system satisfies the property, as they do not have a way to assess which transition probabilities have the greatest contribution to the satisfaction of the property.

Considering different model classes, experiment design is also used in system identification [7]. Recent studies [10,11] have incorporated experiment design to data-driven statistical verification over *dynamical systems* with partly unknown dynamics, controllable inputs and noisy measurements. Similar to our approach, they also compute a confidence estimate on the properties of interest by gathering data through *optimal experiment design*.

Action selection for Markov decision processes, though in our context used for experiment design, is a known problem that in general amounts to synthesising strategies. [15] presents an overview for MDPs with static rewards, and [8] provides solutions for MDPs with non-Markovian rewards. Closer to our approach, [9] synthesises strategies for MDPs online, where an agent learns a state cost only after selecting an action. [13] use inference-based techniques over strategies to pick a strategy that maximises the expected reward for an MDP with arbitrary rewards.

2 Background

We model a fully known system as a Markov decision process [2].

Definition 1. *A discrete-time Markov decision process (MDP)* **M** *is a tuple* $(S, \text{Act}, \mathbb{T}, \iota_{init}, AP, L)$, *where:*

- *S is a finite, non-empty set of states,*
- *Act is a set of actions,*
- *$\mathbb{T} : S \times Act \times S \to [0,1]$ is the transition probability function, such that $\forall s \in S$ and $\forall \alpha \in \text{Act}, \sum_{s' \in S} \mathbb{T}(s, \alpha, s') \in \{0, 1\}$,*
- *$\iota_{init} : S \to [0,1]$ denotes an initial probability distribution over the states S, such that $\sum_{s \in S} \iota_{init}(s) = 1$,*
- *The states in S are labelled with atomic propositions $a \in AP$ via the labelling function $L : S \to 2^{AP}$.*

An action $\alpha \in Act$ is enabled in state s if and only if $\sum_{s' \in S} \mathbb{T}(s, \alpha, s') = 1$. Let $\text{Act}(s)$ denote the set of enabled actions in s. For any state $s \in S$, it is required that $\text{Act}(s) \neq \varnothing$. Each state $s' \in S$ for which $\mathbb{T}(s, \alpha, s') > 0$ is called an α-successor of s. Those states s satisfying the condition $\iota_{init}(s) > 0$ are called initial states.

We assume that the MDP is not known exactly, and instead belongs to the set of MDPs represented by a parametric Markov decision process.

Definition 2. *A discrete-time parametric Markov decision process (pMDP) is a tuple* $\mathbf{M}_{\Theta} = (S, \text{Act}, \mathbb{T}_{\theta}, \iota_{init}, AP, L, \Theta)$, *where $S, \iota_{init}, \text{Act}, AP, L$ are as in Definition 1. The entries in $\mathbb{T}_{\theta}$ are specified in terms of parameters, collected in a parameter vector $\theta \in \Theta$, where Θ is the set of all possible evaluations of θ. Each evaluation gives rise to an induced Markov decision process $\mathbf{M}(\theta)$.*

$\forall s \in S, \forall \alpha \in \text{Act}(s), \forall \theta \in \Theta : \sum_{s' \in S} \mathbb{T}_{\theta}(s, \alpha, s') = 1$, namely any $\theta \in \Theta$ induces an MDP $\mathbf{M}(\theta)$ where the transition function $\mathbb{T}_{\theta}$ can be represented by a *stochastic matrix*. We also assume a prior distribution on the model parameters (to be used in Bayesian inference). We assume all non-parameterised transition probabilities are known exactly.

As in [19], we consider *linearly parameterised* MPDs, where unknown transition probabilities can be linearly related. More precisely, given $\Theta \subseteq [0,1]^n$ and parameter vector $\theta = (\theta_1, \ldots, \theta_n) \in \Theta$ with $\theta_i \in [0,1]$, a pMDP is considered linearly parameterised if all outgoing transition probabilities of state-actions pairs have probability $g_l(\theta)$ or $1 - g_l(\theta)$, where $g_l(\theta) = k_0 + k_1\theta_1 + \ldots + k_n\theta_n$ with $k_i \in [0,1]$ and $\sum k_i \leq 1$. This restriction is due to the transformations presented in [19] necessary to perform Bayesian inference over the model parameters. As before, $\forall s \in S, \forall \alpha \in \text{Act}(s), \forall \theta \in \Theta : \sum_{s' \in S} \mathbb{T}_{\theta}(s, \alpha, s') = 1$.

2.1 Strategies

A strategy for an MDP resolves nondeterminism by choosing an action in each state of the model. In our work experiment design amounts to synthesising a strategy for an MDP, i.e., a sequence of actions, under which we generate data from the system. We focus on *deterministic memoryless* strategies in this paper, i.e., strategies that always pick the same action in any given state, independent of the history of states already visited. Future work will extend to both memory-dependent and randomised strategies.

Definition 3. *A deterministic memoryless strategy for an MDP M is a function* $\pi : S \rightarrow \mathrm{Act}$ *s.t.* $\pi(s) \in \mathrm{Act}(s)\ \forall_{s \in \mathbf{S}}$.

2.2 Properties – Probabilistic Computational Tree Logic

We consider system specifications (aka properties) given in a fragment of Probabilistic Computational Tree Logic (PCTL) [2]. Since we use PRISM [14] for parameter synthesis, we consider *non-nested, unbounded-time "until"* properties expressed in PCTL.

Definition 4. *Let a discrete-time MDP be given. Let* ϕ *be a formula interpreted over states* $s \in S$, *and* φ *be a formula interpreted on paths of the MDP. Also, let* $\bowtie \in \{<, \leq, \geq, >\}$, $n \in \mathbb{N}$, $p \in [0, 1]$, $c \in AP$. *The Syntax of the PCTL fragment we consider is given by:*

$$\phi := \mathrm{true} \mid c \mid \phi \wedge \phi \mid \neg \phi \mid \mathbf{P}_{\bowtie p}(\varphi), \qquad \varphi := \bigcirc \phi \mid \phi \, \mathcal{U} \, \phi.$$

Definition 5. *Consider a PCTL formula* $\phi := \mathbf{P}_{\bowtie p}(\phi_1 \, \mathcal{U} \, \phi_2)$. *Let* $\mathbb{P}_{\mathbf{M}}^{\pi}(s, \varphi)$ *denote the probability associated to the paths of an MDP* $\mathbf{M}$ *starting from* $s \in S$ *satisfying the path formula* φ *under the strategy* π. *Let* $\mathfrak{A}(\mathbf{M})$ *denote all deterministic memoryless strategies for* $\mathbf{M}$. *The satisfaction of the formula* ϕ *by* M *is given by:*

$$\mathbf{M} \models \mathbf{P}_{\bowtie p}(\phi_1 \, \mathcal{U} \, \phi_2) \iff \forall s \in S, \iota_{init}(s) > 0 : \min_{\pi \in \mathfrak{A}(\mathbf{M})} \mathbb{P}_{\mathbf{M}}^{\pi}(s, \phi_1 \, \mathcal{U} \, \phi_2) \bowtie p.$$

We introduce the *feasible set* of parameters, denoted Θ_{ϕ}, which is the set of parameter evaluations for which the property is satisfied.

Definition 6. *Let* $\mathbf{M}(\theta)$ *be an induced MDP of the pMDP* $\mathbf{M}_{\Theta}$, *indexed by parameter vector* $\theta \in \Theta$. *Let* ϕ *be a formula in PCTL. The feasible set* Θ_{ϕ} *is defined as:* $\theta \in \Theta_{\phi} \iff \mathbf{M}(\theta) \models \phi$.

We use $\mathbb{P}(A)$ to denote the probability of an event A, $p(\cdot)$ to represent probability density functions and $\mathbf{P}_{\bowtie p}(\cdot)$ for the probabilistic operator in PCTL.

3   Overview of the Method

Our method is made up of three distinct phases, as shown in Fig. 1.

1. We use a parameter synthesis tool to determine a set of feasible parameters for which the property is satisfied by the system, based on the given parametric Markov decision process, see Sect. 3.1.
2. (a) We synthesise a strategy for collecting data, based on the feasible set and the prior distribution over the parameters, see Sect. 6.
   (b) We collect data from the underlying system using the synthesised strategy, see Sect. 4.
   (c) We use Bayesian inference to infer a distribution over the likely values of the parameters, based on the collected data, and update the respective prior distributions with the new information, see Sect. 4. If we can sequentially collect more data, loop back to step 2 (a).
3. We compute the confidence that the system satisfies the property, based on the data collected, see Sect. 5.

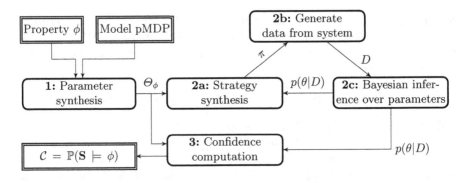

Fig. 1. Overview of the verification procedure.

3.1   Parameter Synthesis

The first phase of the method uses parameter synthesis to find the feasible set of parameters, namely parameter evaluations corresponding to models of the considered pMDP that satisfy the given PCTL property. This step leads to the set of parameters $\Theta_\phi = \{\theta \in \Theta : \mathbf{M}(\theta) \models \phi\}$.

The output of the parameter synthesis procedure is a mapping from hyper-rectangles (which are subsets of parameter evaluations) to truth values, namely "true" if the property is satisfied in the hyper-rectangle and "false" otherwise.

Implementation: We use PRISM [14] for parameter synthesis: the tool computes a rational function of the parameters, which expresses the result obtained from model checking the PCTL property on the parameterised model. Our approach can also make use of Storm [21], which shows potential to be scalable to much larger systems. Storm *lifts* a parametric Markov decision process to a parameter-free Stochastic Game (SG) between two players, and solves the resulting SG via standard value iteration.

4   Bayesian Inference in Parametric Markov Decision Processes

In this work, we collect data from the underlying system and use Bayesian learning to infer a probability distribution over parameters of the pMDP model based on the collected data. Bayesian inference maintains a probability distribution over these parameters and updates the distribution by employing Bayes' rule as more observations are gathered [22]. An initial prior distribution $p(\theta)$ is assumed.

Data. We collect *finite traces* from the underlying system, in the form of a sequence of visited states and actions. We use D to denote a set of finite traces. We split the data into transition counts: D_{s_k,α_1,s_l} denotes the number of times the transition from s_k to s_l under action α_1 appears within the data set D. Each transition count is the outcome of an independent trial in a multinomial distribution[1] with event probabilities given by the transition probabilities.

Assume for now that the transitions are parameterised either with constants or with single parameters of the form θ_i or $1-\theta_i$. We can group transition counts for identically parameterised transitions. We shall denote by D_{θ_j} the transition counts for all transitions with probability given by θ_j.

We wish to obtain posterior distributions for each parameter via *marginal distributions* (which, in this case, are binomial distributions), by applying *parameter-tying* [20] techniques. We thus obtain a number of transition counts for $1-\theta_j$ as the sum of all transitions not parameterised with θ_j, under an action that has a transition parameterised with θ_j, and denote it by $D_{\neg\theta_j}$. Hence D_{θ_j} and $D_{\neg\theta_j}$ are calculated as:

$$D_{\theta_j} = \sum_{s_i \in S, s_l \in S, \alpha_k \in Act} D_{s_i,\alpha_k,s_l} \text{ for } \mathbb{T}(s_i, \alpha_k, s_l) = \theta_j, \text{ and}$$

$$D_{\neg\theta_j} = \sum_{s_i \in S, s_l \in S, \alpha_k \in Act} D_{s_i,\alpha_k,s_l} \text{ for } \mathbb{T}(s_i, \alpha_k, s_l) \neq \theta_j \wedge \exists s_m \in S : \mathbb{T}(s_i, \alpha_k, s_m) = \theta_j.$$

Let $D_{\theta_j,\neg\theta_j}$ denote the pair $(D_{\theta_j}, D_{\neg\theta_j})$. For parameterisations where the transition probabilities are expressed as linear functions of parameters, we obtain $D_{\theta_j,\neg\theta_j}$ by the same procedure that [19] uses. We expand the Markov decision

---

[1] A multinomial distribution is defined by its density function $f(\cdot \mid p, N) \propto \prod_{i=1}^{k} p_i^{n_i}$, for $n_i \in \{0, 1, \ldots, N\}$ and such that $\sum_{i=1}^{k} n_i = N$, where $N \in \mathbb{N}$ is a parameter and p is a discrete distribution over k outcomes.

process, introducing new states and new transitions, allowing us to force all transition probabilities to be expressed as constants, or in the form of θ_j or $1 - \theta_j$, for any parameter $\theta_j \in \theta$. We can then represent the parameter counts over parameters in the new transitions as multinomial distributions. We omit the detail here and refer the reader to the extended version of this paper, and the original work [19].

Bayesian Inference with Data. Consider a parametric Markov decision process $\mathbf{M}_\Theta = (S, \text{Act}, \mathbb{T}_\theta, \iota_{init}, AP, L, \Theta)$ with $\Theta \subseteq [0,1]^n$. Suppose that we have obtained D_{θ_j} and $D_{\neg\theta_j}$ for all $\theta_j \in \theta$, and that we have assumed non-informative, uniform prior distributions for all parameters $\theta_j \in \theta$, denoted by $p(\theta_j)$. The posterior density $p(\theta_j \mid D)$ is given by Bayes' rule:

$$p(\theta_j \mid D) = \frac{\mathbb{P}(D \mid \theta_j) p(\theta_j)}{\mathbb{P}(D)} = \frac{p(\theta_j) \theta_j^{D_{\theta_j}} (1 - \theta_j)^{D_{\neg\theta_j}}}{\mathbb{P}(D_{\theta_j, \neg\theta_j})}.$$

A standard approach [5,17,22] is to consider the prior to be a Dirichlet distribution. The posterior distribution is then updated by adding the event counts to the hyperparameters of the prior. The Dirichlet prior distribution for the pair $(\theta_j, 1 - \theta_j)$ is denoted as $\text{Dir}(\theta_j \mid \mu^{\theta_j})$ with hyperparameters $\mu^{\theta_j} = (\mu_1^{\theta_j}, \mu_2^{\theta_j})$. Thus, the updated posterior distribution for the parameter θ_j is given as: $\theta_j \sim p(\theta_j \mid D) = \text{Dir}(\theta_j \mid D_{\theta_j, \neg\theta_j} + \mu^{\theta_j})$.

The posterior distribution for the entire parameter vector θ, given by $p(\theta \mid D)$ is equal to the product of the posterior distributions for all $\theta_i \in \theta$. This holds due to the independence of each θ_i over independent state-action pairs in the pMDP. Note that, if we have a linearly parameterised MDP, we obtain some of the transition counts in the form of multinomial distributions. We hence obtain realisations of the posterior by a sampling procedure from [19] as explained in the extended version of this paper.

5   Computation of Confidence

We determine a confidence, $\mathcal{C}$, for the satisfaction of a PCTL formula ϕ by a system $\mathbf{S}$ of interest. We first presented this procedure in previous work [19], and we need no extension to this due to the external nondeterminism being factored out in the Bayesian inference calculation given in the previous section.

Definition 7. *Given a PCTL formula ϕ that has a binary satisfaction function, i.e., the property is either satisfied or not, and posterior distributions $p(\theta_i \mid D)$ for all $\theta_i \in \theta$, as obtained in the previous section, the confidence in $\mathbf{S} \models \phi$ can be quantified by Bayesian inference as*

$$\mathcal{C} = \mathbb{P}(\mathbf{S} \models \phi \mid D) = \int_{\Theta_\phi} \prod_{\theta_i \in \theta} p(\theta_i \mid D_{\theta_i, \neg\theta_i}) d\theta, \tag{1}$$

The operation shown in Eq. (1) is equivalent to computing the confidence that each parameter is within its feasible set, and then taking the product of all the parameter confidence values. The integral of a Dirichlet distribution is hard to compute using analytical methods, and so we use Monte Carlo integration. This also allows integration with the calculation of the posterior distribution for pMDPs with linear parameterisations, where we have obtained the posterior distribution by means of sampling, as described in [19].

6   Online Experiment Design

The key contribution in this paper is the design of experiments to generate maximally useful data. We describe in the preceding sections how we use a limited amount of data efficiently to obtain a confidence that the system satisfies the property. In this section, we propose a method for selecting the deterministic memoryless strategy that provides the most useful data to input into our confidence computation in Sect. 5. This allows us to compute the most accurate confidence value for the finite data set of limited size, i.e., the confidence should be high if the underlying system satisfies the property, and low if the underlying system does not satisfy the property.

6.1   Predicted Confidence

We predict the confidence after taking a transition from state s under action α. We define the predicted confidence, $C_{s,\alpha}^{\mathrm{pred}}$, to be the confidence computed using the *expected parameter counts*, after taking a single transition from s under action α: these are denoted by $\mathbb{E}_{s,\alpha}(D_{\theta_i,\neg\theta_i})$ for all $\theta_i \in \theta$. Formally,

$$C_{s,\alpha}^{\mathrm{pred}} = \int_{\Theta_\phi} \prod_{\theta_i \in \theta} p(\theta_i \mid \mathbb{E}_{s,\alpha}(D_{\theta_i,\neg\theta_i}))d\theta,$$

where $p(\theta_i \mid \mathbb{E}_{s,\alpha}(D_{\theta_i,\neg\theta_i}))$ is the predicted posterior distribution obtained by updating the prior, $Dir(\theta_i \mid \mu^{\theta_i})$, with the expected parameter counts, i.e., $\mathrm{Dir}(\theta_i \mid \mu^{\theta_i} + \mathbb{E}_{s,\alpha}(D_{\theta_i,\neg\theta_i}))$.

We first compute the *expected transition counts* for the state-action pair, $\mathbb{E}_{s,\alpha}(D_{s,\alpha})$, from which we extract the *expected parameter counts* using the method in Sect. 4. Consider a state s with an action α, and two transitions with probabilities $\mathbb{T}_\theta(s,\alpha,s') = g_l(\theta) = k_0 + k_1\theta_1 + \ldots + k_n\theta_n$, and $T_\theta(s,\alpha,s) = 1 - g_l(\theta)$. The expected transition counts are given by a multinomial distribution over the outgoing transitions under that action, with event probabilities equal to the *expected transition probabilities*. Note that prior distribution for any parameter $\theta_i \in \theta$ is $Dir(\theta_i \mid \mu^{\theta_i})$. To compute the expected transition probabilities, we require the expected values of the parameters, given by $\mathbb{E}(\theta_i) = \frac{\mu_1^{\theta_i}}{\mu_1^{\theta_i} + \mu_2^{\theta_i}}$ for all $\theta_i \in \theta$. The expected value of the transition probabilities are then given by evaluating $g_l(\mathbb{E}(\theta))$ and $1 - g_l(\mathbb{E}(\theta))$. Hence the expected transition counts $\mathbb{E}_{s,\alpha}(D_{s,\alpha,s'})$ and $\mathbb{E}_{s,\alpha}(D_{s,\alpha,s})$, are equal to the expected transition probabilities for $\mathbb{T}_\theta(s, \alpha, s')$ and $\mathbb{T}_\theta(s, \alpha, s)$. Consider only the transition parameterised with $\mathbb{T}_\theta(s, \alpha, s') = g_l(\theta)$:

$$\mathbb{E}_{s,\alpha}\left(D_{s,\alpha,s'}\right) = \mathbb{E}\left(\mathbb{T}(s, \alpha, s')\right) = g_l(\mathbb{E}\left(\theta\right))$$

$$= k_0 + k_1 \mathbb{E}\left(\theta_1\right) + \ldots + k_n \mathbb{E}\left(\theta_n\right) = k_0 + \sum_{i=1:n} k_i \frac{\mu_1^{\theta_i}}{\mu_1^{\theta_i} + \mu_2^{\theta_i}}.$$

We can extract the parameter counts as described in Sect. 4, to obtain $\mathbb{E}_{s,\alpha}\left(D_{\theta_i, \neg\theta_i}\right)$.

6.2   Optimisation of Predicted Confidence Gain

The underlying system either satisfies or does not satisfy the given property, so we wish to minimise the difference between our confidence value and the closest among 0 or 1, or to maximise the difference between a confidence of 0.5 and our confidence, i.e., the maximum absolute value of $0.5 - \mathcal{C}$. We can therefore define a predicted confidence gain for a state-action pair (s, α), denoted by $\mathbb{G}_{s,\alpha}$, as the maximisation of this difference, i.e., the biggest step towards either 0 or 1.

$$\mathbb{G}_{s,\alpha} = |0.5 - \mathcal{C}_{s,\alpha}^{\mathrm{pred}}| - |0.5 - \mathcal{C}|$$

For a finite trace of length N, we can calculate the optimal predicted confidence gain for state s and discrete time step t, denoted by x_s^t, as

$$x_s^t = \begin{cases} \max_{\alpha \in Act(s)}(\mathbb{G}_{s,\alpha} + \sum(\mathbb{T}(s, \alpha, s') . x_{s'}^{t+1})) & \text{if } 0 < t < N \\ 0 & \text{if } t \geq N. \end{cases}$$

It is important to note that the confidence gain is not a static quantity, because $\mathbb{G}_{s,\alpha}$ depends on the distribution over the relevant component parameters of θ at time t.

6.3   Optimal Confidence Gain: Experiment Design via Strategy Synthesis

Due to memory dependency of the confidence gain, computing an optimal strategy is intractable, and cannot be solved via conventional dynamic programming methods [8]. However, we put forward a few alternatives.

Explicitly Evaluated Memoryless Strategies. The conventional way of solving a MDP with non-Markovian rewards is to translate the model into an equivalent MDP with Markovian rewards, whose states result from augmenting those of the original model with extra information capturing enough history to make the reward Markovian. This is in general computationally expensive [8]. Given that we will be performing strategy synthesis repeatedly in our method (i.e., once each time a new batch of data is sequentially gathered), we compromise and use a straightforward selection method to find the best memoryless strategy.

This reduces the number of possible strategies and allows us to consider each possible strategy individually. We simplify the calculations in Sect. 6.1 to compute the expected transition counts for a full trace of length N, and then compute the predicted confidence gain for the entire memoryless strategy. This method works well for small trace lengths, however computing the expected transition counts for a full trace of length N amounts to performing a matrix multiplication N times, so this can be time consuming for large N.

Alternative Off-Line Method. An alternative approach would be to disregard the memory dependency of the confidence gain. This corresponds to an off-line approach: we compute a strategy on the model frozen at the time we start generating traces, assuming that the prior distributions remains unchanged over the trace horizon N. We assign confidence gains to state-action pairs and treat them as static rewards. This allows us to use classical dynamic programming to find the best memoryless strategy, which would require introducing a discount factor on the rewards, to avoid infinite returns inside *strongly-connected components*. This method may be faster for long trace lengths than explicitly evaluating possible strategies, as done previously; however, the selected strategy may not be the best memoryless strategy when the trace lengths are large, and specifically when the prior distributions, which are assumed to remain unchanged, actually change significantly over time as the trace length is being reached.

Comparison. Consider the small pMDP shown in Fig. 2, parameterised with $\theta = (\theta_1, \theta_2)$, and the property $\mathbf{P}_{\leq 0.5}(\text{true } \mathcal{U} \ s_1)$. Both parameters have the same prior distributions and both contribute equally to the feasible set. Intuitively, choosing action α_2 or α_3 is better than choosing action α_1, because any trace starting with α_1 only contains one parameterised transition. However, it is also intuitive that choosing α_2 is better than α_3 because any trace starting with α_3 only gives us information about θ_1, whereas traces with α_2 give us information about both parameters.

The dynamic programming approach will pick nondeterministically between action α_3 and α_2 for the first trace, because the reward assigned to (s_0, α_3) is the same as the reward assigned to (s_0, α_2) as the initial priors and the feasible sets are the same. The priors will not be updated until after the full trace is collected. Our strategy synthesis approach calculates the expected updates for these priors, and will thus be able to detect a better strategy, which selects action α_2.

In our experimental evaluation, we use the explicitly evaluated memoryless strategy. Henceforth, the explicitly evaluated memoryless strategy will be referred to as the *synthesised strategy*.

7 Results

We experimentally evaluate the research questions posed in the problem statement: *question 1 – given a limited amount of data, can we use it efficiently to*

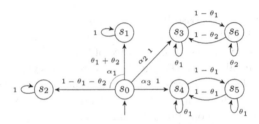

Fig. 2. Example pMDP where offline strategy synthesis may not be optimal

quantify a confidence that our system satisfies a given property? question 2 – can we design experiments that increase the accuracy of this confidence?

Experimental Set-Up. Our approach is implemented in C++. We use PRISM [14] for parameter synthesis, and GSL-2.3 [6] for random number generation.

To answer question 2, we evaluate our *synthesised strategy* approach against two alternatives. The first comparison is against a memoryless strategy, randomly selected from the set of all possible memoryless strategies. We term the resultant strategy as *random static strategy*. The second comparison strategy randomly selects actions at each state as data is collected, and therefore we term it as *no strategy*. All three approaches use the same Bayesian inference framework over parameter counts.

We present the analysis of our approach on the simple pMDP model in Fig. 3 and with the PCTL property $\mathbf{P}_{\geq 0.5}$(true $\mathcal{U}$ complete). We also run our approach on models up to 1000 states, but find the scalability depends on the number of actions in the model. We assign non-informative priors to the parameters. Note that in our model, θ_2 does not contribute to the satisfaction of the property, and having validated that this does not affect the confidence results, we set θ_2 equal to θ_1. We simulate a range of underlying systems, corresponding to models $\mathbf{M}(\theta)$ with different values for θ, which allows us to assess the accuracy of our confidence values against a ground truth, G_{true}. For a simulated system modelled by $\mathbf{M}(\theta)$, this is given by:

$$G_{true} = \begin{cases} 0 \text{ if } \theta_1 \notin [0.369, 0.75], \\ 1 \text{ if } \theta_1 \in [0.369, 0.75]. \end{cases} \tag{2}$$

We collect data from the simulated system in the form of a history of state-action pairs visited. We compute the mean squared error (MSE) between the ground truth from Eq. (2) and the confidence estimate, formally, MSE $= \frac{1}{n} \sum_{i=1}^{n} (G_{true} - G_i)^2$, where n is the number of trials and G_i is the output confidence estimate for the i-th run.

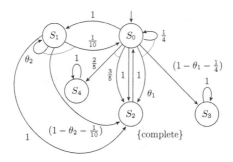

Fig. 3. A simple pMDP for the experimental evaluation.

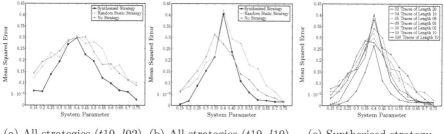

(a) All strategies ($t10$, $l02$)   (b) All strategies ($t10$, $l10$)   (c) Synthesised strategy

Fig. 4. Errors produced by the confidence computation for the three strategies considered. Plots (a) and (b) show the MSE for each type of strategy and for 10 traces of different trace lengths over different simulated systems. Plot (c) presents the MSE for the *synthesised strategy* over different simulated systems and combinations of number of traces with varying trace lengths. We denote by ($t10$, $l02$) a run with 10 traces, each of length 02.

Observations and Discussion. The MSE in the confidence from all three strategies, over a range of underlying systems and varying quantities of data (i.e., for different numbers and lengths of traces), are shown in Fig. 4. The convergence of the confidence outcome is shown in Fig. 5, with box plots showing the interquartile range (IQR), omitting any outliers, and whiskers extending to the most extreme data points not considered to be outliers.

Accuracy of Confidence Results. The confidence for all approaches is low around the lower boundary of Θ_ϕ, and the MSE is high, shown in Fig. 4. This is consistent with the goal of the confidence calculation, where one would need to know the exact value of the system parameter θ if its value is near this edge, to be able to decide whether it falls in Θ_ϕ or not, and hence the calculation has a high sensitivity around this boundary This sensitivity increases as the amount of data increases, as seen by comparing the MSE for $\theta_1 = 0.4$ in Fig. 4a, where the trace length is 2, with Fig. 4b when the trace length has increased up to 10. To explore why this is the case, consider that to compute the confidence we integrate the posterior distribution over the feasible set $\Theta_\phi = [0.369, 0.75]$.

The posterior distribution for $\theta_i = 0.369$ should have a peak centred at 0.369 and half of the probability mass falling in the feasible set, leading to $C = 0.5$. The height and width of the posterior distribution are determined by the amount and spread of data available and for a tall and thin distribution (encompassing a large amount of data), a small change in the position of the peak can move a large percentage of mass of the distribution in or out of the feasible set. This is prominent in Fig. 4b since our approach synthesises a strategy that would yield the highest information gain, i.e., the most useful data. However, as we move away from the edge, increased data effectively places probability mass away from the uncertain regions, thus reducing both variance and MSE. Neither of the other two alternatives has the ability to collect as much useful data and therefore variance is high even at the far ends of the parameter spectrum. The ability of our method to collect more useful data is also illustrated in the convergence graphs shown in Fig. 5, where synthesis approach converges to the ground truth quicker than both comparison strategies.

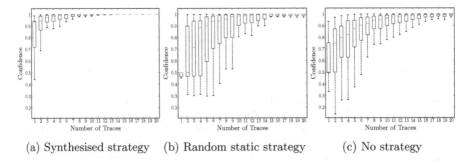

(a) Synthesised strategy   (b) Random static strategy   (c) No strategy

Fig. 5. Convergence of confidence outcomes to the ground truth for a run with 20 traces, each of length 10 over a simulated underlying system with both parameters (θ_1, θ_2) set to 0.7.

We conclude that our strategy synthesis does improve the accuracy of the confidence calculation, unless the parameter value falls close to the boundary of Θ_ϕ, and that away from this boundary the confidence converges to the ground truth and we are able to verify the property over S based on the data collected.

Robustness. We run our implementation with varying lengths of traces, where the total number of transitions in the data remains the same, and the results summarised in Fig. 4c show that our approach, on this case study, is relatively insensitive to this variation (compare Fig. 4a with b). Our method depends on the number of parameterised transitions we visit and so depends on the trace length being long enough to visit some parameterised transitions. This is in contrast to Statistical Model Checking techniques, where the accuracy of the approach depends on the trace length being great enough to satisfy the property, e.g., to reach some desired state. In both cases this will vary depending on the structure of the model.

8   Conclusions

In this paper, we have presented an approach for statistical verification of a fragment of unbounded-time PCTL properties on partially unknown systems, by automating the design of smart experiments that maximise the amount of useful data collected from the underlying system. We validate that our approach increases the accuracy of the confidence that the system satisfies the property, compared to selecting data randomly. We are able to achieve meaningful confidence outcomes with comparably limited amounts of available data.

We are pursuing extensions of this framework for much wider class of probabilistic models, in particular continuous time models, with a broad range of applications.

References

1. Araya-López, M., Buffet, O., Thomas, V., Charpillet, F.: Active learning of MDP models. In: Sanner, S., Hutter, M. (eds.) EWRL 2011. LNCS (LNAI), vol. 7188, pp. 42–53. Springer, Heidelberg (2012). doi:10.1007/978-3-642-29946-9_8
2. Baier, C., Katoen, J.: Principles of Model Checking. MIT Press, Cambridge (2008)
3. Chen, Y., Nielsen, T.D.: Active learning of Markov decision processes for system verification. In: ICMLA, vol. 2, pp. 289–294. IEEE (2012)
4. D'Argenio, P., Legay, A., Sedwards, S., Traonouez, L.: Smart sampling for lightweight verification of Markov decision processes. STTT **17**(4), 469–484 (2015)
5. Friedman, N., Singer, Y.: Efficient Bayesian parameter estimation in large discrete domains. In: NIPS, pp. 417–423. The MIT Press (1998)
6. Galassi, M., Davies, J., Theiler, J., Gough, B., Jungman, G.: GNU Scientific Library - Reference Manual, GSL Version 1.12, 3rd edn. Network Theory Ltd., Bristol (2009)
7. Gevers, M., Bombois, X., Hildebrand, R., Solari, G.: Optimal experiment design for open and closed-loop system identification. Comm. Inf. Syst. **11**(3), 197–224 (2011)
8. Gretton, C., Price, D., Thiébaux, S.: Implementation and comparison of solution methods for decision processes with non-Markovian rewards. In: UAI, pp. 289–296. Morgan Kaufmann (2003)
9. Guan, P., Raginsky, M., Willett, R.M.: Online Markov decision processes with Kullback-Leibler control cost. IEEE Trans. Autom. Control **59**(6), 1423–1438 (2014)
10. Haesaert, S., Van den Hof, P.M.J., Abate, A.: Data-driven property verification of grey-box systems by Bayesian experiment design. In: 2015 American Control Conference (ACC), pp. 1800–1805, July 2015
11. Haesaert, S., Van den Hof, P.M.J., Abate, A.: Experiment design for formal verification via stochastic optimal control. In: ECC, pp. 427–432. IEEE (2016)
12. Henriques, D., Martins, J., Zuliani, P., Platzer, A., Clarke, E.M.: Statistical model checking for Markov decision processes. In: QEST, pp. 84–93. IEEE Computer Society (2012)
13. Hoffman, M.D., de Freitas, N., Doucet, A., Peters, J.: An expectation maximization algorithm for continuous Markov decision processes with arbitrary reward. In: AISTATS, JMLR Proceedings, vol. 5, pp. 232–239. JMLR.org (2009)

14. Kwiatkowska, M.Z., Norman, G., Parker, D.: PRISM 4.0: verification of probabilistic real-time systems. In: Gopalakrishnan, G., Qadeer, S. (eds.) CAV 2011. LNCS, vol. 6806, pp. 585–591. Springer, Heidelberg (2011). doi:10.1007/978-3-642-22110-1_47

15. Kwiatkowska, M.Z., Parker, D.: Automated verification and strategy synthesis for probabilistic systems. In: Van Hung, D., Ogawa, M. (eds.) ATVA 2013. LNCS, vol. 8172, pp. 5–22. Springer, Cham (2013). doi:10.1007/978-3-319-02444-8_2

16. Legay, A., Delahaye, B., Bensalem, S.: Statistical model checking: an overview. In: Barringer, H., et al. (eds.) RV 2010. LNCS, vol. 6418, pp. 122–135. Springer, Heidelberg (2010). doi:10.1007/978-3-642-16612-9_11

17. Pasanisi, A., Fu, S., Bousquet, N.: Estimating discrete Markov models from various incomplete data schemes. Comput. Stat. Data Anal. **56**(9), 2609–2625 (2012)

18. Peter Eichelsbacher, A.G.: Bayesian inference for Markov chains. J. Appl. Probab. **39**(1), 91–99 (2002)

19. Polgreen, E., Wijesuriya, V.B., Haesaert, S., Abate, A.: Data-efficient Bayesian verification of parametric Markov chains. In: Agha, G., Houdt, B. (eds.) QEST 2016. LNCS, vol. 9826, pp. 35–51. Springer, Cham (2016). doi:10.1007/978-3-319-43425-4_3

20. Poupart, P., Vlassis, N.A., Hoey, J., Regan, K.: An analytic solution to discrete Bayesian reinforcement learning. In: ICML. ACM International Conference Proceeding Series, vol. 148, pp. 697–704. ACM (2006)

21. Quatmann, T., Dehnert, C., Jansen, N., Junges, S., Katoen, J.: Parameter synthesis for Markov models: faster than ever. In: Artho, C., Legay, A., Peled, D. (eds.) ATVA 2016. LNCS, vol. 9938, pp. 50–67. Springer, Cham (2016). doi:10.1007/978-3-319-46520-3_4

22. Ross, S., Pineau, J., Chaib-draa, B., Kreitmann, P.: A Bayesian approach for learning and planning in partially observable Markov decision processes. J. Mach. Learn. Res. **12**, 1729–1770 (2011)

23. Sen, K., Viswanathan, M., Agha, G.: Statistical model checking of black-box probabilistic systems. In: Alur, R., Peled, D.A. (eds.) CAV 2004. LNCS, vol. 3114, pp. 202–215. Springer, Heidelberg (2004). doi:10.1007/978-3-540-27813-9_16

24. Younes, H.L.S.: Probabilistic verification for black-box systems. In: Etessami, K., Rajamani, S.K. (eds.) CAV 2005. LNCS, vol. 3576, pp. 253–265. Springer, Heidelberg (2005). doi:10.1007/11513988_25

Data-Driven Model-Based Detection of Malicious Insiders via Physical Access Logs

Carmen Cheh[1(✉)], Binbin Chen[2], William G. Temple[2],
and William H. Sanders[1]

[1] University of Illinois, Urbana, IL 61801, USA
{cheh2,whs}@illinois.edu
[2] Advanced Digital Sciences Center, Singapore, Singapore
{binbin.chen,william.t}@adsc.com.sg

Abstract. The risk posed by insider threats has usually been approached by analyzing the behavior of users solely in the cyber domain. In this paper, we show the viability of using physical movement logs, collected via a building access control system, together with an understanding of the layout of the building housing the system's assets, to detect malicious insider behavior that manifests itself in the physical domain. In particular, we propose a systematic framework that uses contextual knowledge about the system and its users, learned from historical data gathered from a building access control system, to select suitable models for representing movement behavior. We then explore the online usage of the learned models, together with knowledge about the layout of the building being monitored, to detect malicious insider behavior. Finally, we show the effectiveness of the developed framework using real-life data traces of user movement in railway transit stations.

Keywords: Physical access · Physical movement · Cyber-physical Systems · Insider threat · Intrusion detection · User behavior

1 Introduction

Insider threats are a top concern of all organizations because they are common and can have severe consequences. However, insider threats are very difficult to detect, since the adversary already has physical and cyber access to the organization's assets. Much state-of-the-art research [1] and many state-of-the-practice tools [2,3] focus on the cyber aspect of insider attacks by analyzing the user's cyber footprint (e.g., logins and file accesses). However, the strength of an organization's defense mechanisms is only as strong as its weakest link. By failing to consider the physical aspect of users' behavior, an organization not only leaves itself unable to detect precursor physical behavior that could facilitate future cyber attacks, but also opens itself up to less tech-savvy attacks such as vandalism and theft [4].

© Springer International Publishing AG 2017
N. Bertrand and L. Bortolussi (Eds.): QEST 2017, LNCS 10503, pp. 275–291, 2017.
DOI: 10.1007/978-3-319-66335-7_17

Thus, physical security plays a crucial role in an organization's overall defense posture. This is especially true for critical infrastructure systems such as power grids and transportation systems in which a physical breach can have major real-world effects. Building access controls [5] are often used to limit the areas that users can access based on their role in the organization; this is normally achieved through a relatively static assignment of a set of locations to the user's tracking device (e.g., RFID tag or access card). When a user moves between spaces (e.g., swiping a card at a door), information about this movement is logged.

Although building access control restricts the spaces that a user is able to access, it is merely the first step towards physical security. As with other access control solutions, it faces the same problem of being overly permissive [6]. But denying access to rarely accessed rooms is a costly solution, as it places the burden on administrators to grant every access request, which can lead to severe consequences, especially in time-critical situations (e.g., maintenance). Even with a restrictive set of granted permissions, the access control solutions in place do not take into account the context of a user's access.

Thus, we focus on detecting abnormalities in a user's movement within an organization's buildings. Specifically, we explore how physical access logs collected from a railway transit system can be used to develop a more advanced behavior-monitoring capability for the purposes of detecting abnormalities in a user's movement. In particular, we aim to determine (1) the feasibility of characterizing the movement behavior of users in a complex real-world system, (2) the techniques that can be applied to this detection problem, and (3) the ability to integrate real-time detection into physical security.

We provide a systematic approach to tackle these issues in a way that can be generalized to a diverse set of systems. We observe that since an organization consists of users who have a diverse set of roles, the movement patterns of users in different roles may vary vastly because of their job needs. Instead of proposing a single technique to model all users, we construct a methodical approach that selects the appropriate model based on the context of the organizational role and learns that model from historical data. More specifically, we propose metrics to determine the feasibility of modeling the behavior of certain users in a system. We then construct a model that factors in contextual information such as time and location, and show that the model can be used in an online manner. This study is supported by a set of real-life physical access traces that we collected from our industrial collaborator.

In summary, our contributions in this paper are as follows:

- We show that abnormal movement of users can be detected from physical access logs, thus strengthening a system's physical security.
- We define a framework that characterizes a user's physical movement behavior and learns models of the user's behavior using historical data.
- We evaluate our framework using real-world physical access data obtained from railway transit stations. We show that our metric properly differentiates users, allowing us to use appropriate models of user movement behavior to obtain good false positive and false negative detection rates. We also show the feasibility of performing detection in an online manner.

The structure of the paper is as follows. In Sect. 2, we discuss related work in the domain of anomaly detection of physical movement. Section 3 introduces our case study of railway transit systems, and Sect. 4 describes our framework for detecting malicious insiders, applying it to the case study as an example. The evaluation results are presented in Sect. 5. Finally, future work is summarized in Sect. 6, and the conclusion is given in Sect. 7.

2   Related Work

In this section, we discuss the related work spanning domains from physical movement tracking and prediction to anomaly detection of physical movement and cyber events.

There has been a substantial amount of work on use of cyber logs (e.g., network flows and system logs) to profile users and detect events of interest. For example, Kent et al. proposed *authentication graphs* [7] to profile user behavior and detect threats using computer authentication logs in an enterprise network. In contrast, our work focuses on physical access logs, where physical-world factors, like space and time, directly impact the correlation among different access events. Despite these differences, we also observe the importance of distinguishing different user roles.

In the area of physical access control, there has been work in the route anomaly detection area that looked at people or objects moving in a geographical space that was not delineated by rooms [8–10]. Pallotta et al. [8] and Radon et al. [9] both detect deviations in the trajectory of a vessel in the maritime domain. Their approaches use contextual information, such as the speed of the vessel and weather information, in order to predict the next location of a vessel. However, in the maritime domain, the source and destination of the vessel are already known beforehand, and the anomalies are assumed to arise from the differences in trajectories. This is unlike our work, in which we focus on an indoor setting that has unpredictable destinations for each user.

Dash et al. [10] use mobile data to predict the movement of people in a geographical region. They construct multiple Dynamic Bayesian network models, each of which includes different granularities of context (e.g., day of the week vs. time of day). They predict the next visited location by analyzing the results obtained from each of those models. Unlike their completely data-driven approach of applying all models before computing the best result, we propose a more guided approach by first choosing the appropriate model based on an understanding of a person's past movement data.

In contrast to the work described above, we consider the more restrictive setting of indoor location tracking, which reduces the amount of noise in the data and allows us to identify a user's location with more confidence. Because of physical barriers that prevent a user from moving uninhibited from one space to another, the paths that a user can take are also limited.

For indoor physical access, there has been work in both movement prediction and anomaly detection. In the movement prediction domain, Gellert and

Vintan [11] use *Hidden Markov Models* (HMMs) to predict a user's next location. They use real-world physical access data of four users from a single floor of an office building, although the size and topology of the building are very small. Their results show that a simple Markov model of order 1 gives the best performance. Koehler et al. [12] expand on Gellert's work by using ensemble classifiers to predict how long a user will stay at a given location.

In the anomaly detection domain, different techniques to detect differences in a user's movement have been proposed. Graph models have been studied by Eberle and Holder [13] and Davis et al. [14]. Eberle and Holder [13] detect structural anomalies by extracting common subgraph movement patterns [15]. However, they only consider simplified physical layouts and do not distinguish among different user roles. Davis et al. [14] search labeled graphs for both structural and numeric anomalies and apply their approach to physical access logs in an office building.

Other models, ranging from finite state machines to specific rules, have also been studied. Liu et al. [16] model the normal movements of devices as transitions in finite state machines. Unlike us, they focus on the movement of devices (instead of people) in a hospital setting, where their main goal is to detect missing-device events. Biuk-Aghai et al. [17] focus on suspicious behavioral patterns, including temporal, repetitive, displacement, and out-of-sequence patterns. These patterns only involve the time interval between movements and the reachability of locations rather than the sequence of locations that were visited.

Finally, patents from IBM [18] and Honeywell [19] present the general design of using physical access data to detect potential security incidents. However, they do not discuss detailed designs for dealing with complicated building topology and user roles, and do not provide experimental studies on real-world traces.

3   Motivating Use Case

Physical security is of high priority for industrial control facilities and critical infrastructures. Through a project partnership, we have gained deep knowledge about the physical access control challenges faced by railway transit system operators. We will use this real-world use case to motivate our study.

Background. The railway transit system is an important component of a nation's transportation system. The impact of an attack or fault in the system can be very severe, ranging from loss of service and station blackouts to derailment. For example, a Polish teenager rewired a remote control to communicate with the wireless switch junctions, causing derailment of a train and injury of twelve people [20]. Since the track was accessible by the public, the attack was easily performed. However, in our case study, the underground railway system presents a stronger barrier against such an attack. Potential loss of revenue and human life motivates the need for both physical and cyber security of such systems. In particular, the insider threat is of the utmost importance, as can be seen in the 2006 case in which two traffic engineers hacked into a Los Angeles signal system, causing major traffic disruption [21].

System Architecture. A railway station consists of a single building that may house one or multiple railway lines through it. The general public accesses the railway lines by passing through fare gates in the concourse area and moving to the platform. Figure 1 depicts the topology of the railway station in our case study. In addition to the concourse and platform area, the railway station contains many rooms hidden from the public eye that house the equipment necessary to maintain the running of the station and its portion of the railway track. Each room serves a specific function, and there are multiple rooms that share the same function. The rooms are distributed throughout the station on multiple levels. The railway staff can access those spaces only by swiping their access cards at readers on the doors. Although most of the doors inside the staff-only spaces have card readers, there are a number of doors that allow free access. Different stations have different floor plans, and the number of rooms within a station may vary. However, all the stations share the same types of rooms (e.g., power supply room).

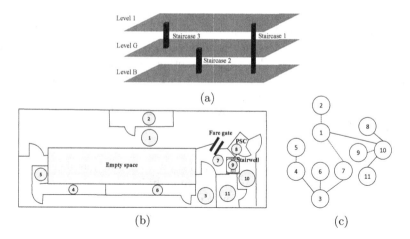

Fig. 1. (a) The different levels of a railway station building with staircases connecting two or more levels. (b) A small sample floor plan of one of the levels. The PSC room represents the Passenger Service Center. (c) Graph representation of (b). Each edge in the graph represents a pair of directed edges between the vertices. Bolded edges imply that a card reader exists on the door bordering the spaces (vertices).

Threat Model. Our threat model focuses on users who have gained physical access to the rooms in a railway station. Those users may be malicious railway staff or an outsider who has gained control over an employee's access control device. Since building access control solutions are in place, we assume the adversary's goal is to tamper with devices in a room to which he or she already has physical access. In a railway station, almost all the rooms house critical assets. Thus, we cannot narrow our focus to any specific portion of the railway station to reduce the space of possible movement trajectories. The level of risk involved in letting an adversary achieve his or her goal is too high. However, restricting a user's access to rooms in a station can also result in severe consequences. Since railway staff require access to rooms in order to conduct maintenance on devices within those rooms, denying them access could cause disruption of service.

Opportunities and Challenges. Unlike an enterprise system for which the office building has a simple, systematic layout across all levels (e.g., a single corridor branching out to multiple rooms), a railway station has a complex non-symmetrical layout. There are multiple paths with varying lengths that a user can take to get from one room to another. This implies that topology is an important factor in determining whether a user's physical movement is anomalous. In addition, the railway transit system consists of diverse user roles (e.g., station operators and power maintenance staff). The job scopes of such users vary in terms of work shifts, responsibilities, and work locations, all of which affect their physical movement behavior. Even users in the same role exhibit different movement behaviors based on their assigned duties and personal habits.

The building access control system that is in place offers a limited view of users' physical movement. Since card readers may fail and certain doors are not outfitted with card readers, we are unable to determine a user's full movement trajectory. A user may also tailgate another user, and thus the access will remain invisible to us. Therefore, it is challenging to detect deviations in a user's movement behavior. We tackle this problem in the next section by integrating knowledge of the system layout and by learning models of users' behaviors from historical physical access data.

Envisioned Monitoring. Currently, a railway system staff member would need to look through the physical access logs manually in order to detect malicious behavior. We aim to reduce the amount of manual effort by automatically presenting a smaller subset of potentially malicious physical accesses in real-time to the staff member. The staff member can then focus his or her attention on the smaller subset, using video surveillance to corroborate evidence of malicious activity. To aid the decision-making, we can also supplement the suspicious accesses with a model of the users' normal behavior.

4   Malicious Insider Detection Framework

In this section, we describe our framework that systematically analyzes users' physical movement logs to detect malicious insiders. The framework dissects the problem into three parts: understanding the characteristics of users' behaviors, learning a suitable model representation, and using the model together with knowledge of the system layout to estimate the probability of an abnormal access.

4.1   Preliminaries and Definitions

We define a system $sys = \{U, Env\}$ (e.g., enterprise organization, critical infrastructure) as the collection of users U who work for it, and the environment Env that contains the system's assets. The environment Env consists of both the physical and cyber aspects of the system. The physical aspect is composed of the building $\mathbf{B}$ and the physical assets $\mathbf{Q}$ within it. The cyber aspect consists of the networked computer system and its digital assets. The cyber and physical aspects are interrelated, but we focus only on the physical aspect in this paper.

We represent the building topology $\mathbf{B}$ as a directed graph $G = (V, E)$ in which the set of vertices represents the spaces S in the building. A directed edge $e(v_1, v_2)$ represents possible movement from v_1 to v_2.[1] For example, the floor plan in Fig. 1b is represented as the graph in Fig. 1c. The set of spaces S can be divided into two covering disjoint subsets: rooms $\mathbf{R}$, and common areas $\mathbf{C}$ (e.g., staircases, corridors). The edges have attributes that represent the access door codes that are associated with user access. We also assign weights to edges based on the spaces to which they are incident.

The state of the system at time t, $State_t(sys)$, is thus defined as the combination of the current location of all users $Loc(u)$ and the state of the environment $State_t(Env)$. The location of the users is defined with respect to the building $\mathbf{B}$, $Loc(U) = \{s | Loc(u) = s \in S, u \in U\}$. The state of the environment $State_t(Env)$ is the condition of the physical and cyber topology and assets (e.g., malfunctioning devices, change in networking access).

4.2   Phase 1: Offline

The framework consists of an offline and an online phase as shown in Fig. 2. The offline phase consists of two stages: characterization of users based on their past movement behavior, and construction of models based on users' characteristics and past movement. The inputs to this phase are the past system states

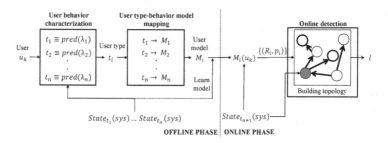

Fig. 2. The framework is divided into offline and online phases, where the offline phase is fed into the online phase.

---

[1] This implies that if $e(v_1, v_2)$ exists, the backward edge $e(v_2, v_1)$ also exists in G.

$State_{Past} = State_{t_1}(sys) \ldots State_{t_n}(sys)$, and the output of this phase is a collection of tuples $(\mathcal{M}, g)$, in which $\mathcal{M}$ is a model representing the movement behavior of a user and $g : State_t(sys) \rightarrow \{R_i, p_i\}$ is a function that takes in the current system state, and uses the model $\mathcal{M}$ to estimate the probabilities p_i of a user's entering a set of rooms R_i.

User Types. The first stage of the offline phase is to distinguish between different users by using their past movement behavior. Typical access control systems assign roles to users based on the sets of rooms that they need to access. However, these roles do not directly reflect the user behavior. Instead, we propose to categorize users according to how they move within a building.

We define the different types of user behavior $\mathbb{T}$ based on the users' "reasons" for movement, where "reason" refers to the context that facilitates users' movement patterns. For each reason or user type $q_i \in \mathbb{T}$, we define a metric λ_i that characterizes the type of behavior that falls under that reason. The metric λ_i takes as input the historical system state pertaining to the user $State_{Past}$ and outputs a real number in $\mathbb{R}$. So we map users to user types, $type(u) = q_i \in \mathbb{T}$, by calculating a predicate function on the output of λ_i, $q_i \equiv pred(\lambda_i)$.

Application: In our railway station case study, there are two main types of user movement behavior. The first type, $q_1 \in \mathbb{T}$, involves users who have a very regular movement behavior, of which the primary members are station operators. Station operators work a fixed set of hours in the station, and, because of their job scope, their movement patterns are fairly consistent. They remain in the *Passenger Service Center* (PSC) to assist the public and monitor the state of the station, visit storerooms and staff rooms, and clock in and out.

The second type of users, $q_2 \in \mathbb{T}$, involves those whose movement is triggered when an event occurs. This applies to maintenance staff who visit rooms to conduct maintenance of the equipment. Different maintenance staff members are in charge of different subsystems (e.g., power supply or signaling), and thus they access different sets of rooms in the station.[2]

In order to categorize a user into q_1 or q_2, we define a metric λ_e based on the approximate entropy of a time series [22] constructed using the collected historical access data. The metric is defined as $\lambda_e = ln(C_m/C_{m+1})$, where C_m is the prevalence of repetitive patterns of length m in the time series. Each subsequence of length m in the sequence is compared to other subsequences. If the number of similar subsequences is high, then C_m is large. This metric is shown to be able to quantify the predictability of user movement [23,24]. We choose m to be 3, which provides a good metric for characterizing our trace as shown in Sect. 5. If the user's entropy value is low, the user belongs to q_1; otherwise, the user belongs to q_2. In other words, $q_1 \equiv (\lambda_e(u) < \mathbf{E})$, where $\mathbf{E}$ is a numerical threshold. The choice of parameter $\mathbf{E}$ is discussed in Sect. 5.

---

[2] This may apply to other systems too. E.g., a security guard doing rotations in a building belongs to q_1, and a technical support staff member who goes to an office when his or her assistance is required belongs to q_2.

User Behavior Models. Next, we construct behavior models for each behavior type $q \in \mathbb{T}$ defined earlier. Since each user is motivated to move within the building for different reasons, it is not possible to specify a single model for all users' behavior. Such a model would be inherently biased towards a certain set of users and perform badly for others.

Instead, for each $q \in \mathbb{T}$, we select an appropriate modeling technique $\mathcal{M} \in \mathbb{U}_{\mathcal{M}}$ from a large set of possible modeling techniques $\mathbb{U}_{\mathcal{M}}$. The model should leverage q's distinct characteristics and provide insight into the likelihood that a user will access a room given the current system state $State_{t_{n+1}}(sys)$.

For each user u that has $type(u) = q \in \mathbb{T}$, we learn the model by analyzing past system states $State_{Past}$ in order to assign probabilities to the rooms in $\mathbf{R}$. Finally, we define the function g that takes the state $State_{t_{n+1}}(sys)$ and use the learned model $\mathcal{M}$ to determine the probabilities associated with the user's entering a set of rooms next. Based on $State_{t_{n+1}}(sys)$, $\mathcal{M}$ will calculate and return a set $\{R_i, p_i\}$, where $R_i \in \mathbf{R}$ is a room in the building and $p_i \in [0, 1]$ is the probability that the user will access R_i next.

Application: Users belonging to q_1 have a low entropy value $\lambda_e(u) < \mathbf{E}$. This implies that their movement patterns are highly predictable and repetitive. Thus, we choose to represent a user's movement behavior with a Markov model[3]. The states in the Markov model are the set of rooms $\mathbf{R}$, and a transition from state i to state j implies that a user visits room R_j after R_i.

Given the previous system states $State_{Past}$, we learn the Markov model of a user u. The system state at any point of time $State_{t_i}(sys)$ contains the physical access records for u. We can reconstruct the full movement sequence $Seq(u) = R_1 \ldots R_n$ as the sequence of rooms that were visited. The sequence $Seq(u)$ can be divided into segments based on the lengths of the time intervals between consecutive physical accesses. Each segment represents a series of movements that occur close together in time. A period of inactivity (more than 3 hours) separates any two segments. The initial probability vector $\pi(0)$ is the normalized frequency with which each room $r \in \mathbf{R}$ appears at the beginning of each segment of $Seq(u)$. The transition probability p_{ij} is the normalized frequency with which the user visits R_i and then R_j.

However, the users belonging to q_2 have less regular movements and may change movement patterns based on events in the system. So we combine the Markov model with additional contextual knowledge about the states of the devices in the rooms. After vetting the accesses through the Markov model, we correlate the remaining suspicious accesses with logs about device state. Intuitively, if a device in room R_d fails and then a physical access into R_d is logged, that physical access is considered non-malicious. Then, given the device failure incidents in $State_{t_{n+1}}(sys)$, the probability p_d associated with device failures in

---

[3] Although the Markov model imposes certain assumptions about the movement behavior, such as the memoryless property, it can be extended to include temporal and spatial correlations. We intend to explore these extensions in future work.

room R_d in the set $\{R_i, p_i\}$ is suitably changed such that any accesses leading to R_d are considered non-malicious.

4.3   Phase 2: Online

The online phase involves determining, based on the behavior models derived from the offline phase, whether a user's access is an abnormality. The inputs to this phase are the tuples $(\mathcal{M}, g)$ from the offline phase and the current state of the system $State_{t_{n+1}}(sys)$. The output of this phase is a real number in $\mathbb{R}$ that indicates the degree of abnormality of the access. The algorithm for this phase is given below in the ONLINEDETECTION function.

The current system state $State_{t_{n+1}}(sys)$ includes the location of the user $Loc(u) = R_1 \in \mathbf{R}$ and the physical access that is being made, $A = S_1 \to S_2$, $S_1, S_2 \in S$. In other words, the user is moving from S_1 to S_2. We update the user's behavior model to reflect the current state of the user in the system by computing $g(State_{t_{n+1}}(sys))$. The output is the set $\{R_i, p_i\}$, where $R_i \in \mathbf{R}$ is a room and $p_i \in [0,1]$ is the probability that the user will access R_i next. Using knowledge of the building topology $\mathbf{B}$, we determine the likelihood that the access A is anomalous based on the paths from the user's current location to the set of rooms R_i.

Given the access A, we want to determine all the rooms that the user is likely to access. We first find all the rooms that are reachable from S_2, i.e., $P_T = \{R_i | \exists path(S_2, R_i)\}$. For all such vertices $R_i \in P_T$, we decide whether the user is likely to access R_i by moving to S_2 from S_1. If it's easier to access R_i from S_2, then we consider R_i as one of the likely rooms. To decide whether R_i is easily accessed, we calculate path lengths using the weighted edges. We calculate the shortest path from S_1 to R_i, $d(S_1, R_i)$, and compare it to the shortest path from S_1 to R_i through S_2, $d(S_1, S_2) + d(S_2, R_i)$. If the shortest path through S_2 is similar in length to the shortest path, then we consider R_i as a possible room that the user wants to access. With the resulting shortlisted set of rooms, we sum up their likelihoods $\sum_{R_i \in P_T} p_i$ to obtain a final score. If the score is below a threshold value Z, access A is deemed anomalous.

Algorithm 1. ONLINEDETECTION algorithm

Require: $(Loc(u), A = S_1 \to S_2) \in State$
  function ONLINEDETECTION$(State, g)$
    $score \leftarrow 0; \{R_i, p_i\} \leftarrow g(State)$
    for all $R_i \in \{R_i, p_i\}$ **do**
      $shortestlen \leftarrow GetShortestPath(S_1 \to R_i)$
      $len \leftarrow GetShortestPath(S_2 \to R_i) + e_w(S_1, S_2)$
      if $len < shortestlen \times k$ **then** $score \leftarrow score + p_i$ **end if**
    end for
    return score
  end function

Application: We keep track of the system state, which is the room that the user has last accessed: $State_{t_{n+1}}(sys) = r_C \in \mathbf{R}$. The function g takes $State_{t_{n+1}}(sys)$

and returns the set of probabilities associated with the next visited room $\{R_i, p_i\}$. In ONLINEDETECTION, we rely on edge weights to calculate path lengths. We assign all edges a weight of 1, with the exception of edges that connect different levels of the building (i.e., staircases, elevators, and escalators). We assume that users prefer to take as few staircases as possible. So we assign a weight of 10 to those edges that connect different levels.

We compare the score returned by ONLINEDETECTION with threshold Z. We choose Z based on the probability distribution p_{r_C}. If the probability distribution has a heavy tail, then there are rooms that the user very rarely visits and may be deemed suspicious. So we choose the threshold value as the 95th percentile. Otherwise, we choose the threshold value as the minimum probability in the distribution. The percentile value can be changed by practitioners based on the system requirements; a higher value reduces the false positives, but potentially malicious movements are missed, while a lower value catches more malicious movements but increases the false positives. Since our results focus more on the trends, the exact value of this threshold is not critical.

5   Evaluation

In this section, we utilize real-world data traces to demonstrate the effectiveness of our framework in our railway transit station case study. First, by evaluating our usage of the entropy metric, we answer the question of whether the movement behavior of users can be characterized effectively in a complex system. Second, we determine the detection capability of our proposed behavior models. Finally, we examine the possibility of detecting malicious movement in an online manner.

5.1   Experiment Setup

We use a real-world data set containing physical card accesses to a railway station in a city. The duration of the accesses is from June to October 2016. The station has 62 rooms, with a total of 32,100 accesses made by 314 users. While we focus on one station in this work, the whole railway line consists of 33 stations, 12 of which are interchange stations. We estimate that the average number of accesses per hour over all the stations is approximately 450, whereas the highest number of accesses per hour is around 1,200. This poses a significant challenge if the associated logs need to be examined manually.

The data set contains the following information regarding physical accesses: (1) date and time, (2) door code, (3) user identification, and (4) result of access (success or failure). When the access is a failure, it implies either that the user's card had expired or that the user did not have permission to access the room. Those failed accesses serve as ground truth for known abnormal accesses.

We simulated malicious movement in order to conduct a more thorough assessment of the detection ability. For each user, we injected accesses into the testing data. With a certain small probability, we replaced a legitimate access $A = S_1 \rightarrow S$ with a series of injected accesses. We randomly selected a target room $R_T \in \mathbf{R}$ and calculated the shortest path from S_1 to R_T as $S_1 e_1 S_2 \ldots S_n e_n R_T$. For each edge $e_i, i \in [1, n]$ that has a door code, we added an injected access $A_i = S_i \rightarrow S_{i+1}$.

We split the data set into 80% training and 20% testing subsets and performed 10-fold cross-validation. We conducted the experiments on a Windows 7 Home Premium machine with a 2.7 GHz CPU core and 4 GB of RAM.

5.2   Results

In this subsection, we present the evaluation results for our approach from Sect. 4 based on the physical card accesses data from the railway station.

Implementation Performance. We evaluated the running time of both the offline and online phases. The average running time of the construction of Markov models in the offline phase was 33 ms, whereas the average running time of the ONLINEDETECTION function in the online phase was 1.3 ms. The offline phase can be conducted sporadically during system downtime, whereas the online phase is fast enough to be executed in a real-time manner.

Detection Capability. Our approach marked 2,975 out of 32,100 accesses as suspicious. Hence, the practitioner's effort would have been reduced by over 90%. Figure 3 shows the number of physical accesses over the ten testing subsets. For each subset, the left bar represents all the accesses, and the right bar represents the accesses marked as malicious. We can see that most of the injected accesses and malicious ground truth data are detected as malicious. The numbers of false positives are also low and fairly constant over the ten subsets. On average, our approach gives a false positive rate of 0.08 and a false negative rate of 0.34.

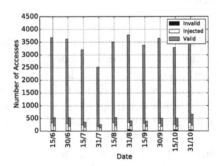

Fig. 3. The number of physical accesses over time. Each tick on the x-axis represents a two-week period; the label indicates the end date of the period. Each bar is divided into three sections representing the valid (or non-malicious) accesses, the malicious ground truth accesses, and injected accesses.

To interpret this result in more detail, we compare our solution with a baseline method that marks any access leading to a previously unvisited room as malicious. It is easy to see that both solutions can identify malicious paths that lead to any previously unvisited room. However, if an attacker carefully selects his or her path by moving only to previously visited rooms, the baseline method will not be able to identify any of those paths (i.e., its false negative rate will be 100%). In comparison, our solution can still raise an alarm if the path covers any unusual transitions among previously visited rooms. The reason that our false negative rate in Fig. 3 is relatively high (i.e., 0.34) is that we randomly choose a destination room and generate the shortest path to that destination; thus, a substantial fraction of the generated malicious paths are indistinguishable from legitimate paths that a user actually traveled before. In other words, since most of the generated paths are short, it becomes impossible in a certain fraction of cases to differentiate anomalous and normal movement behavior.

To study how the length of the attacker's path affects the performance of our approach, we experimented with increasing the length of the malicious path we injected (to consider the case when an attacker wanders around the space to do a site survey and explore potential attack opportunities). Instead of injecting paths that ended at a room, we randomly generated paths that went through a sequence of previously visited rooms. We varied the number of visited rooms in the path; the results are presented in Fig. 4a. The baseline method is still unable to detect any of these malicious paths, regardless of their lengths. In contrast, the probability of our method's detecting the path approaches 100% as the path grows longer.

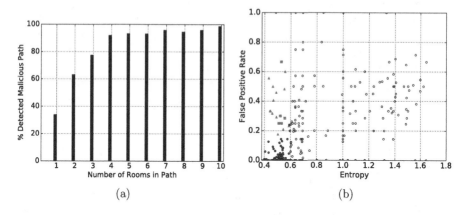

(a)   (b)

Fig. 4. (a) The percentage of detected malicious paths vs. number of rooms in the path. (b) The distributions of false positive rates with respect to the user's entropy.

User Characterization. Next, we evaluate whether the entropy metric defined in Sect. 4 is suitable for characterizing user behaviors. If the entropy metric can differentiate user behaviors, then the Markov models constructed for users with low entropy ($q_1 \in \mathbb{T}$) will have a better detection capability (lower false positive and negative rates) than those constructed for users with high entropy. We plot the entropy vs. false positive rates in Fig. 4b. Each point in Fig. 4b represents a single user in one subset. One user should map to a maximum of 10 points in the plot. We can see that almost all users whose entropy is below 0.6 (filled markers) have low false positive rates. When the entropy is above 0.6 (unfilled markers), the false positive rates are high. So we can set our entropy threshold **E** to 0.6 in order to distinguish between the user types. Then, 15% of the users would belong to q_1 and 85% to q_2. Although fewer users are categorized under q_1, these users account for 79% of the accesses. Thus, having a low false positive rate for these users implies that entropy is a suitable metric for characterizing user behavior.

However, several outliers show a high false positive rate for q_1-type users. We studied each of them individually and found that there were reasons why the accesses were marked as suspicious. The users represented by triangles in Fig. 4b accessed rooms that they had not previously visited, whereas the users represented by squares had a small testing set (<5 accesses), so their false positive rates are disproportionately high. In actual operation, the training data set and real-time accesses will be much larger, so there won't be outliers.

Integration of Device State. In Sect. 4, we proposed to correlate logs about device state with physical access logs in order to decrease the false positives for the q_2-type users. We assume that the timing information in these logs is synchronized with that in the physical access logs. The logs collected for each type of device (e.g., breaker or lights) are different, and thus the amount of information about the device's state that can be extracted varies. However, we only need to know when a device fails, since the failure could trigger entrance of a maintenance staff member (user of type q_2) into the room to repair the device. In particular, there are four rooms in the station for which we could identify the failure of a device from the logs with particularly strong confidence. These rooms contained devices that controlled the environment in the station (e.g., air chiller and water pumps). We extracted the textual description and alarm values that indicated device failure and searched the device logs for failure incidents. We compared the timestamps of the failure incidents to the times of the users' accesses. If the user was not in the room prior to the failure, entered the room after the detected failure, and subsequently left the room when the maintenance was complete, then we consider that physical access to be non-malicious. As a result, we reduced the false positives for a subset of the users by an average of 0.45 for the four rooms. This preliminary result shows that we can use additional logs regarding the system environment to determine whether an access is malicious.

Online Detection. We determined the feasibility of detecting malicious movement in an online manner by studying how early on a malicious path of a certain length can be detected. We considered injected paths with a length of 4, and the false negative rates for the first, second, third, and fourth injected accesses in the path were 0.54, 0.25, 0.09, and 0, respectively, in our experiment. This shows that our approach is able to detect malicious paths (with a certain minimum length) with high confidence, and even before an attacker reaches the destination room.

6   Discussion and Future Work

In this paper, we present the first step towards understanding how physical access logs can be used to enhance the detection capability of a system. In our case study of railway transit stations, we characterize two different types of user movement behavior. The first user type, q_1, performs well in terms of false positive rates. In ongoing work, we added a notion of time into the states of the Markov model and applied it specifically to the set of station operators in q_1. By separating the station operators into a third user type and honing the model, we have obtained encouraging reductions in false positive rates. For the second user type q_2, the false positive rates are much higher than q_1's. We have shown that using knowledge of the device states improves the false positive rates. However, many issues need to be resolved, such as time drifts between the device logs and physical access logs, differences in contextual understanding of diverse device logs, and missing data regarding device state. We intend to address these issues and pursue this line of thought in future work.

We can also enhance our Markov model further by taking into account the amount of time a user spends in a room, and the function of the room (e.g., storeroom vs. power room). The parameters that we use in our approach can be further tuned and targeted to different users for enhanced detection capability.

In this paper, we only create movement models for each user in isolation. Thus, we do not handle colluding insiders who may tailor their movements such that both parties remain within their movement patterns, but they are able to achieve their malicious goal together. We need to have a more comprehensive view of the system and user movements as a whole in order to tackle such adversaries, and we are currently pursuing this direction by using richer models.

This paper also shows favorable results in using online-based detection in a real-world system. If we can detect a malicious physical access early on in a user's movement, we can make suitable responses to prevent a potential breach. For example, an administrator can temporarily remove a user's permissions to certain critical rooms, or place the user under further observation.

7   Conclusion

One way in which organizations address insider threats is through physical security. However, the state of the art in building access control is lacking. In this paper, we study the use of physical access logs for detecting malicious movement within a building. We propose a systematic framework that uses knowledge of the system and its users in order to analyze physical access logs. We characterize users by using a set of metrics that take historical physical access data as input. Each user type is mapped to a behavior model, and the details of the model are learned through use of the user's past physical accesses. Finally, we develop an online detection algorithm that takes the behavior model and the building topology as input, and returns a score indicating the likelihood that the user's access is anomalous. We apply our framework to a real-world data trace of physical accesses in railway stations. The results show that our framework is useful in analyzing physical access logs for the purpose of detecting malicious movement.

Acknowledgements. This work was supported in part by the National Research Foundation (NRF), Prime Minister's Office, Singapore, under its National Cybersecurity R&D Programme (Award No. NRF2014NCR-NCR001-31) and administered by the National Cybersecurity R&D Directorate, and supported in part by the research grant for the Human-Centered Cyber-physical Systems Programme at the Advanced Digital Sciences Center from Singapore's Agency for Science, Technology and Research (A*STAR). This work was partly done when Carmen Cheh was a research intern at ADSC. We also want to thank the experts from SMRT Trains LTD for providing us data and domain knowledge.

References

1. Salem, M., Hershkop, S., Stolfo, S.J.: A survey of insider attack detection research. In: Stolfo, S.J., Bellovin, S.M., Keromytis, A.D., Hershkop, S., Smith, S.W., Sinclair, S. (eds.) Insider Attack and Cyber Security: Beyond the Hacker. AIS, vol. 39, pp. 69–90. Springer, Boston (2008)
2. Alien Vault: Insider threat detection software (2016). https://www.alienvault.com/
3. Insider threat security & detection (2016). http://www.tripwire.com/
4. CERT Insider Threat Center: Insider threat and physical security of organizations (2011). https://insights.sei.cmu.edu/insider-threat/2011/05/insider-threat-and-physical-security-of-organizations.html
5. Luallen, M.E.: Managing insiders in utility control environments. Technical report, SANS Institute (2011)
6. Bauer, L., Cranor, L.F., Reeder, R.W., Reiter, M.K., Vaniea, K.: Real life challenges in access-control management. In: Proceedings of ACM SIGCHI Conference on Human Factors in Computing Systems, pp. 899–908 (2009)
7. Kent, A.D., Liebrock, L.M., Neil, J.C.: Authentication graphs: analyzing user behavior within an enterprise network. Comput. Secur. **48**, 150–166 (2015)
8. Pallotta, G., Jousselme, A.L.: Data-driven detection and context-based classification of maritime anomalies. In: Proceedings of 18th International Conference on Information Fusion, pp. 1152–1159 (2015)
9. Radon, A.N., Wang, K., Glasser, U., Wehn, H., Westwell-Roper, A.: Contextual verification for false alarm reduction in maritime anomaly detection. In: Proceedings of IEEE International Conference on Big Data, pp. 1123–1133 (2015)
10. Dash, M., Koo, K.K., Gomes, J.B., Krishnaswamy, S.P., Rugeles, D., Shi-Nash, A.: Next place prediction by understanding mobility patterns. In: Proceedings of IEEE International Conference on Pervasive Computing and Communication Workshops, pp. 469–474 (2015)
11. Gellert, A., Vintan, L.: Person movement prediction using hidden Markov models. Stud. Inf. Control **15**(1), 17–30 (2006)
12. Koehler, C., Banovic, N., Oakley, I., Mankoff, J., Dey, A.K.: Indoor-ALPS: an adaptive indoor location prediction system. In: Proceedings of ACM International Joint Conference on Pervasive and Ubiquitous Computing, pp. 171–181 (2014)
13. Eberle, W., Holder, L.: Anomaly detection in data represented as graphs. Intell. Data Anal.: Int. J. **11**(6), 663–689 (2007)

14. Davis, M., Liu, W., Miller, P., Redpath, G.: Detecting anomalies in graphs with numeric labels. In: Proceedings of 29th ACM Conference on Information and Knowledge Management, pp. 1197–1202 (2011)
15. Eberle, W., Holder, L., Graves, J.: Detecting employee leaks using badge and network IP traffic. In: IEEE Symposium on Visual Analytics Science and Technology, October 2009
16. Liu, C., Xiong, H., Ge, Y., Geng, W., Perkins, M.: A stochastic model for context-aware anomaly detection in indoor location traces. In: Proceedings of IEEE 12th International Conference on Data Mining, pp. 449–458 (2012)
17. Biuk-Aghai, R.P., Si, Y.W., Fong, S., Yan, P.F.: Individual movement behaviour in secure physical environments: modeling and detection of suspicious activity. In: Cao, L., Yu, P.S. (eds.) Behavior Computing, pp. 241–253. Springer, London (2012)
18. Hoesl, M.J.: Integrated physical access control and information technology security U.S. Patent No. 6641090 B2, granted on 17 June 2014
19. Khurana, H., Guralnik, V., Shanley, R.: System and method for insider threat detection U.S. Patent No. 8793790 B2, granted on 29 July 2014
20. Baker, G.: Schoolboy hacks into city's tram system, 11 January 2008. http://www.telegraph.co.uk/news/worldnews/1575293/Schoolboy-hacks-into-citys-tram-system.html
21. Grad, S.: Engineers who hacked into L.A. traffic signal computer, jamming streets, sentenced, 1 December 2009. http://latimesblogs.latimes.com/lanow/2009/12/engineers-who-hacked-in-la-traffic-signal-computers-jamming-traffic-sentenced.html
22. Pincus, S.M.: Approximate entropy as a measure of system complexity. Proc. Nat. Acad. Sci. 88(6), 2297–2301 (1991)
23. Li, X.: Using complexity measures of movement for automatically detecting movement types of unknown GPS trajectories. Am. J. Geogr. Inf. Syst. 3(2), 63–74 (2014)
24. Song, C., Qu, Z., Blumm, N., Barabási, A.L.: Limits of predictability in human mobility. Science 327(5968), 1018–1021 (2010)

Tools

Tulsa: A Tool for Transforming UML to Layered Queueing Networks for Performance Analysis of Data Intensive Applications

Chen Li[1](✉), Taghreed Altamimi[2], Mana Hassanzadeh Zargari[2],
Giuliano Casale[1], and Dorina Petriu[2]

[1] Imperial College London, London SW7 2AZ, UK
{chen.li1,g.casale}@imperial.ac.uk
[2] Carleton University, Ottawa K1S 5B6, Canada
{taghreedaltamimi,manazargar,petriu}@sce.carleton.ca

Abstract. Motivated by the problem of detecting software performance anti-patterns in data-intensive applications (DIAs), we present a tool, Tulsa, for transforming software architecture models specified through UML into Layered Queueing Networks (LQNs), which are analytical performance models used to capture contention across multiple software layers. In particular, we generalize an existing transformation based on the Epsilon framework to generate LQNs from UML models annotated with the DICE profile, which extends UML to modelling DIAs based on technologies such as Apache Storm.

1 Introduction

The objective of our research is to design tools for iteratively enhancing the quality of data-intensive applications (DIAs) that leverage Big Data technologies hosted in private or public clouds. We consider that the DIAs are developed in a DevOps process, where the developers obtain runtime monitoring information, especially performance metrics, and reflect them back into design time models to reason about system performance improvements. In order to achieve that, a performance model needs to be generated from the DIA architecture model and runtime information. In this work, we use the Unified Modeling Language (UML) to specify the software architecture at the design stage. The architecture model characteristics (see Sect. 3) and its performance attributes are mainly captured by DICE profile [2], a recently proposed UML profile to annotate technology specific aspects of Storm, Hadoop and Spark into UML diagrams. DICE profile extends the standard MARTE profile [5], so it inherits the MARTE stereotypes for non-functional properties and performance attributes [4]. The specific problems we consider in this paper is how to annotate the runtime performance measurements in the UML model, and how to transform the UML model into the performance model for subsequent performance analysis.

This work is partially supported by the European Commission grant no. 644869, DICE.

© Springer International Publishing AG 2017
N. Bertrand and L. Bortolussi (Eds.): QEST 2017, LNCS 10503, pp. 295–299, 2017.
DOI: 10.1007/978-3-319-66335-7_18

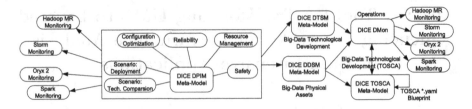

Fig. 1. High level abstract view of DICE profile

Several approaches have been proposed for generating performance models, such as queueing networks [1], stochastic Petri nets [3] and layered queueing networks (LQNs) [6] from architecture models. While these studies remain relevant, the advent of Big Data has popularized technologies such as Apache Storm and Hadoop in the implementation of DIAs. However, there is a shortage of methods for specifying UML models for these DIAs and automatically deriving performance models.

In this paper, we focus on Storm applications and transform the corresponding UML model into a performance LQN model. There are three reasons for choosing LQNs. First, a Storm topology may be seen as a network of buffers and processing elements that exchange messages, so it is quite natural to map them into a queueing network model. Second, the core elements of LQN models are semantically similar to the corresponding elements of UML activity and deployment diagrams. Third, LQN solvers such as LINE[1] or LQNS[2] are available to provide analytical methods to solve the LQN model. This paper proposes a new tool called Tulsa, which leverages DICE profile as a better way of annotating DIA UML models and transforms them to LQN models. Our work extends an existing UML+MARTE-to-LQN transformation based on the Epsilon framework[3] to leverage specific stereotypes of the DICE profile in the generation of LQN models [6].

2   DICE Profile

The DICE profile expresses some familiar model-driven architecture concepts for DIAs. In particular, the DICE profile offers three new models, called DICE Platform Independent Model (DPIM), DICE Technology Specific Model (DTSM), and DICE Deployment Specific Model (DDSM) [2,4]. Figure 1 shows the high level abstract view of DICE profile. DPIM provides an abstract specification of the DIA architecture, allowing the inclusion of computation nodes and storage nodes. At this abstraction layer, DPIMs help the developer to define a high-level topology, the main services exposed by the DIA and their QoS requirements.

---

[1] http://line-solver.sourceforge.net/.

[2] https://github.com/layeredqueuing/V5/tree/master/lqns.

[3] http://www.eclipse.org/epsilon/.

The DTSM layer is a refinement of the DPIM layer that encompasses techno-logical decisions. For example, data processing needs are detailed in a DTSM through configuration requirements for appropriate Big Data technologies, such as Hadoop. Lastly, DDSM enables the designer to specify deployment decisions on cloud infrastructures. In the DICE framework, such decisions can be subse-quently translated into a concrete deployment blueprint based on TOSCA [7]. Our tool mainly focuses on the DTSM and DDSM layers, which are appropriate for performance evaluation.

3   Model Transformation

Comparing with our previous work in [6], Tulsa not only implements model transformation for general distributed systems, but also supports Storm applica-tions by leveraging DICE profile. The underpinning scripts are mainly written in Epsilon Object Language (EOL) and Epsilon Transformation Language (ETL). Tulsa supports complex workflow which is assembled by a set of ANT tasks.

3.1   Model Mapping

The source model accepted by Tulsa considers two types of UML diagrams: deployment and activity diagram. The deployment diagram specifies the sys-tem configuration, e.g., indicating the functional components, assigning key attributes and defining constraints, and the activity diagram defines the behav-ior of the system. Table 1 shows the corresponding mapping from UML model annotated with DICE and MARTE stereotypes to LQN model.

DICE UML model uses *Device* to stand for a VM cluster or a single server. DDSM provides a stereotype ≪*DdsmVMsCluster*≫ to capture characteristics of the VM cluster, e.g., *instances* tag means the number of single server in the cluster. *ExecutionEnvironment* represents the platform which is deployed on the VM. DDSM provides a stereotype ≪*DdsmStormCluster*≫ to annotate the Storm platform. The related single servers are nested in this *ExecutionEnvi-ronment*. Tulsa transforms a single server (i.e., *Device*), which provides services to Storm platform, to a *Processor* in LQN model. An *Artifact* is used or pro-duced by a software development process or deployment and operation of a system, e.g., software component. An *Artifact* can be transformed into a *Task* which stands for the software component in LQN model. In Storm applications, there are two types of *Artifact* called *Spout* and *Bolt*. DTSM defines stereotypes ≪*StormSpout*≫ and ≪*StormBolt*≫ for them respectively. These stereotypes provide tags for specifying the level of parallelism and the execution time, e.g., *parallelism* (i.e., specifying the number of threads).

Due to space limitations, we only describes some core elements and stereo-types which are mainly considered for Storm applications. More details on the elements with MARTE stereotypes can be found in [6].

Tulsa is available at https://github.com/dice-project/DICE-Enhancement-APR. Pre-requirements of running the Tulsa are to install JDK 8, Eclipse 4.6.1,

Table 1. Model mapping: from UML + DICE + MARTE to LQN element

UML model element	DICE + MARTE stereotype	LQN model element
model	None	lqnmodel
Deployment diagram		
Device	GaExecHost, DdsmVMsCluster	Processor
ExecutionEnvironment	DdsmStormCluster	None
Artifact	Scheduler, DdsmBigDataJob, StormSpout, StormBolt	Task
Activity diagram		
AcceptEventAction	GaStep	Entry
InitialNode	GaWorkloadEvent	Entry
OpaqueAction, CallOperationAction, SendSignalAction	GaStep	Activity
DecisionNode, MergeNode, JoinNode, ForkNode	None	Precedence
ControlFlow	StormStreamStep	Precedence, synch-call, asynch-call

```
    lqnDICE.lqnx
  1 <?xml version="1.0" encoding="us-ascii"?>
  2 <lqn-model xmlns:xsi="http://www.w3.org/2001/XMLSchema-instance" xsi:noNamespaceSchemaLocation="/usr/local/share/
  3   <processor name="Server_Spout" multiplicity="2" scheduling="fcfs">
  4     <task name="WikiArticleSpout" multiplicity="8" scheduling="ref" think-time="40.0">
  5       <entry name="StrartPoint" type="PH1PH2">
  6         <entry-phase-activities>
  7           <activity name="OpaqueAction1" host-demand-mean="2.0" phase="1"/>
  8           <activity name="CallOperationAction6" host-demand-mean="0.1" phase="2">
  9             <asynch-call dest="AcceptEventAction3" calls-mean="1.0"/>
 10           </activity>
 11         </entry-phase-activities>
 12       </entry>
 13     </task>
 14   </processor>
 15   <processor name="Server_Link1" multiplicity="2" scheduling="fcfs">
 16     <task name="LinkCounterBolt" multiplicity="2" scheduling="fcfs">
 17       <entry name="AcceptEventAction3" type="PH1PH2">
 18         <entry-phase-activities>
```

Fig. 2. Fragment of a generated LQN model

Table 2. Performance results produced by the LQN solver

Processor	Throughput	Service-time	Utilization
Server_Spout	0.189944	2.117742	0.398882
Server_Link1	0.189944	0.9	0.170949
Server_Link2	0.189944	4.28414	0.60782

the DICE Profile [4] and the Epsilon framework. Figure 2 and Table 2 show a screenshot of a LQN model and the results produced by lqns solver respectively.

4   Conclusion

In this paper we have presented Tulsa, a tool for transforming a UML model annotated with DICE profile into a LQN model. Tulsa is a part of the DICE project, whose objective is to define a quality-driven framework for developing data-intensive applications that leverage Big Data technologies hosted in private or public clouds. Further development of the tool is expected to support other DIAs such as Hadoop and Spark.

References

1. Dubois, D.J., et al.: Model-driven application refactoring to minimize deployment costs in preemptible cloud resources. In: CLOUD 2016. IEEE Press, USA (2016)
2. Casale, G., Ardagna, D., Artac, M., et al.: DICE: quality-driven development of data-intensive cloud applications. In: MiSE 2015, pp. 78–83. IEEE Press (2015)
3. Merseguer, J., Campos, J.: Software performance modeling using uml and petri nets. Perform. Tools Appl. Netw. Syst. **2965**, 265–289 (2004)
4. D2.1 Design and quality abstractions. http://www.dice-h2020.eu/deliverables/
5. UML Profile for MARTE: Modeling and Analysis of Real-Time Embedded Systems, Version 1.1, Object Management Group (2011)
6. Altamimi, T., Zargari, M.H., Petriu, D., Performance analysis roundtrip: automatic generation of performance models and results feedback using cross-model trace links, In: CASCON 2016, Toronto, Canada. ACM Press (2016)
7. D2.3 Deployment abstractions. http://www.dice-h2020.eu/deliverables/

Modelling and Performance Evaluation with TimeNET 4.4

Armin Zimmermann[⊠]

Department of Computer Science and Automation, Systems and
Software Engineering Group, TU Ilmenau, Ilmenau, Germany
armin.zimmermann@tu-ilmenau.de
http://timenet.tu-ilmenau.de

Abstract. The paper presents the current status of the software tool
TimeNET. It supports modeling and performance evaluation of stochastic models, including extended deterministic and stochastic Petri nets,
colored stochastic Petri nets, and Markov chains as well as UML extensions. Among its main characteristics are simulation and analysis modules for stationary and transient evaluation of Petri nets including non-
exponentially distributed delays, as well as a simulation module for complex colored models. Recent enhancements include algorithms for the efficient rare-event simulation of Petri nets, a new multi-trajectory hybrid
simulation/analysis algorithm, and a net class for Markov chains.

Keywords: Modeling tool · TimeNET · Stochastic Petri nets · Colored
Petri nets · Performance evaluation

1 Introduction

TimeNET is a software tool for the modeling and performability evaluation with
several variants of stochastic Petri nets including GSPNs, eDSPNs, and colored
SCPNs (for definitions see e.g. [8]). In comparison to other related tools such
as GreatSPN [1], SPNP [4], Möbius [3] and CPN Tools [5], it supports evaluation of models combining exponential and deterministic as well as more general
non-exponentially distributed firing delays. Numerical analysis and simulation
methods both for transient and steady-state solution have been implemented
as well as structural analysis modules. Moreover, TimeNET supports colored
stochastic Petri nets as well as rare-event simulation algorithms for these model
classes. The token game can be run interactively or automatically to validate
and test eDSPN and SCPN models.

The software architecture contains a Java graphical user interface, shell
scripts controlling analysis processes, and evaluation algorithms implemented
mainly in C++ running as background processes. The tool runs in 32 and
64 Bit Linux and Windows environments. It is available free of charge for
non-commercial use from http://timenet.tu-ilmenau.de/. Successful applications
reported in the literature include communication systems, reliability evaluation,

© Springer International Publishing AG 2017
N. Bertrand and L. Bortolussi (Eds.): QEST 2017, LNCS 10503, pp. 300–303, 2017.
DOI: 10.1007/978-3-319-66335-7_19

manufacturing and transportation, and business as well as logistics processes. Numerous papers including application examples are listed in the tool's web page.

2   New Features in TimeNET 4.4

This paper presents changes in TimeNET since the previous tool description [9], which covered version 4.1 in 2012. More in-depth coverage of history and tool architecture can be found in [8,9].

Among the various changes in the tool since 2012, the scientifically most relevant extensions cover rare-event simulation methods motivated by reliability applications. Such examples will otherwise lead to unacceptably long run times because of the number of events to be simulated until enough samples of interest are generated.

An example application model is shown in Fig. 1, describing a sample network architecture of the Avionic Full-Duplex Ethernet used in modern aircraft [11]. Reliability of such systems is a major concern, and a model-based analysis can show that the required end-to-end message delays are achievable for a certain setup. This is a typical example of industrial systems in which numerical analysis is impossible because of the large state space and concurrent non-Markovian activities, while standard simulation would need exceedingly high run times to compute the results with acceptable statistical accuracy.

While rare-event simulation is a well-known technique for the efficient evaluation of highly reliable systems, the available algorithms require significant

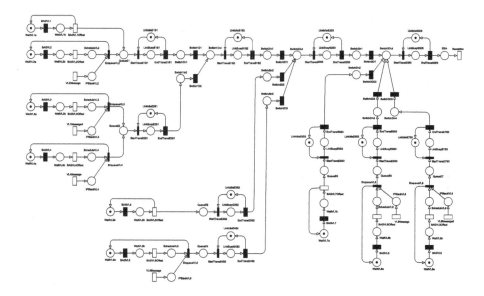

Fig. 1. AFDX network model

background knowledge or apply to quite restricted model classes only. Our goal is to make such methods available for tool users, aiming at semi-automatic algorithm configuration using the available model information.

TimeNET implements a variant of the splitting technique RESTART, which is now extended by automatically deriving an estimation of the result via performance bounds to calculate splitting factors, and an online distance estimation using structural properties analyzed via a linear programming problem [12]. This avoids the user-defined importance function otherwise necessary. In addition to that, an automated importance sampling method for rare-event SPN simulation proposed by Reijsbergen et al. [6] is currently being integrated in the tool.

A new multi-trajectory simulation algorithm for eDSPNs [10] has been implemented recently. It combines elements of simulation and numerical analysis for the first time such that the behavior of the performance evaluation method follows either method just depending on the amount of trajectories (state particles) being stored and followed. This allows a mixed approach, avoiding the pitfalls of simulation (rare events) and numerical analysis (large state spaces).

Another addition is a simulation method for eDSPN models with incompletely known initial state. Automated and distributed optimization of eDSPN and SCPN models [2] have been implemented as a separate add-on. The initial transient phase of steady-state SCPN simulations is now detected and deleted. Modern random number generators for simulations (Mersenne twister) are used instead of standard library implementations now.

Model extensions include a new model class stochastic automata that allows to specify discrete-time and continuous-time Markov chains for teaching purposes. A standard stationary solution has been implemented. UML state charts extended with stochastic elements based on the MARTE profile and energy usage stereotypes have been added as another model class [7]. Such models can be edited and translated into eDSPNs models for the analysis of embedded systems. Model parameters (or definitions) are now unified without a numeric type and may depend on each other in most model classes. Reward definitions for performance measures have been unified for eDSPNs and SCPNS, and extended by transition throughputs (impulse rewards). Color-dependent SCPN model parts were extended and streamlined, including performance measures, arc expressions, and firing delays.

GUI and user interaction have been improved with several details, including a graphical visualization of the reachability graph, and adaptive mouse pointer appearance when adding objects.

User support outside the tool itself is undergoing a major update in 2017: A new integrated support web page will be made available until summer 2017. It contains installation and background information, example models, FAQs, daily and by-version downloads, and user feedback for bug reports etc.

Efficiency and quality of the internal **software development** process are now supported by a build and test server that automatically and regularly checks out the complete source code on different supported systems, compiles the tool, tests it, and makes new builds available for download. The tool is now available as a native 64 Bit application.

The author wishes to thank the numerous students who have contributed to the development of TimeNET over the years.

References

1. Bernardi, S., Bertoncello, C., Donatelli, S., Franceschinis, G., Gaeta, G., Gribaudo, M., Horvàth, A.: GreatSPN in the new millenium. In: International Multiconference on Measurement, Modelling and Evaluation of Computer-Communication Systems, Research Report 760, 2001, Universität Dortmund, Germany: tools of Aachen 2001, pp. 17–23 (2001)
2. Bodenstein, C., Zimmermann, A.: TimeNET optimization environment - batch simulation and heuristic optimization of SCPNs with TimeNET 4.2. In: 8th International Conference on Performance Evaluation Methodologies and Tools (VALUETOOLS 2014), Bratislava, Slovakia, December 2014
3. Courtney, T., Gaonkar, S., Keefe, K., Rozier, E., Sanders, W.: Möbius 2.3: an extensible tool for dependability, security, and performance evaluation of large and complex system models. In: IEEE/IFIP International Conference on Dependable Systems Networks, pp. 353–358 (2009)
4. Hirel, C., Tuffin, B., Trivedi, K.S.: SPNP: stochastic Petri nets. Version 6.0. In: Haverkort, B.R., Bohnenkamp, H.C., Smith, C.U. (eds.) TOOLS 2000. LNCS, vol. 1786, pp. 354–357. Springer, Heidelberg (2000). doi:10.1007/3-540-46429-8_30
5. Jensen, K., Kristensen, K.L., Wells, L.: Coloured Petri nets and CPN tools for modelling and validation of concurrent systems. Int. J. Softw. Tools Technol. Transf. (STTT) 9(3–4), 213–254 (2007)
6. Reijsbergen, D., de Boer, P.-T., Scheinhardt, W., Haverkort, B.: Automated rare event simulation for stochastic Petri nets. In: Joshi, K., Siegle, M., Stoelinga, M., D'Argenio, P.R. (eds.) QEST 2013. LNCS, vol. 8054, pp. 372–388. Springer, Heidelberg (2013). doi:10.1007/978-3-642-40196-1_31
7. Shorin, D., Zimmermann, A.: Extending the software tool TimeNET by power consumption estimation of UML MARTE models. In: Proceeding of the 4th International Conference on Simulation and Modeling Methodologies, Technologies and Applications (SIMULTECH 2014), pp. 83–91, Vienna, Austria, August 2014
8. Zimmermann, A.: Stochastic Discrete Event Systems. Springer, Heidelberg (2007). doi:10.1007/978-3-540-74173-2
9. Zimmermann, A.: Modeling and evaluation of stochastic Petri nets with TimeNET 4.1. In: Proceeding 6th International Conference on Performance Evaluation Methodologies and Tools (VALUETOOLS), pp. 54–63, Corse, France (2012)
10. Zimmermann, A., Hotz, T., Canabal Lavista, A.: A hybrid multi-trajectory simulation algorithm for the performance evaluation of stochastic Petri nets. In: Bertrand, N., Bortolussi, L. (eds.) QEST 2017. LNCS, vol. 10503, pp. 107–122. Springer, Cham (2017)
11. Zimmermann, A., Jäger, S., Geyer, F.: Towards reliability evaluation of AFDX avionic communication systems with rare-event simulation. In: Proceeding Probabilistic Safety Assessment & Management Conference 2014 (PSAM 12), pp. 1–12, Honolulu, Hawaii, USA, June 2014
12. Zimmermann, A., Maciel, P.: Importance function derivation for RESTART simulations of Petri nets. In: 9th International Workshop on Rare Event Simulation (RESIM 2012), pp. 8–15, Trondheim, Norway, June 2012

RODES: A Robust-Design Synthesis Tool for Probabilistic Systems

Radu Calinescu[1], Milan Češka[2], Simos Gerasimou[1(✉)], Marta Kwiatkowska[3], and Nicola Paoletti[4]

[1] Department of Computer Science, University of York, York, UK
simos.gerasimou@york.ac.uk
[2] Faculty of Information Technology, Brno University of Technology, Brno, Czech Republic
[3] Department of Computer Science, University of Oxford, Oxford, UK
[4] Department of Computer Science, Stony Brook University, Stony Brook, USA

Abstract. We introduce RODES – a tool for the synthesis of probabilistic systems that satisfy strict reliability and performance requirements, are Pareto-optimal with respect to a set of optimisation objectives, and are robust to variations in the system parameters. Given the design space of a system (modelled as a parametric continuous-time Markov chain), RODES generates system designs with low sensitivity to required tolerance levels for the system parameters. As such, RODES can be used to identify and compare robust designs across a wide range of Pareto-optimal tradeoffs between the system optimisation objectives.

1 Introduction

Quantitative verification is an effective technique for analysing the quality attributes (e.g. performance and reliability) of alternative system designs from the early stages of the development lifecycle [5]. The quality attributes of interest are formalised as probabilistic temporal logic properties, and are evaluated over Markov models of different system designs. The model that achieves the best tradeoff between the quality attributes is then used as a basis for the implementation of the system. However, if this implementation cannot precisely match the parameters of the selected model, the quality attributes of the system may differ significantly from the values predicted by the quantitative verification of its model. This limits the applicability of recently proposed approaches for the automated synthesis of probabilistic system designs [4,7].

Our RObust DEsign Synthesis (RODES) tool addresses this limitation by generating parametric continuous-time Markov chains (pCTMCs) whose transition rates are allowed to vary within small bounded intervals that correspond to user-specified tolerances for the parameters of the system. RODES implements our theoretical results from [1], which combine probabilistic model synthesis [7] and precise parameter synthesis [2] to generate Pareto-optimal sets of

This work has been supported by the Czech Grant Agency grant No. GA16-17538S.

© Springer International Publishing AG 2017
N. Bertrand and L. Bortolussi (Eds.): QEST 2017, LNCS 10503, pp. 304–308, 2017.
DOI: 10.1007/978-3-319-66335-7_20

pCTMCs (i.e. designs) using a *sensitivity-aware Pareto dominance relation*. This relation [1] acts as a tradeoff between optimality and robustness, and enables adding robust but suboptimal designs into the Pareto-optimal sets. To this end, the relation takes into account both a set of optimisation objectives (requiring the minimisation or maximisation of certain quality attributes) and the benefit of selecting *robust designs*, i.e., designs with quality attributes insensitive to the tolerance-induced variations in the pCTMC transition rates.

The rest of the paper presents RODES and its extensible architecture (Sect. 2), and the tool scalability and applicability to systems from different domains (Sect. 3). The RODES code, supplementary case study material, and full experimental results are available at https://github.com/gerasimou/RODES.

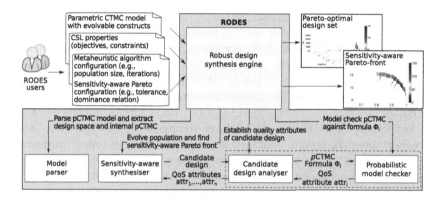

Fig. 1. High-level RODES architecture

2   RODES Functionality and Architecture

RODES (Fig. 1) is a Java-based tool with the inputs described below.

(1) A pCTMC model of the entire design space, expressed in the modelling language of the model checker PRISM [10] extended with the constructs

$$\text{evolve double } k \ [k_{min}..k_{max}] \tag{1}$$

$$\text{evolve int } d \ [d_{min}..d_{max}] \tag{2}$$

$$\text{evolve module } ComponentName \tag{3}$$

which are used to specify ranges for the continuous and discrete parameters of the system, and alternative component designs, respectively. A RODES design is also a pCTMC, obtained from the design-space pCTMC by constraining its continuous parameters (1) to small bounded intervals

$$[k_0 - \delta, k_0 + \delta] \subset [k_{min}, k_{max}], \tag{4}$$

fixing the values of its discrete parameters (2), and selecting one of the alternative designs (3) for each distinct *ComponentName* value.

(2) Continuous stochastic logic (CSL) properties specifying the optimisation objectives and constraints for the quality attributes of the system.

(3) Configuration parameters for the design-search metaheuristic algorithm, and the following parameters of the sensitivity-aware Pareto dominance relation:
- a small *tolerance* $\gamma > 0$ for each continuous parameter (1) such that the allowed parameter-value variation δ from (4) is $\delta = \gamma(k_{max} - k_{min})$;
- a small *sensitivity coefficient* $\epsilon \geq 0$ such that a design needs to have $(1+\epsilon)$ times better quality attributes to dominate a more robust design.

The operation of RODES is managed by a *Robust-design synthesis engine* (Fig. 1). First, a *Model parser* (built using the Antlr parser generator, www. antlr.org) preprocesses the design-space *p*CTMC. Next, a *Sensitivity-aware synthesiser* employs the jMetal Java framework for multi-objective optimisation with metaheuristics (http://jmetal.github.io/jMetal) to evolve an initially random population of *candidate designs*, generating a close approximation of the sensitivity-aware Pareto front. This involves using a *Candidate design analyser*, which invokes the probabilistic model checker PRISM-PSY [3] to obtain the ranges of values for the relevant quality attributes of candidate designs through precise parameter synthesis. The Pareto front and corresponding Pareto-optimal set of designs are then plotted using MATLAB/Octave scripts, as shown in Fig. 2.

A key feature of RODES is its modular architecture. The Sensitivity-aware synthesiser supports several metaheuristics algorithms, including variants of genetic algorithms and swarm optimisers. Further, the sensitivity-aware Pareto dominance relation can be adapted to match better the needs of the system under development (e.g., by comparing designs based on the worst, best or average quality attribute values). Finally, different solvers could be plugged in the probabilistic model checker component, including e.g. the GPU-accelerated version of PRISM-PSY [3], or parameter synthesis tools for DTMCs [6].

3   Case Studies and Experimental Results

We evaluate RODES in three case studies: a variant of the producer-consumer problem;[1] a workstation cluster [9]; and a replicated file system used by Google's search engine [8]. Runtimes (Table 1) depend on the number of evaluations (using more typically improves the quality of the Pareto fronts) and by the time required to analyse a candidate design. These runtimes were obtained using the sequential version of PRISM-PSY, but we are currently integrating the GPU-accelerated version, which will significantly improve the scalability of the tool [3].

Figure 2 shows Pareto fronts and designs obtained for a producer-consumer model comprising a slow high-capacity buffer and a fast buffer of small capacity. The design space has two continuous parameters—overall production rate, prod_rate, and probability of using the fast buffer, p_send_fast; and a discrete parameter that selects between two alternative designs (3) so either packets stay

---

[1] E.W. Dijkstra. "Information Streams Sharing a Finite Buffer" Inf. Proc. Letters, 1972. The model can be found at https://github.com/gerasimou/RODES/wiki.

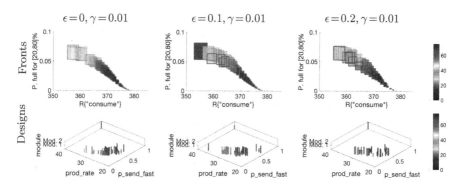

Fig. 2. Sensitivity-aware Pareto fronts (top) for the producer-consumer model, and corresponding synthesised Pareto-optimal designs (bottom). Boxes represent quality-attribute regions, coloured by sensitivity (red: sensitive, blue: robust). Red-bordered boxes indicate sub-optimal robust designs. Designs are compared based on the worst-case quality attribute value (i.e. lower-left corner of each box). (Color figure online)

in the designated buffers, or packets in the slow buffer are redirected to the fast buffer with probability proportional to the slow buffer occupancy. We aim to maximise two objectives: the expected system throughput (x-axis), and the probability that both buffers are utilised at between 20–80% of their capacity (y-axis). Figure 2 shows results for tolerance $\gamma = 0.01$ and for several values of the sensitivity coefficient ϵ (cf. Sect. 2). As expected, the number of slightly sub-optimal but more robust solutions increases with ϵ. The sensitivity-aware Pareto fronts provide unique insights into the system behaviour, and facilitates the selection of designs with a wide range of robustness levels, making RODES an effective tool for the synthesis of robust designs from multi-objective specifications.

Table 1. Time (mean ± SD) for the synthesis using 10,000 evaluations. **variant**: values of scenario parameters. **#states** (**#trans.**): number of states (transitions) of the underlying pCTMC. $|K|$: number of continuous parameters.

Model/variant:	Google file system ($|K|$=2)			Workstation cluster ($|K|$=2)			Prod.-cons. ($|K|$=2)
	S = 5000	S = 10000	S = 20000	N = 9	N = 12	N = 15	
#states	1323	1893	2406	3440	5876	8960	5632
#trans.	7825	11843	15545	18656	32204	49424	21884
Time (m)	104 ± 4	149 ± 4	180 ± 9	185 ± 19	191 ± 27	205 ± 42	29 ± 1

References

1. Calinescu, R., Češka, M., Gerasimou, S., Kwiatkowska, M., Paoletti, N.: Designing robust software systems through parametric Markov chain synthesis. In: ICSA 2017, pp. 131–140 (2017)

2. Češka, M., Dannenberg, F., Paoletti, N., et al.: Precise parameter synthesis for stochastic biochemical systems. In: Acta Informatica, pp. 1–35 (2016)
3. Češka, M., Pilař, P., Paoletti, N., Brim, L., Kwiatkowska, M.: PRISM- PSY: precise GPU-accelerated parameter synthesis for stochastic systems. In: TACAS 2016, pp. 367–384 (2016)
4. Chen, T., Hahn, E.M., Han, T., et al.: Model repair for Markov decision processes. In: TASE 2013, pp. 85–92 (2013)
5. Cheung, L., Roshandel, R., Medvidovic, N., Golubchik, L.: Early pre-diction of software component reliability. In: ICSE 2008, pp. 111–120 (2008)
6. Dehnert, C., Junges, S., Jansen, N., et al.: PROPhESY: A PRObabilistic ParamEter SYnthesis Tool. In: CAV 2015, pp. 214–231 (2015)
7. Gerasimou, S., Tamburrelli, G., Calinescu, R.: Search-based synthesis of probabilistic models for QoS software engineering. In: ASE (2015)
8. Ghemawat, S., Gobioff, H., Leung, S.-T.: The Google file system. In: SOSP 2003, pp. 29–43 (2003)
9. Haverkort, B.R., Hermanns, H., Katoen, J.-P.: On the use of model checking techniques for dependability evaluation. In: SRDS (2000)
10. Kwiatkowska, M., Norman, G., Parker, D.: PRISM 4.0: verification of probabilistic real-time systems. In: Gopalakrishnan, G., Qadeer, S. (eds.) CAV 2011. LNCS, vol. 6806, pp. 585–591. Springer, Heidelberg (2011). doi:10.1007/978-3-642-22110-1_47

QUEST: A Tool for State-Space Quantization-Free Synthesis of Symbolic Controllers

Pushpak Jagtap$^{(\boxtimes)}$ and Majid Zamani

Technical University of Munich, Munich, Germany
{pushpak.jagtap,zamani}@tum.de

Abstract. In this paper, we introduce QUEST, a new tool for automated controller synthesis of incrementally input-to-state stable nonlinear control systems. This tool accepts ordinary differential equations as the descriptions of the nonlinear control systems and constructs their symbolic models using state-space quantization-free approach which can potentially alleviate the issue of so-called *curse of dimensionality* while computing discrete abstractions of the systems with high-dimensional state-space. The tool supports computation of both minimal and maximal fixed points and thus provides natively algorithms to synthesize controllers enforcing safety and reachability specifications. All the computations are done in C++. Finally, we illustrate the performance of the tool on a 10-room building temperature control. The tool together with user manual and some examples are available for download at www.hcs. ei.tum.de/software.

1 Introduction and Motivation

Controller synthesis techniques using so-called discrete abstractions have gained considerable attention in the past few years. They provide tools for automated, correct-by-construction controller synthesis for various systems to enforce complex specifications (usually given in linear temporal logic (LTL) formulae).

There have been recently various software tools on the symbolic controller synthesis for various classes of nonlinear control systems including Pessoa [3], CoSyMa [4], and SCOTS [5]. However, the discrete abstractions obtained in these results and corresponding tools are based on state-space quantization. The need for state space quantization results in an exponential increase in computational complexity with the dimension of state space in the concrete system, and, hence, these techniques suffer severely from the issue of so-called curse of dimensionality especially for the systems with high-dimensional state-space.

In [2,7,8], it has been shown that one can construct discrete abstractions without state-space quantization which are approximately bisimilar to incrementally input-to-state stable nonlinear control systems. The technique uses fixed

This work was supported in part by the German Research Foundation (DFG) through the grant ZA 873/1-1 and the TUM International Graduate School of Science and Engineering (IGSSE).

© Springer International Publishing AG 2017
N. Bertrand and L. Bortolussi (Eds.): QEST 2017, LNCS 10503, pp. 309–313, 2017.
DOI: 10.1007/978-3-319-66335-7_21

length of quantized input sequence as a symbolic state of the abstraction which helps to alleviate the curse of dimensionality. The length of input sequences, referred as temporal horizon N, is used as a parameter to adjust the abstraction precision; a larger value of N results in a higher precision of the abstraction, and, consequently, in a larger abstraction in terms of the number of states.

In this paper, we introduce QUEST, an open-source software tool implementing the synthesis of symbolic controllers based on state-space quantization-free approach proposed in [2,7,8]. QUEST provides algorithms for the construction of discrete abstractions which are approximately bisimilar to the original incrementally stable dynamics without the need to discretize the state space. Moreover, it provides algorithms for synthesizing controllers enforcing some classes of LTL specifications over concrete systems using fixed point computations.

2   Tool Details

QUEST is implemented in C++ and employs binary decision diagrams (BDDs) as underlying data structure to store and manipulate boolean functions representing symbolic abstractions and controllers. QUEST provides two fixed point algorithms for maximal and minimal fixed point computation as described in [6] and thus, natively, provides algorithms to synthesize controllers for safety and reachability specifications. Moreover, one can use combinations of these fixed point algorithms for synthesizing controllers enforcing customized specifications such as reach and stay.

Inputs: QUEST accepts the description of the dynamics of incrementally input-to-state stable nonlinear control systems in the form of an ordinary differential equation; see [1] for a characterization of incremental stability in terms of Lyapunov functions. Additionally, the user needs to provide an input set, an input set quantization parameter η, a source state x_s, a sampling time τ, and a temporal horizon N; see [8] for the role of those parameters. The computation of parameter N for a given desired abstraction precision ε is provided in [8].

Output: QUEST synthesizes controllers with the help of fixed point computations and stores them in the form of BDD. QUEST also provides an option to simulate the closed-loop system equipped with the synthesized controller.

Installation and Usage: The detailed discussion about installation and usage of QUEST along with sample examples are provided in user manual available at www.hcs.ei.tum.de/software.

3   Example

To demonstrate QUEST, we synthesize a controller regulating temperatures in a ten-room building shown schematically in Fig. 1(a). QUEST accepts the dynamic given as an ordinary differential equation as shown below.

```
const int sDIM = 10; /* System dimension */
const double T = 25; /* Sampling time */
size_t N = 12; /* Temporal Horizon */
typedef std::array<double,sDIM> state_type;
auto system_post = [](state_type &x, double* u) -> void {
auto rhs=[u](state_type &dx, const state_type &x) -> void {
    const double a=0.05, ae2=0.005, ae5=0.005, ae=0.0033, ah=0.0036;
    const double te=12; /* External temperature */
    const double th=100; /* Heater temperature */
    dx[0]=(-a-ae)*x[0]+a*x[1]+ae*te;
    dx[1]=(-4*a-ae2-ah*u[0])*x[1]+a*x[0]+a*x[6]+a*x[8]+a*x[2]+ae2*te+ah*th*u[0];
    dx[2]=(-2*a-ae)*x[2]+a*x[1]+a*x[3]+ae*te;
    dx[3]=(-2*a-ae)*x[3]+a*x[2]+a*x[4]+ae*te;
    dx[4]=(-4*a-ae5-ah*u[1])*x[4]+a*x[3]+a*x[7]+a*x[5]+a*x[9]+ae5*te+ah*th*u[1];
    dx[5]=(-a-ae)*x[5]+a*x[4]+ae*te; dx[6]=(-a-ae)*x[6]+a*x[1]+ae*te;
    dx[7]=(-a-ae)*x[7]+a*x[4]+ae*te; dx[8]=(-a-ae)*x[8]+a*x[1]+ae*te;
    dx[9]=(-a-ae)*x[9]+a*x[4]+ae*te; };
size_t nint = 5; /* no. of time step for ode solving */
ode_solver(rhs,x,nint,h); /* Runga Kutte solver */ }
```

In this example, we consider that the control inputs u[0] and u[1] corresponding to heaters H_1 and H_2 are equal to 1 if the corresponding heaters are on and equal to 0 if the corresponding heaters are off. Here, we assume that at most one heater is on at each time instance. Thus, the input set of the system is given as

```
const int iDIM = 2; /* Input dimension */
const size_t P = 3; /* Number of elements in the input set*/
double ud[P][iDIM]={{0,0},{0,1},{1,0}};
```

For this example, we consider the objective to synthesize a controller enforcing all the temperatures to stay within $W = [18, 21.5]^{10}$. This corresponds to the LTL specification $\Box W$ (i.e. safety specification) and is given to the tool as

```
auto setBounds = [](state_type y) -> bool {
    double ul=21.8, ll=18; /*upper and lower bound on the temperature in each room*/
    bool s = true;
        for(int j = 0; j < sDIM; j++){
                if( y[j] >= ul || y[j] <= ll ) {s = false; break;}}
    return s;}
```

We use temporal horizon $N = 12$, sampling time $\tau = 25$ time units, and source state $x_s = [17, 17, 17, 17, 17, 17, 17, 17, 17, 17]^T$ which result in precision $\varepsilon = 0.1$ for the discrete abstraction [8]. The computation of discrete abstraction and controller synthesis have been performed using QUEST on a windows computer with CPU 3.5GHz Intel Core i7. Figure 1(b) shows the evolution of the temperatures ξ in all rooms starting from initial condition $x_0 = [18.9, 19, 19.1, 19.5, 20.8, 19.7, 19.2, 19.9, 19, 19.8]^T$. Figure 1(c) illustrates the corresponding input trajectories υ_{H_1} and υ_{H_2}. Note that, the figures are generated using MATLAB by simulating system dynamics with the control inputs generated by QUEST. In Table 1, we show the effect of N on the size of the abstraction (given by the number of transitions), computation times, and precision ε. Remark that, due to the large dimension of the state-space, the existing tools such as Pessoa [3], CoSyMa [4], and SCOTS [5] fail to synthesize any controller.

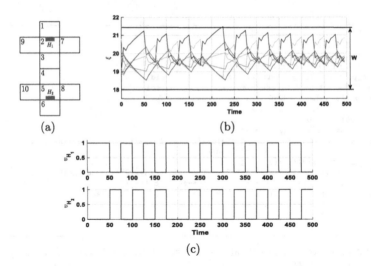

Fig. 1. (a) A schematic of ten-room building, (b) Evolution of temperatures in all rooms under synthesized controller, (c) Input trajectories given by synthesized controller.

Table 1. Performance comparison for different values of N.

N	13	12	11	10	9
Number of transitions in the abstraction	4782969	1594323	531441	177147	59049
Number of transitions in the controller	173980	55808	17888	5582	1722
Abstraction computation time (sec)	5.87	1.6	0.4	0.072	0.014
Controller computation time (sec)	319.63	96.89	29.56	9.12	2.67
Precision ε	0.05	0.1	0.22	0.5	1.4

Acknowledgments. The authors would like to thank S Sairam Akhil for some of the implementation.

References

1. Angeli, D.: A Lyapunov approach to incremental stability properties. IEEE Trans. Autom. Control **47**(3), 410–21 (2002)
2. Le Corronc, E., Girard, A., Goessler, G.: Mode sequences as symbolic states in abstractions of incrementally stable switched systems. In: 52nd IEEE Conference on Decision and Control (CDC). pp. 3225–3230. IEEE (2013)
3. Mazo, M., Davitian, A., Tabuada, P.: PESSOA: a tool for embedded controller synthesis. In: Touili, T., Cook, B., Jackson, P. (eds.) CAV 2010. LNCS, vol. 6174, pp. 566–569. Springer, Heidelberg (2010). doi:10.1007/978-3-642-14295-6_49
4. Mouelhi, S., Girard, A., Gössler, G.: CoSyMa: a tool for controller synthesis using multi-scale abstractions. In: Proceedings of the 16th International Conference on Hybrid Systems: Computation and Control, pp. 83–88. ACM (2013)

5. Rungger, M., Zamani, M.: SCOTS: a tool for the synthesis of symbolic controllers. In: Proceedings of the 19th International Conference on Hybrid Systems: Computation and Control, pp. 99–104. ACM (2016)
6. Tabuada, P.: Verification and Control of Hybrid Systems: A Symbolic Approach. Springer Science & Business Media, Heidelberg (2009)
7. Zamani, M., Abate, A., Girard, A.: Symbolic models for stochastic switched systems: a discretization and a discretization-free approach. Automatica **55**, 183–196 (2015)
8. Zamani, M., Tkachev, I., Abate, A.: Towards scalable synthesis of stochastic control systems. Discrete Event Dyn. Syst. **27**(2), 341–369 (2017)

Statistical Model Checking

Three-Valued Spatio-Temporal Logic: A Further Analysis on Spatio-Temporal Properties of Stochastic Systems

Ludovica Luisa Vissat$^{1(\boxtimes)}$, Michele Loreti^2, Laura Nenzi^3, Jane Hillston^1, and Glenn Marion^4

^{1} School of Informatics, University of Edinburgh, Edinburgh, UK
L.Luisa-Vissat@sms.ed.ac.uk, jane.hillston@ed.ac.uk
^{2} DiSIA, University of Firenze, Florence, Italy
^{3} Faculty of Informatics, Vienna University of Technology, Vienna, Austria
^{4} Biomathematics and Statistics Scotland, Edinburgh, UK

Abstract. In this paper we present Three-Valued Spatio-Temporal Logic (TSTL), which enriches the available spatio-temporal analysis of properties expressed in Signal Spatio-Temporal Logic (SSTL), to give further insight into the dynamic behaviour of systems. Our novel analysis starts from the estimation of satisfaction probabilities of given SSTL properties and allows the analysis of their temporal and spatial evolution. Moreover, in our verification procedure, we use a three-valued approach to include the intrinsic and unavoidable uncertainty related to the simulation-based statistical evaluation of the estimates; this can be also used to assess the appropriate number of simulations to use depending on the analysis needs. We present the syntax and three-valued semantics of TSTL and a specific extended monitoring algorithm to check the validity of TSTL formulas. We conclude with two case studies that demonstrate how TSTL broadens the application of spatio-temporal logics in realistic scenarios, enabling analysis of threat monitoring and control programmes based on spatial stochastic population models.

1 Introduction

In many case studies, considering spatial structure is of key importance to better understand and predict the evolution of the system under study. For example, dispersive processes such as spread of disease, invasive species or fire spread have an intrinsic and fundamental spatial dimension that has to be included in the model. Spatial stochastic models provide a good representation of such system dynamics, typically studied through simulations. Correspondingly, the formal analysis of these spatial stochastic models has to also accommodate spatial and temporal modalities to be able to describe and verify properties about the spatio-temporal evolution of the specific systems.

Suitable analysis is provided by spatio-temporal logics and model checking. In most cases a statistical approach [1] is needed to estimate satisfaction

© Springer International Publishing AG 2017
N. Bertrand and L. Bortolussi (Eds.): QEST 2017, LNCS 10503, pp. 317–332, 2017.
DOI: 10.1007/978-3-319-66335-7_22

probabilities of given properties, expressed using logical formulas. Simulation trajectories alone make it difficult to fully analyse the dynamic behaviour and to compare different systems, and the exhaustive exploration of all possible spatio-temporal trajectories is computationally infeasible. Using current simulation-based approaches the outcome is summary information about the satisfaction of properties over the spatial domain.

In our work we seek to add value to such information by providing a novel logic, called Three-Valued Spatio-Temporal Logic, to reason about spatial and temporal evolution of the satisfaction of these properties, giving further insight into the dynamic behaviour of the system under study. For example, in the analysis of the efficacy of a control measure for disease spread, we can verify whether the spread in a specific area will happen with probability under a given threshold over time. We can also identify the locations at highest risk, being surrounded by locations with high probability of becoming infected. The new TSTL atomic propositions are inequalities on the estimated satisfaction probabilities of given logical formulas (in the spatio-temporal logic SSTL [2] in this case, which formally describes and verifies properties of spatio-temporal trajectories), which are estimated using statistical model checking. This simulation-based evaluation has an intrinsic and unavoidable uncertainty, but frequently it is the only computationally feasible approach, requiring just an executable model. We use a three-valued approach to keep track of the associated uncertainty in the results of our model checking and we interpret the inequalities with different degrees of truth, using *true*, *false* and a third value *unknown*. This extension can be also used to give an indication of when more simulations are needed to make the evaluation of atomic propositions more precise and thus allowing stronger conclusions to be drawn. Conversely this enables initial explorations with relatively few simulations and assessment of whether they result in sufficient precision. We implemented the monitoring algorithms for the TSTL logical operators, to evaluate the satisfaction function of TSTL properties. The operators and the procedures are defined in a similar way to SSTL but on a different domain, dealing with three truth values.

Related Work. Several existing logics can be used to describe spatial properties of systems and estimate satisfaction probability values. Much of this work is based on topological models [3], looking at properties of subsets of points of topological spaces, whilst we take a more concrete representation of space. Other literature concerns spatial logics for process algebra with locations [4], used to study the mobility of concurrent systems; here space is represented as a tree and locations are nested. Based on a graph structure, there are logics such as the Multiprocess Network Logic [5], which can express spatio-temporal properties in discrete time. Considering stochastic systems, there are existing logics for expressing properties on probabilities, such as Probabilistic Computation Tree Logic (PCTL) [6] and Continuous Stochastic Logic (CSL) [7]. In these cases, the analysis is limited to temporal aspects, without spatial modalities, while our novel approach considers both. Three-valued logics, such as ours, with just one additional truth value, are a simple case in the field of multi-valued logics [8]. The initial concept was created by Łukasiewicz [9] and developed further by different logicians, such as Kleene [10], introducing the concept of "undefined" dealing with partial recursive functions. The three-valued approach is used in [11], for the definition of a new abstraction method for fully probabilistic systems and in [12], for model checking of Discrete-Time Markov Chains. We are not aware of any current use of a three-valued logic approach in the field of spatio-temporal analysis of stochastic systems.

Paper Structure. The paper is structured as follows: Sect. 2 introduces notation and background work on SSTL while Sect. 3 presents the novel logic TSTL. Section 4 introduces the process algebra MELA we used to perform stochastic simulations, the monitor jSSTL we used to verify SSTL properties and how we linked all these aspects together to verify TSTL properties. Sections 5 and 6 present two different case studies and applications of TSTL. Section 7 reports discussion and future directions for investigation while conclusions are reported in Sect. 8.

2   Background

In this section we introduce some fundamental concepts and notation that we will use in this paper aligned with the syntax and semantics of the existing spatio-temporal logic SSTL.

Notation. We define a *spatial population model*, on a discrete representation of space; it describes a large number of different agents that can perform actions, take different states, interact and move between different locations. More formally, a *spatial population model* $\mathcal{M}$ is defined as a tuple $\mathcal{M} = (\mathcal{S}, G, \mathbf{X}, \mathbf{X}_0, Tr)$ where:

- $\mathcal{S} = \{1, \ldots, n\}$ is the set of states that the population agents can take.
- $G = (L, E, w)$, a *finite weighted undirected graph* that represents the current choice of underlying spatial structure of the spatial population model:
  - L is the finite set of locations (nodes)
  - $E \subseteq L \times L$ is the set of connections (edges)
  - $w : E \rightarrow \mathbb{R}_{\geq 0}$ is the function *cost* (weights). We extend w to E^*, the transitive closure of E (set containing all the pairs of connected nodes). w gives the sum of costs of the shortest path between two different nodes, where this shortest path is the one that minimizes the sum of the costs.
- $\mathbf{X} : L \rightarrow \mathbb{R}^n$, where $\mathbf{X}(l) = (X_1, \ldots, X_n) \in \mathbb{R}^n$ is the state vector, that represents the state of the population in each location. The entries of the vector $\mathbf{X}(l)$ represent the number of agents in location l in the ith state; therefore, to be more specific, these *counting variables* are $X_i \in \mathbb{N}_0$.
- $\mathbf{X}_0 : L \rightarrow \mathbb{R}^n$, where $\mathbf{X}_0(l)$ is the initial state of the state vector, for each location.
- Tr is the set of transitions, $\tau_i = (\alpha_i, v_i, r_i)$, describing the events that change the global state of the system. Each transition consists of a label α_i in the label set $\mathcal{L}$, an update vector, $v_i : L \rightarrow \mathbb{R}^n$ recording the change to each

counting variable in each location due to the transition, and a rate function r_i, which may depend on the global state of the system.

We can interpret the dynamical evolution of these models either stochastically as a Markov chain or deterministically as a system of Ordinary Differential Equations (ODEs); in this work we focus on stochastic spatio-temporal systems. We can describe the temporal evolution of our spatial population models using:

- σ, a *spatio-temporal trajectory* of $\mathcal{M}$. $\sigma : L \times \mathbb{R}_{\geq 0} \to \mathbb{R}^n$ gives the state of the population vector for each location $l \in L$ and each time $t \in \mathbb{R}_{\geq 0}$
- Σ, a *set of spatio-temporal trajectories*, that will be used in the analysis.

SSTL Syntax. Signal Spatio-Temporal Logic (SSTL) [2] is a spatial extension of Signal Temporal Logic (STL) [13], a temporal logic suitable for describing properties of real-valued signals. The syntax of SSTL is given by:

$$\varphi ::= \mu \mid \neg\varphi \mid \varphi_1 \vee \varphi_2 \mid \varphi_1 \, \mathcal{U}^{[t_1, t_2]} \, \varphi_2 \mid \Diamond_{[w_1, w_2]} \varphi \mid \varphi_1 \mathcal{S}_{[w_1, w_2]} \varphi_2$$

The SSTL *atomic proposition* μ is of the form $\mu \equiv (f \geq 0)$, $f : \mathbb{R}^n \to \mathbb{R}$, an inequality on expressions with population counts, given in the spatio-temporal trajectory. *Negation* $\neg$ and *disjunction* $\vee$ are the standard boolean operators and $\mathcal{U}$ is the *bounded until* operator. This temporal operator $\mathcal{U}$ is used to verify that the property φ_2 will be satisfied at some time instant in the interval $[t_1, t_2]$ and that at all preceding time instants φ_1 holds. SSTL introduces two spatial operators: the *bounded somewhere* operator $\Diamond_{[w_1, w_2]}$ and the *bounded surround* operator $\mathcal{S}_{[w_1, w_2]}$, with w_1, w_2 real values, $w_1 \leq w_2$. The *bounded somewhere* operator requires that the property φ holds in a location reachable from the current one, with a cost w, $w \in [w_1, w_2]$. The operator *bounded surround* describes the property of being surrounded by a φ_2-region, while being in a φ_1-region: the formula $\varphi_1 \mathcal{S}_{[w_1, w_2]} \varphi_2$ is true in a location l, if l belongs to a set of locations A where φ_1 holds, such that its external boundary $B^+(A)$ contains only locations satisfying φ_2. The external boundary of a subset of locations A is defined as $B^+(A) := \{l \in L \mid l \notin A \ \wedge \ \exists l' \in A \text{ s.t. } (l, l') \in E\}$. Moreover, the locations in the $B^+(A)$ have to be reached from location l with a cost w, $w \in [w_1, w_2]$. Examples of SSTL formulas will be provided throughout the paper.

SSTL Boolean Semantics. SSTL presents a *boolean* semantics that returns the value true/false ($\mathbb{B} = \{T, F\}$) depending on whether the observed trajectory satisfies the defined SSTL formula or not. The *boolean semantics* of a SSTL formula φ is interpreted over a spatio-temporal trajectory σ of $\mathcal{M}$, for each location $l \in L$ and at time $t \in \mathbb{R}_{\geq 0}$, given values in the set $\mathbb{B}$:

$$\beta(\mathcal{M}, \sigma, l, t, \varphi) \in \mathbb{B}$$

The satisfaction function β is defined as follows:

$$\beta(\mathcal{M}, \sigma, l, t, \mu) = \mu(\sigma(l, t))$$
$$\beta(\mathcal{M}, \sigma, l, t, \neg\varphi) = \neg\beta(\mathcal{M}, \sigma, l, t, \varphi)$$
$$\beta(\mathcal{M}, \sigma, l, t, \varphi_1 \vee \varphi_2) = \beta(\mathcal{M}, \sigma, l, t, \varphi_1) \vee \beta(\mathcal{M}, \sigma, l, t, \varphi_2)$$

$$\beta(\mathcal{M},\sigma,l,t,\varphi_1\,\mathcal{U}^{[t_1,t_2]}\,\varphi_2) = \bigvee_{t'\in[t+t_1,t+t_2]} (\beta(\mathcal{M},\sigma,l,t',\varphi_2)\ \wedge$$

$$\bigwedge_{t''\in[t,t')} \beta(\mathcal{M},\sigma,l,t'',\varphi_1))$$

$$\beta(\mathcal{M},\sigma,l,t,\Diamond_{[w_1,w_2]}\varphi) = \bigvee_{l'\in L, w(l,l')\in[w_1,w_2]} \beta(\mathcal{M},\sigma,l',t,\varphi)$$

$$\beta(\mathcal{M},\sigma,l,t,\varphi_1 S_{[w_1,w_2]}\varphi_2) = \bigvee_{A\in SR_l^{[w_1,w_2]}} (\bigwedge_{l'\in A} \beta(\mathcal{M},\sigma,l',t,\varphi_1)\ \wedge$$

$$\bigwedge_{l''\in B^+(A)} \beta(\mathcal{M},\sigma,l'',t,\varphi_2))$$

where the surrounding region $SR_l^{[w_1,w_2]} = \{A \subseteq L \mid \forall l' \in A : 0 \leq w(l,l') \leq w_2 \wedge \forall l'' \in B^+(A) : w_1 \leq w(l,l'') \leq w_2\}$.

Monitoring algorithms have been defined to evaluate the validity of SSTL properties, given a spatio-temporal trajectory, working inductively bottom-up on the parse tree of the formula. To make the verification procedure tractably computable, the time-domain has to be discretised, giving as output a piece-wise constant approximation of the result. For this reason, in the analysis we talk about time-steps, although we start from discrete-event continuous-time simulations.

As discussed previously, in the study of stochastic systems we are generally interested in evaluating the probability that given properties are satisfied; a commonly used approach consists of estimating these values using statistical methods on a set of trajectories. Therefore, given a SSTL property φ, we shift the analysis from a single trajectory σ to a set of trajectories Σ, assigning to each trajectory a truth value, according to the boolean semantics. After this step we can estimate the satisfaction probability p^* of the formula φ, provided with a confidence interval. We define $\mathcal{P}_\beta$ over the set of trajectories Σ, in terms of β:

$$\mathcal{P}_\beta(\mathcal{M},\Sigma,l,t,\varphi) = (p^*,\delta) \tag{1}$$

where $(p^*,\delta) \in [0,1] \times [0,1]$ and represents the interval $[p^* - \delta, p^* + \delta]$.

$$p^* = \frac{|\Sigma_T|}{|\Sigma|} \quad \text{and} \quad \delta = f_\delta(|\Sigma|,|\Sigma_T|,\epsilon) \tag{2}$$

where $|\Sigma_T| = \{\sigma \in \Sigma \mid \beta(\mathcal{M},\sigma,l,t,\varphi) = T\}$ and δ is calculated with a given confidence level ϵ, according to a suitable function f_δ. There are several approaches to compute the confidence interval. For the sake of simplicity in our presentation we assume this interval to be symmetric. Given the boolean nature of the observations, SSTL uses the binomial proportion confidence interval and the most common choice for the calculation presupposes that the error distribution is approximated by a normal distribution. From this point on, all the results of SSTL monitoring are given at 95% confidence.

3 Three-Valued Spatio-Temporal Logic

We now present the novelty of our research, introducing the syntax and three-valued semantics of TSTL, providing also derived operators and a specific monitoring algorithm.

TSTL Syntax. With the existing SSTL we are able to verify spatio-temporal properties of stochastic systems and estimate the satisfaction probabilities of given formulas. After this initial analysis we use our proposed extension to perform spatio-temporal analysis of these estimated values. The syntax of Three-Valued Spatio-Temporal Logic (TSTL) is given by:

$$\psi ::= \mathcal{P}_{<p}(\varphi) \mid \tilde{\neg}\psi \mid \psi_1 \tilde{\vee} \psi_2 \mid \psi_1 \tilde{\mathcal{U}}^{[t_1,t_2]} \psi_2 \mid \tilde{\Diamond}_{[w_1,w_2]} \psi \mid \psi_1 \tilde{\mathcal{S}}_{[w_1,w_2]} \psi_2$$

where $p \in [0,1]$ and φ is a given SSTL formula. The atomic TSTL formula $\mathcal{P}_{<p}(\varphi)$ expresses an inequality on the estimated satisfaction probability of the SSTL formula φ, checking if it is below the given threshold p. The logical TSTL operators link the TSTL propositions in a similar way to the SSTL ones, but working with estimated values and on a three-valued domain, as explained in the next section. We have *negation* $\tilde{\neg}$ and *disjunction* $\tilde{\vee}$ operators, *bounded until* $\tilde{\mathcal{U}}$, *bounded somewhere* $\tilde{\Diamond}_{[w_1,w_2]}$ and *bounded surround* $\tilde{\mathcal{S}}_{[w_1,w_2]}$. Conceptually all these operators are identical to the SSTL operators, but they operate on a different domain, reasoning about estimated satisfaction probabilities and not population counts. In the remainder we will show examples and differences between the two spatio-temporal logics; we will use the letter φ for SSTL formulas and ψ for TSTL ones.

Three-Valued Semantics. TSTL presents a three-valued semantics that returns a truth values in $\mathbb{T} = \{T, U, F\}$ (true/unknown/false). The truth tables for TSTL negation $\tilde{\neg}$, disjunction $\tilde{\vee}$ and conjunction $\tilde{\wedge}$ (that can be defined in terms of negation and disjunction) are given by:

$\tilde{\neg}$	T	U	F
	F	U	T

$\tilde{\vee}$		ψ_2	
	T	U	F
ψ_1 T	T	T	T
U	U	T	U	U
F	T	U	F

$\tilde{\wedge}$		ψ_2	
	T	U	F
ψ_1 T	T	U	F
U	U	U	U	F
F	F	F	F

as for Kleene's logic of indeterminacy K3 [10]. The three-valued satisfaction function τ for the atomic TSTL proposition $\mathcal{P}_{<p}(\varphi)$ will return a value in $\mathbb{T}$:

$$\tau(\mathcal{M}, \Sigma, l, t, \mathcal{P}_{<p}(\varphi)) = [\![p^* <_\delta p]\!] \in \mathbb{T}$$

that is evaluated starting from the resulting (p^*, δ) given by $\mathcal{P}_\beta(\mathcal{M}, \Sigma, l, t, \varphi)$, as shown in the Eqs. (1), (2). The associated truth value will be:

$$[\![p^* <_\delta p]\!] = \begin{cases} T & \text{if } p > p^* + \delta \\ U & \text{if } p \in [p^* - \delta, p^* + \delta] \\ F & \text{otherwise} \end{cases}$$

The three-valued satisfaction function τ for the TSTL operators is defined as follows, in an analogous manner as SSTL:

$$\tau(\mathcal{M}, \Sigma, l, t, \tilde{\neg}\psi) = \tilde{\neg}\tau(\mathcal{M}, \Sigma, l, t, \psi)$$

$$\tau(\mathcal{M}, \Sigma, l, t, \psi_1 \tilde{\vee} \psi_2) = \tau(\mathcal{M}, \Sigma, l, t, \psi_1) \tilde{\vee} \tau(\mathcal{M}, \Sigma, l, t, \psi_2)$$

$$\tau(\mathcal{M}, \Sigma, l, t, \psi_1 \tilde{\mathcal{U}}^{[t_1,t_2]} \psi_2) = \bigvee_{t' \in [t+t_1, t+t_2]}^{\sim} (\tau(\mathcal{M}, \Sigma, l, t', \psi_2) \tilde{\wedge}$$

$$\bigwedge_{t'' \in [t,t')}^{\sim} \tau(\mathcal{M}, \Sigma, l, t'', \psi_1))$$

$$\tau(\mathcal{M}, \Sigma, l, t, \tilde{\diamondsuit}_{[w_1,w_2]} \psi) = \bigvee_{l' \in L, w(l,l') \in [w_1,w_2]}^{\sim} \tau(\mathcal{M}, \Sigma, l', t, \psi)$$

$$\tau(\mathcal{M}, \Sigma, l, t, \psi_1 \tilde{\mathcal{S}}_{[w_1,w_2]} \psi_2) = \bigvee_{A \in SR_l^{[w_1,w_2]}}^{\sim} (\bigwedge_{l' \in A}^{\sim} \tau(\mathcal{M}, \Sigma, l', t, \psi_1) \tilde{\wedge}$$

$$\bigwedge_{l'' \in B^+(A)}^{\sim} \tau(\mathcal{M}, \Sigma, l'', t, \psi_2))$$

Note the similarity between the structure of β and τ, with operators that refer to SSTL and TSTL respectively. We want to clarify that SSTL results are provided performing SSTL monitoring with a given confidence level. Therefore, we are not talking about confidence level of TSTL results, but about TSTL results, given the confidence level for the SSTL monitoring. With the current definition of TSTL we can derive more operators. We can obtain the operator $\mathcal{P}_{>p}(\varphi)$ as:

$$\mathcal{P}_{>p}(\varphi) := \mathcal{P}_{<1-p}(\neg\varphi)$$

Moreover, the *everywhere* spatial operator $\tilde{\boxdot}_{[w_1,w_2]}$ can be defined as:

$$\tilde{\boxdot}_{[w_1,w_2]} \psi := \neg \tilde{\diamondsuit}_{[w_1,w_2]} \neg\psi$$

This requires ψ to hold in all the locations reachable from the current one with a total cost between w_1 and w_2. The *eventually* $\tilde{\mathcal{F}}^{[t_1,t_2]}$ and the *globally* $\tilde{\mathcal{G}}^{[t_1,t_2]}$ operators are defined as usual:

$$\tilde{\mathcal{F}}^{[t_1,t_2]} \psi := T \tilde{\mathcal{U}}^{[t_1,t_2]} \psi \qquad\qquad \tilde{\mathcal{G}}^{[t_1,t_2]} \psi := \neg\tilde{\mathcal{F}}^{[t_1,t_2]} \tilde{\neg}\psi$$

The *eventually* formula holds if ψ becomes true within t_1 and t_2 time units from the current one, while the *globally* formula requires ψ to be satisfied for each time unit in the relative interval $[t_1, t_2]$. As we already presented, TSTL provides an additional level of analysis for evaluation of spatio-temporal properties of estimated satisfaction probabilities of SSTL properties. Hence, there is a crucial difference between both the analysis and the logical operators used in SSTL and TSTL. For example, the following two TSTL properties ψ_1 and ψ_2:

$$\psi_1 := \mathcal{P}_{<p}(\varphi_1 \wedge \varphi_2) \qquad \psi_2 := \mathcal{P}_{<p}(\varphi_1) \, \tilde{\wedge} \, \mathcal{P}_{<p}(\varphi_2)$$

are intrinsically different and therefore they can take on different truth values. For example, let us assume that we are working with a disease spread model and we have the following SSTL properties on the number of infected agents I:

$$\varphi_1 := I > 5 \qquad \varphi_2 := I > 10$$

Let assume that, for a given disease probability threshold p:

$$\tau(\mathcal{M}, \Sigma, l, t, \mathcal{P}_{<p}(\varphi_1)) = F \qquad \tau(\mathcal{M}, \Sigma, l, t, \mathcal{P}_{<p}(\varphi_2)) = T$$

This can happen if we choose the value of p between the two estimates, outside their respective intervals. Since $\varphi_1 \wedge \varphi_2 \equiv \varphi_2$ then:

$$\tau(\mathcal{M}, \Sigma, l, t, \mathcal{P}_{<p}(\varphi_1 \wedge \varphi_2)) = \tau(\mathcal{M}, \Sigma, l, t, \mathcal{P}_{<p}(\varphi_2)) = T$$

while:

$$\tau(\mathcal{M}, \Sigma, l, t, \mathcal{P}_{<p}(\varphi_1)) \, \tilde{\wedge} \, \tau(\mathcal{M}, \Sigma, l, t, \mathcal{P}_{<p}(\varphi_2)) = F$$

Moreover, the first could perhaps be derived empirically from observations, but the second is only expressible with the new logical operator $\tilde{\wedge}$ and the domain $\mathbb{T}$.

Monitoring the Three-Valued Semantics of the Bounded Surround. To evaluate the validity of TSTL formulas we implemented monitoring algorithms for each logical operator, structured in a similar way to SSTL monitoring [2]. We illustrate now the monitoring algorithm for the TSTL *bounded surround* operator, which is more elaborate than the other procedures. Given a location l and a TSTL bounded surround formula $\psi = \psi_1 \tilde{\mathcal{S}}_{[w_1, w_2]} \psi_2$, the algorithm returns the piecewise constant approximation $s_{\psi, \hat{l}}$ of the function that maps each time t with $\tau(\mathcal{M}, \Sigma, \hat{l}, t, \psi_1 \tilde{\mathcal{S}}_{[w_1, w_2]} \psi_2)$, in the discrete time set $\mathcal{T}$. The cardinality of this set $\mathcal{T}$ depends on the given SSTL and TSTL formulas; it is the shortest finite sequence of time-steps for which we have the values of the satisfaction function of all the formulas involved[1]. As shown in Algorithm 1, as the first step of the algorithm, we compute the value $s_{\psi_1, l}$ for all the locations $l : 0 \leq w(\hat{l}, l) \leq w_2$ and the value $s_{\psi_2, l}$ for all the locations $l : w_1 \leq w(\hat{l}, l) \leq w_2$. These values are obtained by recursive invocation of the monitoring algorithm on the TSTL sub-formulas ψ_1 and ψ_2. We set these values for the other locations to be F, $\forall t \in \mathcal{T}$.

---

[1] We need to take into account that a temporal formula looks T_f time units into the future, hence the domain $[0, T]$ becomes $[0, T - T_f]$.

After this initial step, we iteratively compute a fixed-point function, on the set of locations satisfying the cost bounds, to get the value of the bounded surround formula, for each time step in the discrete time set $\mathcal{T}$. This fixed-point coincides with the limit of the sequence $(\chi_i)_{i \in \mathbb{N}}$, $\chi_i : L \to \mathbb{T}$, defined as follows:

1. $\chi_0(l) = s_{\psi_1,l}(t)$
2. $\chi_{i+1}(l) = \chi_i(l) \,\widetilde{\wedge}\, (\bigwedge\limits_{l':(l,l')\in E} (\chi_i(l') \,\widetilde{\vee}\, s_{\psi_2,l'}(t)))$

where i indicates the iteration. The upper bound on the number of iterations of the algorithm is given by the diameter d_G of the graph; given $\chi(l)$ the fixed point of $\chi_i(l)$, then $\chi(l) = \chi_{d_G+1}(l)$, $\forall l \in L$. The proof of the correctness of the method follows that of the SSTL monitoring. The cost of this computation for each location is $O(d_G|L||\mathcal{T}|)$; therefore, the cost for all locations is $O(d_G|L|^2|\mathcal{T}|)$. For more details, see [2], where a similar approach is used.

Algorithm 1:
Three-Valued Spatio-Temporal Logic: bounded surround operator

---

input: $\hat{l}, \psi = \psi_1 S_{[w_1,w_2]}\psi_2, \mathcal{T}$
for all $l \in L$ **do**
$\quad$ **if** $0 \leq w(\hat{l}, l) \leq w_2$ **then**
$\quad\quad |$ compute $s_{\psi_1,l}$;
$\quad$ **else**
$\quad\quad$ **if** $w(\hat{l}, l) \geq w_1$ **then**
$\quad\quad\quad |$ compute $s_{\psi_2,l}$;
$\quad\quad$ **else**
$\quad\quad\quad \llcorner s_{\psi_2,l} = F$
$\quad\quad \llcorner s_{\psi_1,l} = F; s_{\psi_2,l} = F$

for all $t \in \mathcal{T}$ **do**
$\quad$ **for all** $l \in L$ **do**
$\quad\quad \chi_{prec}(l) = T$
$\quad\quad \llcorner \chi(l) = s_{\psi_1,l}$
$\quad$ **while** $\exists l \in L : \chi_{prec}(l) \neq \chi(l)$ **do**
$\quad\quad \chi_{prec} = \chi$
$\quad\quad$ **for all** $l \in L$ **do**
$\quad\quad\quad \llcorner \chi(l) = \chi_{prec}(l) \,\widetilde{\wedge}\, (\bigwedge\limits_{l':(l,l')\in E} (\chi_{prec}(l') \,\widetilde{\vee}\, s_{\psi_2,l'}))$
$\quad \llcorner s_{\psi,\hat{l}}(t) = \chi(\hat{l})$
return $s_{\psi,\hat{l}}$

---

4   Modelling and Monitoring: MELA and jSSTL

We used the process algebra MELA [14] to formally describe spatial population models and to perform stochastic simulations, in order to produce spatio-

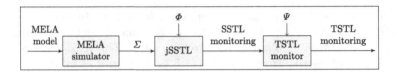

Fig. 1. Σ is the set of spatio-temporal trajectories, Φ the set of SSTL formulas and Ψ the set of TSTL formulas

temporal trajectories for the SSTL monitoring. This process algebra MELA has been developed to build spatial population models of ecological systems, since consideration of the spatial aspect has been recognized as of key importance in ecology. MELA allows one to build models on different discrete spatial structures, to define agent behaviours with spatial constraints on their interactions and probability for these interactions to be effective. Agents can perform different types of actions, that might change their state, their location, or their number in the system. The components in the MELA model generate the states of the underlying stochastic model, a Continuous Time Markov Chain (CTMC) and we perform stochastic simulations using Gillespie's Stochastic Simulation Algorithm (SSA) [15], extracting initial configuration, model structure and parameter values directly from the MELA model description. We chose to use MELA to facilitate the creation of spatial population models since it presents features that fit perfectly with SSTL monitoring settings, such as discrete representation of space and focus on spatial population models. Accordingly, it has been used in order to produce spatio-temporal trajectories, used as input for jSSTL [16], a Java library developed to support monitoring of SSTL properties. Since SSTL works with a discrete space, in particular with weighted graphs, the grid spatial structures in MELA are mapped to a weighted graph structure, to fit with the SSTL framework, with all the weights equal to 1. The results of jSSTL monitoring are used as input to verify TSTL properties. The structure of our spatio-temporal analysis is shown in Fig. 1.

5   Case Study: Defining Safety Zones

We now present two case studies, related to fire propagation, using TSTL properties for the identification of safe zones and exit routes. The actual MELA models and more details about the spatio-temporal analysis can be found in https://ludovicalv.github.io/TSTL. For the first case study we build a MELA model of forest fire: the spatial structure in this model is a 2D grid, 25×25, with Von Neumann neighbourhood of range 1 and absorbing boundaries. The considered grid is crossed by a road (R), that has a high probability of causing fire (B) in its neighbouring forest. We have zones of particular interest (P) for which we wish to provide strong protection (e.g. picnic areas, houses, regional parks) and zones of safety (S) that will never burn. We want to identify safe areas during the spread of fire. The spread can initiate from the danger zone and it can expand to the neighbouring cells.

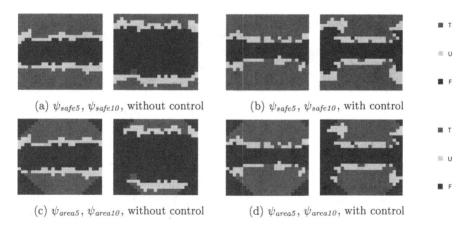

(a) ψ_{safe5}, ψ_{safe10}, without control   (b) ψ_{safe5}, ψ_{safe10}, with control

(c) ψ_{area5}, ψ_{area10}, without control   (d) ψ_{area5}, ψ_{area10}, with control

Fig. 2. (a,b): validity of TSTL formulas ψ_{safe5}, ψ_{safe10} at $t = 0$ (low risk) (c,d): validity of TSTL formulas ψ_{area5}, ψ_{area10} at $t = 0$ (safe zones)

– Zones of interest (picnic areas P, brown sign)
– Zones of safety (fire assembly points S, green sign)
– Road (R, black line) and danger zone (D, neighbouring area)
– Green area: vegetation ($25\,\text{km}^2$)

To reach our goal we start with two SSTL properties: the SSTL property φ_{pos}, a "static" property, that predicates about respective position of locations, and the property φ_{fire5}, related to eventually burning (B) over a given time horizon, here for example 5 time steps. With the formula φ_{pos} we identify the locations that connect P and S with a bounded cost, where the cost bounds are chosen given the distance between the two areas, and that are not part of the road. We verify these properties at time 0.

$$\varphi_{pos} := (\Diamondblack_{[0,17]}(S > 0)) \wedge (\Diamondblack_{[0,17]}(P > 0)) \wedge (\neg(R > 0))$$

$$\varphi_{fire5} := \mathcal{F}^{[0,5]}(B > 0)$$

We find the safe zones using the TSTL property ψ_{area5}, that identifies the locations satisfying the position requirements and with low probability of burning.

$$\psi_{pos} := \mathcal{P}_{>0.01}(\varphi_{pos}) \qquad \psi_{safe5} := \mathcal{P}_{<0.2}(\varphi_{fire5}) \qquad \psi_{area5} := \psi_{pos} \,\widetilde{\wedge}\, \psi_{safe5}$$

The results of this analysis are shown in Fig. 2a and c. Using TSTL we can also identify the zone of higher risk: we use the operator *everywhere* to identify the locations for which all the closest neighbours have high probability of being on fire, within the first 10 time steps (ψ_{risk}).

$$\varphi_{fire10} := \mathcal{F}^{[0,10]}(B > 0) \qquad \psi_{fire10} := \mathcal{P}_{>0.8}(\varphi_{fire10}) \qquad \psi_{risk} := \widetilde{\boxdot}_{[1,1]} \psi_{fire10}$$

The results are shown in Fig. 3. We can observe that the locations neighbouring the road and the safe zones do not satisfy ψ_{risk}: in fact, not all their neighbouring

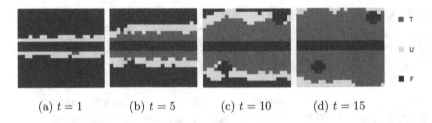

(a) $t = 1$ (b) $t = 5$ (c) $t = 10$ (d) $t = 15$

Fig. 3. TSTL property ψ_{risk}: high risk zones, for $t = 1, 5, 10, 15$ time steps

locations will burn, since both the road and the safe zones will never catch fire. We can also introduce a control measure, like a firebreak [17], to protect the areas between the zone of interest and the safe zones. Firebreaks work as a barrier and they usually consist of a gap in vegetation that slows down the fire spreads. For this reason, we add in our model fire detectors that, once the fire is detected at a given distance, will activate the control measure. These control actions will reduce the probability of fire spread, e.g. cutting the vegetation in the neighbouring area. Using MELA we can also estimate the expense (accumulated performance cost) associated with the control measures, such as the cumulative reward for Continuous Stochastic Reward Logic [18], and use these estimations to balance between effectiveness and expense of the control actions. The difference between the two models, without/with control, is shown in Fig. 2, where we can observe a wider safe zone in the model with the control measure. We verify the TSTL properties ψ_{pos} (for position) and ψ_{safe5}, ψ_{safe10} (for low risk of fire spread in 5 and 10 time steps, respectively), while ψ_{area5} and ψ_{area10} are used to define the safe zones in the different scenarios:

$$\psi_{safe5} := \mathcal{P}_{<0.2}(\varphi_{fire5}) \qquad \psi_{safe10} := \mathcal{P}_{<0.2}(\varphi_{fire10})$$

$$\psi_{area5} := \psi_{pos} \overset{\sim}{\wedge} \psi_{safe5} \qquad \psi_{area10} := \psi_{pos} \overset{\sim}{\wedge} \psi_{safe10}$$

We can expand our analysis using TSTL to identify the limited areas of risk, in situations where we have isolated danger spots spread in the area, using the TSTL *bounded surround* operator and the formula:

$$\psi_{riskArea} := (\mathcal{P}_{>0.8}(B > 0)) \overset{\sim}{\mathcal{S}}_{[w_1, w_2]} (\mathcal{P}_{<0.2}(B > 0))$$

We can also analyse the effectiveness of the control measures, verifying that the risk probabilities do not exceed a given threshold over time.

$$\psi_{lowRisk} := \mathcal{G}^{\sim[0,t]} \mathcal{P}_{<0.2}(B > 0)$$

In both case studies, for each property we run 30 simulations to perform TSTL verification. With these relatively few runs we have an overall insight about the dynamics and about the differences between distinct models, with a simple representation of complex systems and properties. Nevertheless, this analysis takes into account the uncertainty, that can be used to determine the need of more runs for more precise results. We will expand our current framework to automatically provide additional simulations to refine the analysis of the atomic propositions, when a more precise result is needed.

6   Case Study: Emergency Evacuation Route

For the second case study we introduce another fire related model: we want to identify the most appropriate fire exit in a situation where the evacuation routes are already defined. We aim to identify the safe evacuation routes from the centre of the grid to the assembly points, located in the corners, having a fire starting in the location in the lower left corner, that can spread to the neighbouring locations. We build a MELA model on a 2D grid (25×25) where the fire can spread everywhere, apart from the assembly points (safe zones). In the parts of the grid defined as a route we have two agents in parallel, one that identifies the presence of people and the other one that identifies the presence of fire.

- Fire spread model (on a grid, 25 km^2)
- Agents at the center
- Different exit routes (grey lines) to safe zones (located in the corners)
- For each route: ($fire \parallel people$)

In this example the movement of people and fire are modelled separately, they do not influence each other in the model. We will gather both types of information in the study of TSTL properties. This model focusses more on the concrete safe path and on actual movement, while in the previous one we were defining the different safe areas in time, without specifying the route. In the inflammable area the $fire$ agent can be on fire (B, burning) or not (I, inflammable) while the exit route locations can be empty (EM), occupied (Occ) or passed (P); P represents a cell that was occupied but empty again. To identify the safe evacuation routes, we use TSTL to identify cells that have low probability of being on fire (ψ_{fire}) and non-zero probability of being occupied (ψ_{occ}), given the agent movement in the model.

$$\varphi_{occ} := Occ > 0 \qquad \varphi_{fire} := B > 0$$

$$\psi_{occ} := \mathcal{P}_{>0.01}(\varphi_{occ}) \qquad \psi_{fire} := \mathcal{P}_{<0.2}(\varphi_{fire}) \qquad \psi_{safe} := \psi_{occ} \overset{\sim}{\wedge} \psi_{fire}$$

The verification output of TSTL property ψ_{safe} shows the routes that will lead to the assembly point safely, as shown in Fig. 4. To be able to identify the safe evacuation routes from the beginning, instead of observing their temporal evolution, we can check the TSTL property $\psi_{safeRoute}$ at $t = 0$. We want to identify the route that, with probability higher than 0.8, will not be on fire if occupied, in this case in the first 10 time-steps.

$$\varphi_{route} := (EM > 0) \vee (Occ > 0) \vee (P > 0)$$

$$\varphi_{notFire} := \neg((Occ > 0) \wedge (B > 0)) \equiv \neg(Occ > 0) \vee \neg(B > 0)$$

$$\varphi_{Gsafe} := \mathcal{G}^{[0,10]}(\varphi_{route} \wedge \varphi_{notFire}) \qquad \psi_{safeRoute} := \mathcal{P}_{>0.8}(\varphi_{GSafe})$$

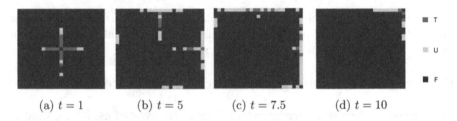

(a) $t = 1$ (b) $t = 5$ (c) $t = 7.5$ (d) $t = 10$

Fig. 4. Temporal evolution of safe evacuation routes: TSTL property ψ_{safe}, with movement rate of the agents equal to 2.0

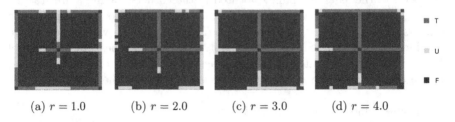

(a) $r = 1.0$ (b) $r = 2.0$ (c) $r = 3.0$ (d) $r = 4.0$

Fig. 5. Safe evacuation routes: TSTL property $\psi_{safeRoute}$, with different movement rate of the agents

We will check this TSTL property changing the rate of agent movement in the MELA model, as shown in Fig. 5. We can observe that if the rate of movement is not high enough, there are not safe options to reach the assembly points. As further analysis, we examine the number of unknown values over time, given TSTL properties and changing the quantity of spatio-temporal trajectories to analyse. We study the percentage of locations having unknown as truth value for different formulas, in both case studies, for different numbers of simulation runs. We observed that the percentage decreases with the increase of the number of runs. Since the width of the confidence intervals depends to a large extent on this value, with an increase in the number of runs we tend to give a more precise estimation of the satisfaction probability. Therefore we have narrower confidence intervals as input for TSTL monitoring and a consequent smaller percentage of unknown values. The three-valued approach is useful to discriminate among TSTL properties in the process of acquiring spatio-temporal trajectories, until the satisfaction set is large enough.

7   Discussion

In the current framework development we use verification of SSTL formulas as input for TSTL monitoring and the starting point for the spatio-temporal analysis. We want to point out that TSTL can be used to predicate on estimated satisfaction probabilities of formulas specified with other logics and also on more general uncertain values with an estimated confidence, as long as the required format is maintained (estimated value for each location at each time point).

As future case studies we will apply our framework to model the spread of invasive species, in particular giant hogweed [19]: we will analyse the effectiveness of different control measures to protect areas of interest, such as regional parks, taking into account also the suitability of the different locations for plant colonisation. In particular, we will analyse the difference between prevention (control outside the boundaries of the area) and direct action (eradication when the invasive species are detected inside the area), considering also the expense associated with the different measures.

As a future extension for TSTL we will define and implement the operator *bounded reachable* $\tilde{\mathcal{R}}$. This operator can be seen as a spatial until with direction and associated with a path. We will be able to verify properties related with locations reachable within a given cost range and satisfying defined TSTL properties, and the existence of a connecting path formed only by locations satisfying a given set of TSTL properties. In the case studies we presented, the use of this new operator would have allowed us to identify safe paths without having to mimic the actual movement, detecting different possible solutions. Using this new TSTL operator we could verify if *there is a safe location (assembly point S) that we can reach passing only through locations with low probability of burning, with a cost w, $w \in [w_1, w_2]$:*

$$\psi := (\mathcal{P}_{<0.2}(B > 0)) \; \tilde{\mathcal{R}}_{[w_1, w_2]} \; (\mathcal{P}_{>0.01}(S > 0))$$

8   Conclusions

In this paper we presented Three-Valued Spatio-Temporal Logic (TSTL), an extension of Signal Spatio-Temporal Logic (SSTL) that allows us to widen the analysis of spatio-temporal properties of stochastic systems. We have shown how this extension is used to study the spatio-temporal evolution of the estimated satisfaction probabilities of given SSTL formulas. We implemented the monitoring algorithms for each TSTL operator and used them in the case studies to perform the novel analysis, checking the validity of different TSTL formulas. We used TSTL to identify the zones that have high risk of catching fire during a fire spread and to find the safest evacuation routes, checking the ones that have high probability to be safe over time. We provide the novel spatio-temporal logic with a three-valued semantics to handle the intrinsic uncertainty related to the statistical methods used to estimate the satisfaction probabilities. The three-valued approach allows us to perform initial analysis with a relatively small set of spatio-temporal trajectories, taking into account the uncertainty; on the other hand, it also provides a decision tool on the number of simulations needed for drawing stronger conclusions.

Acknowledgement. This work was supported by Microsoft Research Cambridge through its PhD Scholarship Programme and by the EU project QUANTICOL 600708. Glenn Marion was funded by the Scottish Government Rural and Environment Science and Analytical Services Division (RESAS).

References

1. Legay, A., Delahaye, B., Bensalem, S.: Statistical model checking: an overview. In: Barringer, H., Falcone, Y., Finkbeiner, B., Havelund, K., Lee, I., Pace, G., Roşu, G., Sokolsky, O., Tillmann, N. (eds.) RV 2010. LNCS, vol. 6418, pp. 122–135. Springer, Heidelberg (2010). doi:10.1007/978-3-642-16612-9_11

2. Nenzi, L., Bortolussi, L., Ciancia, V., Loreti, M., Massink, M.: Qualitative and quantitative monitoring of spatio-temporal properties. In: Bartocci, E., Majumdar, R. (eds.) RV 2015. LNCS, vol. 9333, pp. 21–37. Springer, Cham (2015). doi:10.1007/978-3-319-23820-3_2

3. Aiello, M., Pratt-Hartmann, I., Van Benthem, J.: Handbook of Spatial Logics. Springer, Netherlands (2007)

4. Cardelli, L., Gordon, A.D.: Anytime, anywhere: modal logics for mobile ambients. In: POPL 2000, pp. 365–377. ACM (2000)

5. Reif, J., Sistla, A.: A multiprocess network logic with temporal and spatial modalities. J. Comput. Syst. Sci. **30**(1), 41–53 (1985)

6. Hansson, H., Jonsson, B.: A logic for reasoning about time and reliability. Formal Aspects Comput. **6**(5), 512–535 (1994)

7. Baier, C., Haverkort, B., Hermanns, H., Katoen, J.P.: Model-checking algorithms for continuous-time Markov chains. IEEE TSE **29**, 524–541 (2003)

8. Gottwald, S.: Many-valued logic. In: Stanford Encyclopedia of Philosophy (2008)

9. Łukasiewicz, J.: Selected Works. North-Holland Publishing Company, Amsterdam (1970)

10. Kleene, S.C.: On notation for ordinal numbers. JSL **3**(4), 150–155 (1938)

11. Katoen, J.-P., Klink, D., Leucker, M., Wolf, V.: Three-valued abstraction for probabilistic systems. JLAP **81**(4), 356–389 (2012)

12. Sen, K., Viswanathan, M., Agha, G.: Model-checking Markov chains in the presence of uncertainties. In: Hermanns, H., Palsberg, J. (eds.) TACAS 2006. LNCS, vol. 3920, pp. 394–410. Springer, Heidelberg (2006). doi:10.1007/11691372_26

13. Maler, O., Nickovic, D.: Monitoring temporal properties of continuous signals. In: Lakhnech, Y., Yovine, S. (eds.) FORMATS/FTRTFT -2004. LNCS, vol. 3253, pp. 152–166. Springer, Heidelberg (2004). doi:10.1007/978-3-540-30206-3_12

14. Luisa Vissat, L., Hillston, J., Marion, G., Smith, M.J.: Mela: modelling in ecology with location attributes. In: EPTCS, vol. 227, pp. 82–97 (2016)

15. Gillespie, D.T.: Exact stochastic simulation of coupled chemical reactions. J. Phys. Chem. **81**(25), 2340–2361 (1977)

16. Nenzi, L., Bortolussi, L., Loreti, M.: jSSTL - a tool to monitor spatio-temporal properties. ValueTools (2016)

17. Cerotti, D., Gribaudo, M., Bobbio, A., Calafate, C.T., Manzoni, P.: A Markovian agent model for fire propagation in outdoor environments. In: Aldini, A., Bernardo, M., Bononi, L., Cortellessa, V. (eds.) EPEW 2010. LNCS, vol. 6342, pp. 131–146. Springer, Heidelberg (2010). doi:10.1007/978-3-642-15784-4_9

18. Baier, C., Haverkort, B., Hermanns, H., Katoen, J.-P.: On the logical characterisation of performability properties. In: Montanari, U., Rolim, J.D.P., Welzl, E. (eds.) ICALP 2000. LNCS, vol. 1853, pp. 780–792. Springer, Heidelberg (2000). doi:10.1007/3-540-45022-X_65

19. Catterall, S., Cook, A.R., Marion, G., Butler, A., Hulme, P.E.: Accounting for uncertainty in colonisation times: a novel approach to modelling the spatio-temporal dynamics of alien invasions using distribution data. Ecography **35**(10), 901–911 (2012)

Sequential Schemes for Frequentist Estimation of Properties in Statistical Model Checking

Cyrille Jegourel[1]($\boxtimes$), Jun Sun[1], and Jin Song Dong[2]

[1] Singapore University of Technology and Design, Singapore, Singapore
cyrille.jegourel@gmail.com, sunjunhqq@gmail.com
[2] Griffith University, Brisbane, Australia
dongjs1@gmail.com

Abstract. Statistical Model Checking (SMC) is an approximate verification method that overcomes the state space explosion problem for probabilistic systems by Monte Carlo simulations. Simulations might be however costly if many samples are required. It is thus necessary to implement efficient algorithms to reduce the sample size while preserving precision and accuracy. In the literature, some sequential schemes have been provided for the estimation of property occurrence based on predefined confidence and *absolute* or *relative* error. Nevertheless, these algorithms remain conservative and may result in huge sample sizes if the required precision standards are demanding. In this article, we compare some useful bounds and some sequential methods based on frequentist estimations. We propose outperforming and rigorous alternative schemes, based on Massart bounds and robust confidence intervals. Our theoretical and empirical analysis show that our proposal reduces the sample size while providing guarantees on error bounds.

1 Introduction

Probabilistic Model Checking (PMC) [16] is a formal verification method to analyse quantitative properties of probabilistic systems. PMC algorithms perform an exhaustive traversal of the state space of the system. However, real-world applications often involve multiple interacting components and the resulting state space becomes intractable. This limitation has led to the development of alternative methods like discrete event simulation and Statistical (Probabilistic) Model Checking (SMC) [22]. These simulation-based approaches require the use of an executable model of the system and then estimate the probability of a property based on simulations. SMC provides rigorous bounds of the error of the estimated results, based on robust statistical techniques (e.g., [4,18,20]). For real-world complex systems, SMC has a lot of potential as it requires little memory and remains very efficient for large systems. Finally, SMC is sometimes the only option for verifying many realistic models.

SMC also faces some specific problems. For example, simulations may be costly and time consuming. Moreover, the specifications of critical or important events are in practice tight. SMC must thus focus on additional statistical aspects

© Springer International Publishing AG 2017
N. Bertrand and L. Bortolussi (Eds.): QEST 2017, LNCS 10503, pp. 333–350, 2017.
DOI: 10.1007/978-3-319-66335-7_23

to provide optimised sampling schemes while guaranteeing a rigorous confidence of the estimation. The need of rigorous sampling schemes have been addressed from the early days in SMC [11,22] to more recent [8,10] just to cite a few. A key feature in designing a sampling procedure is to determine the number of simulations necessary to generate an estimation within acceptable margins of error and confidence. Bayesian SMC may be used to address this problem. However, in this approach, the probability to estimate must be given by a prior random variable whose density is based on previous experiments and knowledge about the system [24]. This limitation motivates the alternative frequentist estimation approaches. The scope of this article is restricted to this class of methods.

In [11], the authors discussed the notion of *absolute* and *relative* margin of error for SMC. To guarantee that the absolute error is bounded, they introduced a procedure relying on the Okamoto bound[1] that, given fixed confidence and error parameters, determines *a priori* the number of Bernoulli samples required, which is independent of the probability to estimate. Supporting relative errors (i.e., errors which depend on the probability to quantify) is more difficult, although theoretical bounds exist. Dagum et al. [7] proposed an *approximate* algorithm based on Bernstein's inequalities.

Approximate algorithms work by rough parameter estimations that are then reused in a stopping rule to update the number of simulations achieving the desired precision task. More recently, Watanabe proposed a sequential algorithm for bounding the relative error [21] based on a simpler stopping rule. The procedures described in [11] have been at least partially implemented in statistical model checkers like PRISM [16], PLASMA [14], APMC [12] and UPPAAL-SMC [8]. These sampling schemes are however very conservative notably when the probability to estimate is close to 0 or 1. Moreover, Dagum's algorithm was initially used to estimate the mean value of any random variable distributed in [0, 1] and is thus not optimised for Bernoulli random variables.

In this article, our main goal is to provide better performing sampling schemes that *rigorously* fulfil absolute and relative error specifications. The key idea of our schemes is to define sequentially confidence intervals (CI) of the probability and then to apply Massart bounds, sharper than the Chernoff bounds, over the worst value of the CI to decide whether enough traces have been sampled or not. For this purpose, we also aim to clarify the two-sided "Chernoff" bounds for absolute and relative error specifications, to promote Massart bounds and last but not least, to give proofs of all these bounds. Indeed, the original theorems are one-sided and the two-sided versions must be clearly stated. The proofs are sometimes straightforward, at least for Theorems 3 and 5, but sometimes require more arguments. In particular, we could not find clear wordings and proofs of Theorems 2 and 4 in the literature. Finally, as far as we know, Theorems 6 and 7 are original as well as the algorithms using them. The proofs of the bounds can be found in the extended version available online[2].

---

[1] The Okamoto bound is sometimes called the Chernoff bound in the literature.

[2] At least here: http://sav.sutd.edu.sg/publications/.

In Sect. 2, we formally state the absolute and relative specifications which we want to fulfil. We also recall the basics of Monte Carlo estimation and some subtleties concerning coverage and CI. In Sect. 3, we introduce the Massart bounds. So far, they seem to suffer from a lack of recognition. For that reason, we present a comparison with the Chernoff bounds. We then describe some existing sampling schemes related to our problem in Sect. 4. In Sect. 5, we propose alternative sequential algorithms based on two inequalities, previously proven, which depend on the *coverage* of the probability. Finally, we show in Sect. 6 that these new schemes outperform the current approaches for the absolute and relative error problems by reducing significantly the sampling size. Section 7 concludes the article and leaves open questions for future work.

2 Background

In the following, a stochastic system $\mathcal{S}$ is interpreted as a set of interacting components in which the state is determined randomly with respect to a global probability distribution. Let $(\Omega, \mathcal{F}, \mu)$ be the probability space induced by the system with Ω a set of finite paths with respect to system's property ϕ, $\mathcal{F}$ a σ-algebra of Ω and μ the probability distribution defined over $\mathcal{F}$. Before going further, it is worth mentioning that SMC initially addressed the problem of verifying whether a property probability exceeds a threshold or not. This problem can be solved by using the sequential probability ratio test in hypothesis testing [22]. Other issues have been considered since, notably the estimation of the probability that a system property holds. In spite of similarities, both problems are different and in what follows, we focus on the estimation problem.

2.1 Statement of the Problem

Given a probabilistic system $\mathcal{S}$, a property ϕ and a probability γ, we write $\mathcal{S} \models Pr(\phi) = \gamma$ if and only if the probability that a random execution of $\mathcal{S}$ satisfies ϕ is equal to γ. In principle, if γ is unknown, we can apply analytical methods to determine this value. However, due for example to numerical imprecisions, we often relax the constraints over γ and introduce the following notations:

$$\mathcal{S} \models_{\epsilon}^{a} Pr(\phi) = \gamma \quad \text{and} \quad \mathcal{S} \models_{\epsilon}^{r} Pr(\phi) = \gamma \tag{1}$$

The left formula means that a random execution of $\mathcal{S}$ satisfies ϕ with probability γ plus or minus an absolute error ϵ, i.e. $Pr(\phi) \in [\gamma - \epsilon, \gamma + \epsilon]$. The right formula means that a random execution of $\mathcal{S}$ satisfies ϕ with probability γ up to some relative error ϵ, i.e. $Pr(\phi) \in [(1 - \epsilon)\gamma, (1 + \epsilon)\gamma]$.

SMC applies on an executable system $\mathcal{S}$ and a property ϕ that is verified in finite time. In SMC, the satisfaction of property ϕ is quantified by a Bernoulli random variable of unknown mean γ. This mean is then approximated using a Monte Carlo estimation scheme. The output of the scheme is thus not an exact but an approximate value, given within certain error bounds and a confidence parameter δ that is the probability of outputting a false estimate. SMC thus requires a sampling scheme which outputs, after n samples, an estimate $\hat{\gamma}_n$ close to γ up to some absolute or relative ϵ-based error with probability greater or equal than $1 - \delta$. Formally, we write:

$$\mathcal{S} \models_{\epsilon,\delta}^a Pr(\phi) = \hat{\gamma}_n \quad \text{or} \quad \mathcal{S} \models_{\epsilon,\delta}^r Pr(\phi) = \hat{\gamma}_n \tag{2}$$

if and only if an algorithm outputs estimators while guaranteeing:

$$Pr(|\hat{\gamma}_n - \gamma| > \epsilon) \le \delta \tag{3}$$

or respectively:

$$Pr(|\hat{\gamma}_n - \gamma| > \epsilon\gamma) \le \delta. \tag{4}$$

We call (3) the absolute error specification and (4) the relative error specification. The goal of the article is thus to equip SMC with sampling algorithms that fulfil Specification (3) or (4) with as few samples as possible.

2.2   Monte Carlo Estimation

Let ω be a path sampled from space Ω with respect to distribution μ; z be a function from Ω to $\{0,1\}$ assigning 1 if ω satisfies property ϕ and 0 otherwise and γ be the probability that an arbitrary path of the system satisfies ϕ. In SMC, the behaviour of function z is interpreted as a Bernoulli random variable Z with mean parameter γ. By definition, the average value γ is the integral of function z with respect to distribution μ over space Ω: $\gamma = \mathrm{E}_\mu[Z] = \int_\Omega z(\omega)\,d\mu(\omega)$ and an estimator $\hat{\gamma}_n$ is given by the Monte Carlo method by drawing n independent samples $\omega_i \sim \mu$, $i \in \{1, \ldots, n\}$, as follows:

$$\hat{\gamma}_n = \frac{1}{n} \sum_{i=1}^n z(\omega_i) \approx \mathrm{E}_\mu[Z] \tag{5}$$

Let $m = \sum_{i=1}^n z(\omega_i)$ be the number of successes and $\sigma^2 = \gamma(1 - \gamma)$ the variance of Z. In what follows, for sake of simplicity, we use both notations $\hat{\gamma}_n$ and m/n to denote the estimate.

Confidence Intervals and Coverage. An estimator is given in general within a CI. However, in order to make use of the theorems presented in Sect. 5, we need to distinguish the notion of coverage and approximate CI.

Definition 1. *Given probability γ and a CI I, we call $C(\gamma, I) = Pr(\gamma \in I)$ the coverage of γ (by I).*

Denoting $\Phi(.)$ the standard normal distribution function and $z_{\delta/2} = \Phi^{-1}(1 - \delta/2)$ the $(1 - \delta/2)$th quantile of the normal distribution, the notional $(1 - \delta)$-CI for γ is given by $I = \left[\hat{\gamma}_n - z_{\delta/2}\frac{\sigma}{\sqrt{n}}, \hat{\gamma}_n + z_{\delta/2}\frac{\sigma}{\sqrt{n}}\right]$ in virtue of the central limit theorem. However, in practice, σ^2 is replaced by a sample approximation $\hat{\sigma}_n^2 = \hat{\gamma}_n(1-\hat{\gamma}_n)/n$ (and if n is small, $z_{\delta/2}$ by $t_{\delta/2,n-1}$ the quantile of the Student's

t-distribution with $n-1$ degrees of freedom). Then, an approximate $(1-\delta)$-CI $\tilde{I}$ is given by:

$$\tilde{I} = \left[\hat{\gamma}_n - z_{\delta/2}\hat{\sigma}_n, \hat{\gamma}_n + z_{\delta/2}\hat{\sigma}_n\right] \tag{6}$$

Unfortunately, the coverage of γ by an approximate CI $\tilde{I}$, may be significantly below the (desired) notional coverage: $C(\gamma, \tilde{I}) < C(\gamma, I) = 1-\delta$. More details about this topic are available in the extended version and in [2].

Exact Clopper-Pearson CI. The algorithms proposed in Sect. 5 require an iterative computation of CI to evaluate a rigorous coverage of γ. For that purpose, we use the Clopper-Pearson $(1-\delta)$-CI [6]. This CI guarantees that the actual coverage is always equal to or above the nominal confidence level. In others words, a $(1-\delta)$-Clopper-Pearson CI J guarantees that $C(\gamma, J) \geq 1-\delta$ and its closed-form expression is easily computed: $J = \left[\beta^{-1}\left(\frac{\delta}{2}, m, n-m+1\right), \beta^{-1}(1-\frac{\delta}{2}, m+1, n-m)\right]$ with $\beta^{-1}(\delta, u, v)$ being the δ-th quantile of a Beta distribution parametrised by u and v.

Agresti-Coull CI. As γ decreases, the Clopper-Pearson CI becomes more conservative. The Agresti-Coull CI consists in replacing the number of samples n by $n + z_{\delta}^2$ and the number of successes m by $m + z_{\delta}^2/2$ in the binomial CI (6). The CI is only approximate but still presents a good coverage close to the boundaries and may represent a good compromise between exactness and conservativeness (see [2] for more details).

3 Chernoff-Hoeffding-Okamato and Massart Bounds

In the literature, the Chernoff bounds [4] refer to exponential decreasing bounds, in the number of simulations, of the probability of deviation between a Monte Carlo estimate and its mean. However, they exist under various forms, additive or multiplicative, one or two-sided, more or less "simplified". Moreover, tighter bounds have been established, notably in [17], but they still suffer from a lack of recognition. In this section, we intend to clear up confusion on the bounds by presenting a brief survey of the two-sided bounds and show the improvements achieved by the Massart bounds to give them the attention they deserve.

3.1 Absolute Error Bounds

Though the seminal work is due to Chernoff [4], the two-sided absolute error bound has been first stated for binomial distributions by Okamoto in [19].

Theorem 1 (Okamoto bound). *For any ϵ, $0 < \epsilon < 1$, we have the following inequality:*

$$Pr(|\hat{\gamma}_n - \gamma| > \epsilon) \leq 2\exp(-2n\epsilon^2) \tag{7}$$

Given ϵ, δ, writing out $\delta = 2\exp(-2n\epsilon^2)$, the Okamato bound can be used to determine a minimal number n of simulations to perform a Monte Carlo plan fulfilling the absolute error specification (3). The main advantage of this bound is that it is independent of the value to estimate. However, the bound is very conservative and in many cases, a much lower sample size would achieve the same absolute error specification. Hoeffding provided a one-sided tighter exponential bound in [13]. We present below a two-sided version of his bound.

Theorem 2 (Absolute Error Hoeffding bound). *For any ϵ such that $0 < \epsilon < 1$ and γ such that $0 < \gamma < 1$, we have the following inequality:*

$$Pr(|\hat{\gamma}_n - \gamma| > \epsilon) \le 2\exp\left(-n\epsilon^2 f(\gamma)\right) \tag{8}$$

where $f(\gamma) = \begin{cases} 1/(1-2\gamma)\log((1-\gamma)/\gamma) & \text{if } \gamma \ne 1/2 \\ 2 & \text{if } \gamma = 1/2 \end{cases}$

Surprisingly, we could not find a clear statement and a proof of this result in the literature. We thus present a proof in the extended version.

In this article, the Hoeffding bound is only presented because of its repute. Indeed, Massart established in [17] a sharper bound that holds if the absolute error ϵ is lower than probabilities γ and $1 - \gamma$. In what follows, we use the two-sided absolute and relative error versions of Massart bounds.

Theorem 3 (Absolute Error Massart bound). *For all γ such that $0 < \gamma < 1$ and any ϵ such that $0 < \epsilon < \min(\gamma, 1 - \gamma)$, we have the following inequality:*

$$Pr(|\hat{\gamma}_n - \gamma| > \epsilon) \le 2\exp\left(-n\epsilon^2 h_a(\gamma, \epsilon)\right) \tag{9}$$

where $h_a(\gamma, \epsilon) = \begin{cases} 9/2\left((3\gamma + \epsilon)(3(1-\gamma) - \epsilon)\right)^{-1} & \text{if } 0 < \gamma < 1/2 \\ 9/2\left((3(1-\gamma) + \epsilon)(3\gamma + \epsilon)\right)^{-1} & \text{if } 1/2 \le \gamma < 1 \end{cases}$

3.2 Relative Error Bounds

In practice, the absolute error is set independently of γ. However, it could be that the approximation is meaningless, especially if the absolute error is large with respect to γ. In this case, setting a relative error that remains 'small' with respect of γ may be more adequate. The literature mentions a Chernoff-Hoeffding bound with relative error (e.g. [1]). This bound is known under multiple forms, more or less sharp and one or two-sided. For sake of consistency, we here provide a two-sided bound. As the existing literature adopts slightly different results, sometimes without providing their proof, we give a complete proof in the extended version adapted from two online references[3].

---

[3] http://crypto.stanford.edu/~blynn/pr/chernoff.html and www.cs.princeton.edu/courses/archive/fall09/cos521/Handouts/probabilityandcomputing.pdf.

Theorem 4 (Relative Error Hoeffding bound). *For any ϵ, $0 < \epsilon < 1$ and γ, $0 < \gamma < 1$, we have the following inequality:*

$$Pr\left(|\hat{\gamma}_n - \gamma| > \epsilon\gamma\right) \leq 2\exp\left(-\frac{n\epsilon^2\gamma}{2+\epsilon}\right) \tag{10}$$

Finally, the Massart bound has a two-sided relative form.

Theorem 5 (Relative Error Massart bound). *For γ, $0 < \gamma < 1$ and any ϵ, $0 < \epsilon < (1-\gamma)/\gamma$, we have the following inequality:*

$$Pr(|\hat{\gamma}_n - \gamma| \geq \epsilon\gamma) \leq 2\exp\left(-n\epsilon^2 h_r(\gamma,\epsilon)\right) \tag{11}$$

$$with \quad h_r(\gamma,\epsilon) = \begin{cases} 9\gamma/2\left((3+\epsilon)(3-\gamma(3+\epsilon))\right)^{-1} & if\ 0 < \gamma < 1/2 \\ 9\gamma/2\left((3-\epsilon)(3-\gamma(3-\epsilon))\right)^{-1} & if\ 1/2 \leq \gamma < 1 \end{cases}$$

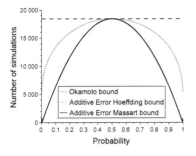

Fig. 1. Okamoto (dash), Hoeffding (dot) and Massart (plain) bounds with absolute error $\epsilon = 0.01$ and confidence parameter $\delta = 0.05$.

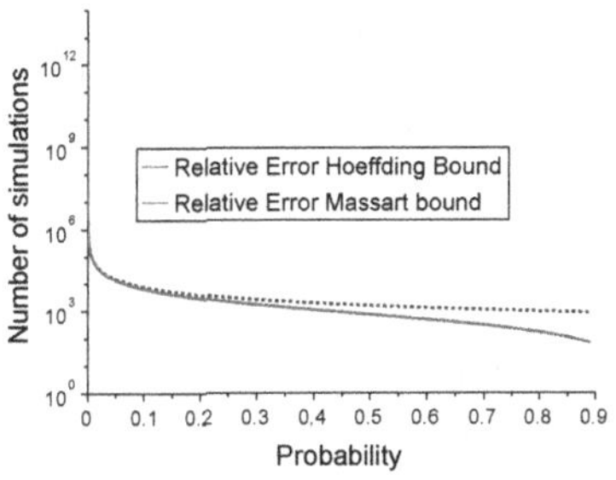

Fig. 2. Hoeffding (dot) and Massart (plain) bounds with relative error $\epsilon = 0.1$ and confidence parameter $\delta = 0.05$.

Notional Sample Size. If we let δ be equal to any of the right side expression of the inequalities given in Theorems (1)–(5), we can deduce a notional sample size n such that specification (3) or (4) is fulfilled. For example, using Theorem 5 given ϵ and δ, we only need to set $n > 1/(h_r(\gamma,\epsilon)\epsilon^2)\log(2/\delta)$ to satisfy the relative error specification (4). However, Hoeffding and Massart inequalities are not directly applicable because they depend on γ, in contrast to the Okamoto bound. But, they still have a theoretical interest: Fig. 1 indicates for any notional γ the number of simulations necessary to produce an (ϵ, δ)-estimator according to the Okamoto, Hoeffding and Massart bounds. Though the bounds are approximately equivalent when γ is $1/2$, the bounds are far apart when γ is away from $1/2$. Given $\epsilon = 0.01$, $\delta = 0.05$ and $\gamma = 0.05$ for example, the absolute error specification would be fulfilled with $n \geq 3283$ simulations according to the Massart bound instead of $n \geq 11276$ or $n \geq 18445$ for the respective Hoeffding and

Okamoto bounds. Similarly for the relative error specification, Fig. 2 shows that the Massart sample size is always lower than the Chernoff-Hoeffding sample size. The gain in sample size is more important when γ is high. With $\epsilon = 0.1$, $\delta = 0.05$, the ratio between Hoeffding and Massart sample sizes tends to decrease to 1.086 when γ tends to zero, that may still be non-negligible if sampling is time-costly.

4   Related Work

In this section, we give a brief summary of existing sequential methods based on frequentis estimations to address specification (3) or (4). Some of them have already been implemented in SMC. We also recall that the specifications can be alternatively addressed by Bayesian SMC, not explored in this article, when beliefs and knowledge about the system are exploitable [24].

4.1   Schemes for the Absolute Error Specification

Given ϵ and δ, the standard method to satisfy specification (3) is to compute a sample size n independently of probability γ using the Okamoto bound. Since there does not exist a bound independent of γ in the relative error case, the sequential schemes are mostly used to address specification (4) but they are not limited to it.

Simple Scheme. A simple idea could be to sample and update a $(1-\delta)$-CI until it is included into an interval $\hat{\gamma}_n \pm \epsilon$. This frequentist approach is implemented in UPPAAL-SMC [8]. However, though this technique may work more often if the CI are computed according to the Clopper-Pearson method, this scheme does not guarantee in general specification (3) for any δ, ϵ and γ (see for example [9]). For sake of understanding, we added a brief but technical explanation in the extended version. It is however possible to pre-compute a value δ^* that guarantees a final coverage greater than $1 - \delta$ (see [9]).

Chen's Scheme [3]. A promising sequential scheme which may work in practice, at least for some common values of ϵ and δ, is the work proposed by Chen in [3]. Chen's scheme also takes advantage of the Massart bounds. The idea is to sample while $n < 2\frac{\log(2/\delta)}{\epsilon^2} \left[1/4 - (|\hat{\gamma}_n - 1/2| - 2/3\epsilon)^2 \right]$. Unfortunately, this rule only guarantees to produce an estimation which does not exceed the error bound ϵ on one side. So far, showing the other half of the bound has not been proven and was conjectured by the authors after some experiments.

4.2   Schemes for the Relative Error Specification

In [11], the relative error specification is addressed by Dagum's algorithm.

Dagum's Scheme [7]. is a three-step procedure to perform an estimation of the mean of a general $[0, 1]$-valued random variable X given relative error ϵ and confidence δ. The two first steps consist in providing a coarse estimation $\hat{\gamma}_k$ and a dispersion parameter $\hat{\rho}_l$. Finally, the third step provides the final estimation $\hat{\gamma}_n$ using $\hat{\gamma}_k$ and $\hat{\rho}_l$. The three steps are independent and depend on three different stopping rules, omitted here for sake of simplicity (see [7] for more details). The final sample size is thus given by $k+l+n$. Nevertheless, Dagum's scheme is based on coarser bounds than the Chernoff bounds. Moreover, this algorithm is used to estimate the mean of any random variable with support in $[0, 1]$. Consequently, the scheme has a very general use but is not optimised for Bernoulli random variables.

Watanabe's Scheme [21]. In order to guarantee the relative error specification, Watanabe proposed to sample until the number of successes is greater than $\frac{3(1+\epsilon)}{\epsilon^2} \log \frac{2}{\delta}$. The main advantage is that this simple scheme does not require to perform pre-samples as in the first two steps of Dagum's algorithm. As far as we know, this scheme, more recent than Dagum's, is not implemented in SMC.

5   A Sequential Scheme Involving Coverage

In this section, we present our sequential scheme for the absolute and relative error specification. Our scheme performs better than Watanabe and Dagum's scheme in the relative error case and, unlike the simple and Chen's schemes, is guaranteed to bound the error on both sides while strictly maintaining a coverage greater than $1 - \delta$. Apart from the Okamoto bound, the inequalities presented in Sect. 3 require the knowledge of γ and they are thus not directly applicable. However, one may still exploit some information about probability γ. For example, depending on the problem, one may know or numerically evaluate with certainty a rough interval in which γ evolves. We present in the first subsection two theorems and the underlying sample sizes and, in the second subsection, our sampling schemes.

5.1   Bounds with Coverage

The following theorems make use of the Massart bounds presented in Theorems 3 and 5 as they are sharper than the Chernoff-Hoeffding bounds.

Theorem 6 (Absolute Error Massart Bound with coverage). *Let a and b be the extrema of CI $I \in \mathcal{B}([0, 1])$ and I^c be the complement of I in $[0, 1]$:*

$$Pr\left(|\hat{\gamma}_n - \gamma| > \epsilon\right) \leq 2 \exp\left(-n\epsilon^2 h_a(x, \epsilon)\right) + C(\gamma, I^c) \tag{12}$$

where function h_a is defined in Theorem 3 and $x = a$ if $b < 1/2$, $x = b$ if $a > 1/2$ and $x = 1/2$ if $1/2 \in I$.

By default, $a = 0$, $b = 1$, $C(\gamma, [0,1]^c) = 0$ and the theorem is consistent with the Okamoto bound. We remark that even if an accurate estimation of γ is not feasible to obtain within a reasonable time, Theorem 6 can exploit coarse but exact bounds a, b calculated analytically. In that case, we would have $C(\gamma, [a,b]^c) = 0$. Finally, a similar theorem involving relative error can be established.

Theorem 7 (Relative Error Massart Bound with coverage). *Let a be a (random) element of $[0,1]$ and h_r defined as in Theorem 5.*

$$Pr(|\hat{\gamma}_n - \gamma| > \epsilon\gamma) \leq 2\exp(-n\epsilon^2 h_r(a,\epsilon)) + C(\gamma, [0,a]) \tag{13}$$

Both theorems state that the probability of absolute or relative error is bounded by the respective Massart bound applied over the most pessimistic value of a CI plus the probability that the CI does not contain γ. We deduce from both theorems the following sample-size result:

Theorem 8. *Let $\delta' < \delta$ such that $C(\gamma, I^c) < \delta'$. (i) Under the conditions of Theorem 6, a Monte Carlo algorithm $\mathcal{A}$ that outputs an estimate $\hat{\gamma}_n$ fulfils Specification (3) if $n > \frac{1}{\min(h_a(a,\epsilon),h_a(b,\epsilon))\epsilon^2} \log \frac{2}{\delta-\delta'}$.*

(ii) Similarly, under the conditions of Theorem 7, a Monte Carlo algorithm $\mathcal{A}$ that outputs an estimate $\hat{\gamma}_n$ fulfils Specification (4) if $n > \frac{1}{h_r(a,\epsilon)\epsilon^2} \log \frac{2}{\delta-\delta'}$.

The proof is immediate in both cases once we set $\delta = 2s + \delta'$ with s being the respective exponential expressions of Theorems 6 or 7.

The bounds of Theorem 8 are more conservative than the bounds induced by Theorems 3 and 5 because the Massart bounds are evaluated in the most pessimistic value of CI $[a,b]$. In addition, our bound also takes into account the probability that γ is not in I, that implies an additional number of samples in the final sample size. In the absolute error case, if a CI I containing $1/2$ is determined, applying the previous theorem is unnecessary because the sample size is simply bounded with respect to the Okamoto bound. Similarly, if a (or b) is lower-bounded (or respectively upper-bounded) by $1/2$ but still close to $1/2$, the Okamoto bound is likely better. However, if γ is closer to 0 or 1, the logarithmic extra number of samples is largely compensated by the evaluation of the Massart bound in a or b.

5.2   Sequential Algorithms

In the following, we present two new sampling schemes. Both of them require three inputs: an error parameter ϵ, and two confidence parameters δ and δ' such that $\delta' < \delta$. After each sample, we update a Monte Carlo estimator and a $(1 - \delta')$-CI for γ. Then, the most pessimistic bound of the CI is used in the Massart function to compute a new minimal sample size n that satisfies Theorem 8. The process is repeated until the calculated sample size is lower than or equal to the current number of runs. We provide the pseudo-code of our Algorithms (1) and (2). Keywords GENERATE corresponds to a sample path generation and

---

Algorithm 1. Absolute Error Sequential Algorithm

Data:

$\epsilon, \delta, \delta'$: the original parameters

$M = \lceil \frac{1}{2\epsilon^2} \log \frac{2}{\delta} \rceil$: the Okamoto bound

$k = 0$

$m = 0$: the number of successes

$n_k = M$

$I_k = [a_k, b_k] \, [0, 1]$: the initial CI to which γ is known to belong

1 **while** $k < n_k$ **do**

2 $\quad$ $k \leftarrow k + 1$

3 $\quad$ GENERATE $\omega^{(k)}$

4 $\quad$ $z(\omega^{(k)}) = \mathbb{1}(\omega^{(k)} \models \phi)$

5 $\quad$ $m \leftarrow m + z(\omega^{(k)})$

6 $\quad$ DETERMINE I_k

7 $\quad$ **if** $1/2 \in I_k$ **then**

8 $\quad\quad$ $n_k = M$

9 $\quad$ **else if** $b_k < 1/2$ **then**

10 $\quad\quad$ $n_k = \lceil \frac{2}{h_a(b_k, \epsilon)\epsilon^2} \log \frac{2}{\delta - \delta'} \rceil$

11 $\quad$ **else**

12 $\quad\quad$ $n_k = \lceil \frac{2}{h_a(a_k, \epsilon)\epsilon^2} \log \frac{2}{\delta - \delta'} \rceil$

13 $\quad$ $n_k \leftarrow \min(n_k, M)$

Output: $\hat{\gamma}_k = m/k$

DETERMINE to the evaluation of the CI, slightly different in both schemes. Theorems 6 and 7 guarantee the correctness of our schemes since, for any tuple (m, n), if we are able to compute a $(1 - \delta')$-CI I and its exact coverage, the deviation probability is bounded by δ defined as the sum of the coverage and the Massart function at n, ϵ and the most pessimistic value of I.

Absolute Error Sequential Algorithm. We initiate the algorithm with a CI I_0 in which γ belongs (by default, $I_0 = [0, 1]$) and a worst-case (ϵ, δ)-sample size $n_0 = M$ with $M = \lceil \frac{1}{2\epsilon^2} \log \frac{2}{\delta} \rceil$ determined by the Okamoto bound ($\lceil . \rceil$ denotes the ceiling function). Once a trace $\omega^{(k)}$ is generated and monitored, the number of successes with respect to property ϕ and the total number of traces are updated. Then, an exact $(1 - \delta')$-CI I_k is evaluated. Iteration after iteration, the CI width tends to shorten and becomes more and more accurate. Theorem 8-i is applied to determine a new sample size n_k, bounded from above by M if necessary. These steps are repeated until $k \geq n_k$ at which specification (3) is rigorously fulfilled.

Relative Error Sequential Algorithm We first assume the existence, in a practical case study, of a threshold γ_{min}, supposedly low, corresponding to a tolerated precision error (e.g. a floating-point approximation). Estimating a

---

Algorithm 2. Relative Error Sequential Algorithm

Data:

$\epsilon, \delta, \delta', \gamma_{min}$: the original parameters

$M = \lceil \frac{1}{\epsilon^2 h_r(\gamma_{min},\epsilon)} \log \frac{2}{\delta} \rceil$

$k = 0$

$n_k = M$

$I_k = [a_k, 1] = [\gamma_{min}, 1]$: the initial CI in which γ is supposed to belong

1 **while** $k < n_k$ **do**

2      $k \leftarrow k + 1$

3   GENERATE $\omega^{(k)}$

4      $z(\omega^{(k)}) = \mathbb{1}(\omega^{(k)} \models \phi)$

5      $m \leftarrow m + z(\omega^{(k)})$

6   DETERMINE I_k

7   **if** $\gamma_{min} \geq a_k$ **then**

8         $n_k = M$

9   **else**

10        $n_k = \lceil \frac{1}{\epsilon^2 h_r(a_k,\epsilon)} \log \frac{2}{\delta - \delta'} \rceil$

11     $n_k \leftarrow \min(n_k, M)$

Output: $\hat{\gamma}_k = m/k$

---

value below γ_{min} is then unnecessary. The maximal number of simulations is consequently bounded by $M = \lceil \frac{1}{\epsilon^2 h_r(\gamma_{min},\epsilon)} \log \frac{2}{\delta} \rceil$. The relative error scheme is similar to the absolute error scheme. Note however that it is only necessary to determine a lower bound of I_k since h_r is a decreasing function in γ. Then, we determine a one-sided Clopper-Pearson $(1 - \delta')$-CI of shape $[a_k, 1]$ with $a_k = \beta^{-1}(\delta', m, n - m + 1)$. Theorem 8-ii is applied to determine a new sample size n_k, upper bounded by M if $a_k < \gamma_{min}$ and the steps are repeated until $k \geq n_k$. If the final output $\hat{\gamma}_k \geq \gamma_{min}$, Specification (4) is rigorously fulfilled. Otherwise, we can still output that γ is lower than γ_{min} with probability greater that $1 - \delta$.

6   Experiment Results

Our methods significantly reduce the sampling size while rigorously guaranteeing the specifications when probability γ gets away from $1/2$ in the absolute error case and for any γ in the relative error case, in comparison to the methods that have been documented for SMC in [11]. Both methods can be easily used to improve existing SMC tools. To give a glimpse of their efficiency, we give the gain in sampling size obtained with our methods in Table 1 over 3 standard Prism benchmarks described in [23]: the tandem queueing network in which queue capacities are equal to 3, the 10-station symmetric polling system and the 20-dependable workstation cluster. We refer to the extended version and [23] for more details concerning the models and the properties. In Prism, the Okamoto sampling size can be computed with the APMC method. For a given ϵ and δ, we report in column "(AE) Gain" the ratio between the Okamoto sampling size and our sampling size (average based on 5 experiments). For example, the property of the cluster model has probability $\gamma = 5.160834 \times 10^{-4}$ to occur. Given absolute error $\epsilon = 10^{-4}$ and confidence parameter $\delta = 0.05$, it requires 184443973 paths to guarantee Specification (3) when our method only requires 462077 paths to guarantee the same specification, which is 399 fewer samples. Similarly, given relative error ϵ and confidence parameters in column "Dagum (ϵ, δ)", "(RE) Gain" corresponds to the ratio of the 5-experiment average sampling sizes obtained by Dagum's algorithm and our method, necessary to fulfil Specification (4). The sampling sizes of these examples are given in the extended version. Our methods are general and the class of probabilistic systems on which the sampling schemes can be applied does not really matter as long as the systems are executable and the executions can be monitored. In what follows, we evaluate our sampling schemes on a small benchmark, available in the extended version, that can be easily investigated using model checker Prism [16] to corroborate our results.

Table 1. Sampling size gains over standard Prism benchmarks

	γ	APMC (ϵ, δ)	(AE) Gain	Dagum (ϵ, δ)	(RE) Gain
Tandem	0.155132	$(0.01, 0.001)$	1.7	$(0.05, 0.001)$	5.18
Polling	0.540786	$(0.001, 0.01)$	1	$(0.01, 0.01)$	3.65
Cluster	5.160834×10^{-4}	$(10^{-4}, 0.05)$	399	$(0.2, 0.05)$	9

6.1 Absolute Error Scheme Results

We compare our algorithm with the simple and Chen's schemes. To guarantee specification (3) in the simple scheme, one can use the algorithm proposed by Frey in [9]. This procedure pre-computes a value δ^* that guarantees a final coverage greater than $1 - \delta$ when the CI are computed according to the Clopper-Pearson method. For each couple of successes and trials (m, n) where n is smaller than the Okamoto bound M, the algorithm computes the number of sequences of observations $h(m, n, \epsilon)$ that lead to the output m/n. Unfortunately, we were unable to get results for ϵ smaller than 0.1 due to overflows of values $h(m, n, \epsilon) > 10^{309}$ in addition to an excessive amount of time required by this recursive computation. Thus, we used the default $\delta^* = \delta$.

We repeated each set of experiments 200 times with the three schemes for several values of γ, ϵ and δ. We estimated the empirical coverage by the number of times the specification (3) is fulfilled divided by 200 and computed the average, the standard deviation and the extrema values of the sample size and of the estimations $\hat{\gamma}$. For sake of clarity, as our results are consistent for all ϵ, δ and are symmetric with respect to $\gamma = 1/2$, we summarize the most relevant results for $\epsilon = 0.01$, $\delta = 0.05$ and $0 < \gamma \leq 1/2$ in Table 2. More details are provided for every scheme and set of experiments in the extended version. For every ϵ and δ, the sampling size is significantly lower for the simple scheme than for Chen and our schemes. However, the empirical coverage is below $1 - \delta$ for some γ

Table 2. Results of the absolute error scheme with $\epsilon = 0.01$ and $\delta = 0.05$

γ	0.005	0.01	0.02	0.05	0.1	0.3	0.5
Coverage (simple)	1	0.965	0.94	0.96	0.965	0.975	0.945
$\hat{\gamma}$ min (simple)	0	0	0.007	0.036	0.087	0.288	0.484
$\hat{\gamma}$ max (simple)	0.013	0.021	0.029	0.062	0.113	0.316	0.513
N mean (simple)	518	729	1107	2172	3777	8278	9703
Coverage (Chen)	1	0.98	1	0.995	1	0.995	0.995
$\hat{\gamma}$ min (Chen)	0	0	0.011	0.04	0.091	0.292	0.492
$\hat{\gamma}$ max (Chen)	0.01	0.017	0.028	0.059	0.107	0.31	0.511
N mean (Chen)	810	1171	1900	3946	7035	15684	18444
Coverage (new)	1	0.99	995	0.995	0.995	1	1
$\hat{\gamma}$ min (new)	0	0	0.01	0.039	0.089	0.291	0.491
$\hat{\gamma}$ max (new)	0.011	0.019	0.027	0.059	0.106	0.309	0.51
N mean (new)	831	1229	2064	4474	8161	18434	18445

(in bold and red in the table). For example, Table (2) indicates an empirical coverage of 0.94 for $\epsilon = 0.01$, $\delta = 0.05$, and $\gamma = 0.02$. Moreover, we remark that for every set of experiments, the simple scheme outputs at least one estimation that exceeds $\gamma \pm 1.25\epsilon$ (in bold and red in the table). This indicates that the difference between the estimation and γ exceeds the absolute error ϵ by more than 25%, that may consequently lead to important analysis errors. In comparison, the difference between $\hat{\gamma}$ and γ never exceeds ϵ by more than 10% in both other schemes. We thus do not recommend to use the simple scheme if specification (3) is rigorously prescribed. The theoretical expectations of Chen and our schemes are empirically confirmed: the coverage is significantly above $1 - \delta$ in each case (>0.95 in Table 2). Specification (3) is thus strictly satisfied. Chen's scheme shows a slightly better performance than our algorithm in terms of sampling size. However, we recall that Chen only guarantees that the estimation does not exceed the error bound ϵ on one side. For that reason, we recommend to use our algorithm that seems to be reasonably more conservative.

Figure 3a shows an empirical plot of the sample size as a function of probability γ. In this experiment, we let the sample size be greater than the Okamoto bound (dotted blue line) to illustrate the gap between the empirical and the notional bounds. With the sampling Algorithm 1 described in Sect. 5, the sample size would be bounded in virtue of Okamoto's inequality between 0.3 and 0.7. Note that the empirical plot has no particular meaning but is a guide to the eye that illustrates the behaviour of our algorithm. As expected, the gain is larger close to 0 and 1. For $\gamma = 0.02$, the Okamoto sample size (18445) is divided in average by 9. The empirical sample size is always maintained above the notional Massart sample size, indicating that the sample size has not been mistakenly minimised due to a wrong CI.

6.2   Relative Error Scheme Results

We repeated 200 times Dagum's, Watanabe's and our relative error schemes for eight values of γ with several ϵ and δ. We reported the results for $(\epsilon, \delta) = (0.1, 0.01)$ in Table 3. More values are provided in the extended version as well as more detailed tables, containing the descriptive statistics of the sample sizes.

Table 3. Sample size average of the relative error scheme for $(\epsilon, \delta) = (0.1, 0.01)$.

γ	0.9	0.7	0.5	0.3	0.1	0.05	0.01	0.001
N mean Dagum	1871	4402	9056	19703	74064	152757	803572	8124356
N mean W	1942	2498	3501	5836	17479	35006	175092	1746713
N mean New	202	623	1373	3043	11365	23812	122426	1236491

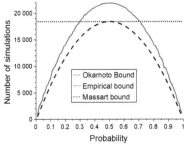

(a) Notional Okamoto (dot) and Massart (dash) bounds versus empirical results (absolute error $\epsilon = 0.01$ and confidence parameter $\delta = 0.05$).

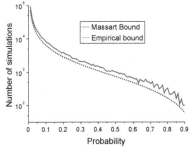

(b) Massart bounds (dot) versus empirical bounds (relative error $\epsilon = 0.1$ and confidence parameter $\delta = 0.05$).

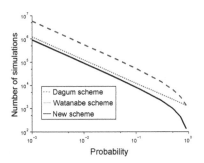

(c) Comparison between Dagum-and-al. (above), Watanabe (middle) and new (below) relative error schemes ($\epsilon = 0.1$ and $\delta = 0.05$).

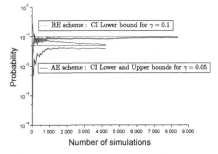

(d) Evolution of the CI bounds for the Absolute and Relative Error schemes, respectively in blue and red with $\gamma = 0.05$ and 0.1.

Fig. 3. Experimental results (Color figure online)

The average sample sizes are drawn for each scheme on Fig. 3c. We did not report the coverage of the sampling schemes because specification (4) was largely satisfied in the three cases. However, Dagum scheme is very conservative in sample size as Table 3 and Fig. 3c illustrate. We observe that our scheme is better than Watanabe's for all values of γ, especially when γ tends to 1. As γ decreases, the Clopper-Pearson CI becomes more conservative but our algorithm still presents better performances. However, when γ is below 0.05, the conservativeness of the Clopper-Pearson CI becomes too significant and exponentially impacts the sample size. Once the number of simulations k exceeds 1000 and $\hat{\gamma}_k \in [1/k, 0.04]$, we thus replaced the evaluation of the Clopper-Pearson CI by the Agresti-Coull CI. The results in the last two columns of Table 3 are obtained using the Agresti-Coull CI. The approximation maintains the highest performance and seems to be a good alternative to exact CI. A deeper investigation is left to future work, but even if the lower bound of the exact CI is below the lower bound of the Agresti-Coull CI, their difference is likely tight and reusing a slightly too optimistic value in the Massart bound is unlikely to pose problem.

We compare in Fig. 3b the empirical plot of the sample size as a function of probability γ with the notional bound. As for the absolute error case, the empirical plot is always maintained above the notional Massart sample size. Figure 3d shows the typical evolution of the CI bounds for the absolute and relative error problem with respectively, $\gamma = 0.05$ and $\gamma = 0.1$ and shows their accuracy and reliability over time.

7 Conclusion

The focus of this paper was to minimise using sequential schemes based on frequentist estimations the sampling size necessary to estimate a property with absolute or relative error in comparison to the standard methods in SMC. To build estimators that fulfil Specifications (3) or (4), we presented two sequential algorithms based on Massart bounds and coverage of probability γ. The comparison with the standard SMC schemes showed significant improvements. Finally, it is worth recalling that all the Monte Carlo sampling schemes anyway require a lot of samples for rare event estimation. Though the problem of designing sampling schemes for Binomial estimators is well-documented, the lack of exact concentration inequalities for importance sampling [5] and splitting estimators [15] in SMC makes the design of robust sampling procedures challenging in the rare event context.

Acknowledgements. Cyrille Jegourel and Jun Sun are partially supported by NRF grant GNRF1501 and Jin Song Dong by the project: Reliable Prototyping Framework for Daily Living Assistance of Frail Ageing People (RELIANCE).

References

1. Angluin, D., Valiant, L.: Fast probabilistic algorithms for Hamiltonian circuits and matchings. J. Comput. Syst. Sci. **18**(2), 155–193 (1979)
2. Brown, L., Cai, T., DasGupta, A.: Interval estimation for a binomial proportion. Stat. Sci. **16**(2), 101–133 (2001)
3. Chen, J.: Properties of a new adaptive sampling method with applications to scalable learning. In: WI, pp. 9–15, Atlanta (2013)
4. Chernoff, H.: A measure of asymptotic efficiency for tests of a hypothesis based on the sum of observations. Ann. Math. Stat. **23**(4), 493–507 (1952)
5. Clarke, E.M., Zuliani, P.: Statistical model checking for cyber-physical systems. In: Bultan, T., Hsiung, P.-A. (eds.) ATVA 2011. LNCS, vol. 6996, pp. 1–12. Springer, Heidelberg (2011). doi:10.1007/978-3-642-24372-1_1
6. Clopper, C.J., Pearson, E.S.: The use of confidence or fiducial limits illustrated in the case of the binomial. Biometrika **26**, 404–413 (1934)
7. Dagum, P., Karp, R., Luby, M., Ross, S.: An optimal algorithm for Monte Carlo estimation. SIAM J. Comput. **29**(5), 1484–1496 (2000)
8. David, A., Larsen, K.G., Legay, A., Mikucionis, M., Poulsen, D.B.: Uppaal SMC tutorial. STTT **17**(4), 397–415 (2015)
9. Frey, J.: Fixed-width sequential confidence intervals for a proportion. Am. Stat. **64**(3), 242–249 (2010)
10. Grosu, R., Peled, D., Ramakrishnan, C.R., Smolka, S.A., Stoller, S.D., Yang, J.: Using statistical model checking for measuring systems. In: Margaria, T., Steffen, B. (eds.) ISoLA 2014. LNCS, vol. 8803, pp. 223–238. Springer, Heidelberg (2014). doi:10.1007/978-3-662-45231-8_16
11. Hérault, T., Lassaigne, R., Magniette, F., Peyronnet, S.: Approximate probabilistic model checking. In: Steffen, B., Levi, G. (eds.) VMCAI 2004. LNCS, vol. 2937, pp. 73–84. Springer, Heidelberg (2004). doi:10.1007/978-3-540-24622-0_8
12. Hérault, T., Lassaigne, R., Peyronnet, S.: APMC 3.0: approximate verification of discrete and continuous time Markov chains. In: QEST, pp. 129–130 (2006)
13. Hoeffding, W.: Probability inequalities for sums of bounded random variables. J. Am. Stat. Assoc. **58**(301), 13–30 (1963)
14. Jegourel, C., Legay, A., Sedwards, S.: A platform for high performance statistical model checking – PLASMA. In: Flanagan, C., König, B. (eds.) TACAS 2012. LNCS, vol. 7214, pp. 498–503. Springer, Heidelberg (2012). doi:10.1007/978-3-642-28756-5_37
15. Jegourel, C., Legay, A., Sedwards, S.: Importance splitting for statistical model checking rare properties. In: Sharygina, N., Veith, H. (eds.) CAV 2013. LNCS, vol. 8044, pp. 576–591. Springer, Heidelberg (2013). doi:10.1007/978-3-642-39799-8_38
16. Kwiatkowska, M.Z., Norman, G., Parker, D.: PRISM 2.0: a tool for probabilistic model checking. In: QEST, pp. 322–323. IEEE (2004)
17. Massart, P.: The tight constant in the Dvoretzky-Kiefer-Wolfowitz inequality. Ann. Prob. **18**, 1269–1283 (1990)
18. Metropolis, N., Ulam, S.: The Monte Carlo method. J. Am. Stat. Assoc. **44**(247), 335–341 (1949)
19. Okamoto, M.: Some inequalities relating to the partial sum of binomial probabilities. Ann. Inst. Statis. Math. **10**, 29–35 (1958)
20. Wald, A.: Sequential tests of statistical hypotheses. Ann. Math. Stat. **16**(2), 117–186 (1945)

21. Watanabe, O.: Sequential sampling techniques for algorithmic learning theory. Theoret. Comput. Sci. **348**, 3–14 (2005)
22. Younes, H.: Verification and planning for stochastic processes with asynchronous events. Ph.D. thesis, Carnegie Mellon University (2004)
23. Younes, H.L.S., Kwiatkowska, M., Norman, G., Parker, D.: Numerical vs. statistical probabilistic model checking: an empirical study. In: Jensen, K., Podelski, A. (eds.) TACAS 2004. LNCS, vol. 2988, pp. 46–60. Springer, Heidelberg (2004). doi:10.1007/978-3-540-24730-2_4
24. Zuliani, P., Platzer, A., Clarke, E.M.: Bayesian statistical model checking with application to stateflow/simulink verification. FMSD **43**(2), 338–367 (2013)

Multilevel Monte Carlo Method for Statistical Model Checking of Hybrid Systems

Sadegh Esmaeil Zadeh Soudjani[1]($\boxtimes$), Rupak Majumdar[1],
and Tigran Nagapetyan[2]

[1] Max Planck Institute for Software Systems, Kaiserslautern, Germany
{Sadegh,Rupak}@mpi-sws.org
[2] Department of Statistics, University of Oxford, Oxford, UK
Tigran.Nagapetyan@stats.ox.ac.uk

Abstract. We study statistical model checking of continuous-time stochastic hybrid systems. The challenge in applying statistical model checking to these systems is that one cannot simulate such systems exactly. We employ the multilevel Monte Carlo method (MLMC) and work on a sequence of discrete-time stochastic processes whose executions approximate and converge weakly to that of the original continuous-time stochastic hybrid system with respect to satisfaction of the property of interest. With focus on bounded-horizon reachability, we recast the model checking problem as the computation of the distribution of the exit time, which is in turn formulated as the expectation of an indicator function. This latter computation involves estimating discontinuous functionals, which reduces the bound on the convergence rate of the Monte Carlo algorithm. We propose a smoothing step with tunable precision and formally quantify the error of the MLMC approach in the mean-square sense, which is composed of smoothing error, bias, and variance. We formulate a general adaptive algorithm which balances these error terms. Finally, we describe an application of our technique to verify a model of thermostatically controlled loads.

Keywords: Statistical model checking · Hybrid systems · Multilevel Monte Carlo · Formal verification · Continuous-time stochastic processes

1 Introduction

Continuous-time stochastic hybrid systems (ct-SHS) are a natural model for cyber-physical systems operating under uncertainty [5,7]. A ct-SHS has a hybrid state space consisting of discrete modes and, for each mode, a set of continuous states (called the invariant). In each mode, the continuous state evolves according to a stochastic differential equation (SDE) in continuous time. The discrete mode may change once the continuous state reaches the boundaries of the invariant.

We consider quantitative analysis of temporal properties of ct-SHS [2,3]. The fundamental analysis problem, called *probabilistic reachability*, consists in computing the probability that the state of a ct-SHS exits a given safe set within

© Springer International Publishing AG 2017
N. Bertrand and L. Bortolussi (Eds.): QEST 2017, LNCS 10503, pp. 351–367, 2017.
DOI: 10.1007/978-3-319-66335-7_24

a given bounded time horizon. Since analytic solutions are not available, there are two common approaches. The first approach is *numerical* model checking that relies on the exact or approximate computation of the measure of the executions satisfying the temporal property. The second approach, called *statistical* model checking, relies on finitely many sample executions of the system, and employs hypothesis testing to provide confidence intervals for the estimate of the probability.

Statistical model checking has proven to be computationally more efficient than numerical model checking as it only requires the system to be executable. Thus, it can be applied to larger classes of systems and of specifications [24]. The main underlying assumption in all statistical model checking techniques is the ability to sample from the space of executions of the system. Unfortunately, we cannot compute exact simulations for the general class of ct-SHS due to the process evolution being continuous in both time and space. In this paper, we describe a statistical model checking approach to ct-SHS using the *multilevel Monte Carlo (MLMC)* method [17,20], which does not require exact executions of the system.

Our procedure works as follows. First, we formulate the quantitative analysis problem as computing the distribution of the first exit time of the system from the given safe set. Then, we build a sequence of approximate models whose executions converge weakly (or in expectation) to the execution of the concrete system. Although these approximate models can be used separately in the classical setting of statistical model checking in order to compute estimates of the exit time, the MLMC method can take advantage of coupling between approximate executions with different time resolutions to provide better convergence rates.

An important challenge in applying the MLMC technique to the quantitative analysis of ct-SHS is that a discontinuous function is applied to the first exit time. While MLMC can be applied to discontinuous functions, the convergence rates we can guarantee are poor. We propose a smoothing step that replaces the discontinuous function with a continuous approximation and show that the replacement decreases the overall computation cost.

Finally, we analyze the asymptotic computational cost of the MLMC approach for a given error bound. We propose an adaptive algorithm which balances errors due to bias, variance, and smoothing, and tunes the hyperparameters of the algorithm on the fly. We illustrate our technique on an example model of thermostatically controlled loads.

Related Work. Formal definitions of various classes of *continuous-time* probabilistic hybrid models are presented in [26], together with a comparison. Over such models, [4] has formalized the notion of probabilistic reachability, [27] has proposed a computational technique based on convex optimization, [14] has provided discretization techniques with formal error bounds, and [15] has developed an approach based on satisfiability modulo theory. For *discrete-time* stochastic hybrid models probabilistic reachability (and safety) has been fully characterized in [1] and computed via software tools [10,11] that use finite abstractions. The methods can be extended to more general probabilistic temporal logics

[28]. These approaches generally suffer from curse of dimensionality and are not applicable to large dimensional models.

An overview of statistical model checking techniques can be found in [22–24]. The paper [9] employs statistical model checking for verifying unbounded temporal properties. The paper [8] has discussed the use of importance sampling to address the issue of rare events in statistical verification of cyber-physical systems. A distributed implementation of statistical model checking is proposed in [6] and a set-oriented method for statistical verification of dynamical systems is presented in [29].

A detailed overview of applications of MLMC can be found in [18]. The MLMC for estimating distribution functions is described in the recent paper [16] and is adapted to our setting.

The article is structured as follows. In Sect. 2, we define the ct-SHS model and the probabilistic reachability problem. In Sects. 3 and 4, we discuss the standard Monte Carlo technique and the MLMC method, respectively, and compare their convergence rates. We then discuss two technical modifications: applying a smoothing operator to the discontinuous function of exit time (Sect. 5) and an adaptive MLMC algorithm (Sect. 6). In Sect. 7, we provide simulation results for an example.

2   Model Definition

We study statistical model checking for the rich class of continuous-time stochastic hybrid systems (ct-SHS).

2.1   Continuous-Time Stochastic Hybrid Systems

Definition 1. *A continuous-time stochastic hybrid system (ct-SHS) is a tuple* $\mathcal{H} = (Q, \mathcal{X}, b, \sigma, x_0, r)$ *where the components are defined as follows.*

States Q *is a countable set of discrete states (modes) and* $\mathcal{X} : Q \to \mathcal{P}(\mathbb{R}^n)$ *maps each mode* $q \in Q$ *to an open set* $\mathcal{X}(q) \subseteq \mathbb{R}^n$, *called the* invariant *for the mode* q. *A state* (q, z) *with* $q \in Q$ *and* $z \in \mathcal{X}(q)$ *is called a* hybrid state. *The hybrid state space* X *is defined as* $X = \{(q, z) \mid q \in Q, z \in \mathcal{X}(q)\}$. *We write* ∂Z *for the boundary of a set* Z *and* $\partial X := \{(q, z) \mid q \in Q, z \in \partial \mathcal{X}(q)\}$.

Evolution $b : X \to \mathbb{R}^n$ *is a vector field and* $\sigma : X \to \mathbb{R}^{n \times m}$ *is a matrix-valued function, with* $n, m \in \mathbb{N}_0$. *For each* $q \in Q$, *define the following SDE:*

$$dz(t) = b(q, z(t))dt + \sigma(q, z(t))dW_t, \tag{1}$$

where $(W_t, \ t \geq 0)$ *is an* m-*dimensional standard Wiener process in a complete probability space. We assume functions* $b(q, \cdot) : \mathcal{X}(q) \to \mathbb{R}^n$ *and* $\sigma(q, \cdot) : \mathcal{X}(q) \to \mathbb{R}^{n \times m}$ *are bounded and Lipschitz continuous for all* $q \in Q$. *The assumption ensures the existence and uniqueness of the solution of (1).*

Initial State $x_0 \in X$ *is the initial state of the system;*

Transition Kernel $r : \partial X \times Q \to [0, 1]$ *is a discrete stochastic kernel which governs the switching between the SDEs defined in* (1). *That is, for all* $q \in Q$, *we assume* $r(\cdot, q)$ *is measurable and, for all* $x \in \partial X$, *the function* $r(x, \cdot)$ *is a discrete probability measure.*

Intuitively, an execution of a ct-SHS starts in the initial state x_0, and evolves according to the solution of the diffusion process (1) for the current mode until it hits the boundary of the invariant of the current mode for the first time. At this point, a new mode q' is chosen according to the kernel r and the execution proceeds according to the solution of the diffusion process for q', and so on.

Let $z^q(t)$, $q \in Q$ be the solution of diffusion process (1) starting from $z^q(0) \in \mathcal{X}(q)$. Define $t^*(q)$ as the first exit time of $z^q(t)$ from the set $\mathcal{X}(q)$,

$$t^*(q) := \inf\{t \in \mathbb{R}_{>0} \cup \{\infty\}, \text{ such that } z^q(t) \in \partial\mathcal{X}(q)\}. \qquad (2)$$

A stochastic hybrid process, describing the evolution of a ct-SHS, is obtained by the concatenation of diffusion processes $\{z^q(t), q \in Q\}$ together with a jumping mechanism given by a family of first exit times $t^*(q)$; we make this formal in Definition 2[1].

Definition 2. *A stochastic process* $x(t) = (q(t), z(t))$ *is called an execution of ct-SHS* $\mathcal{H}$ *if there exists a sequence of stopping times* $T_0 = 0 < T_1 < T_2 < \ldots$ *such that for all* $k \in \mathbb{N}_0$:

- $x(0) = (q_0, z_0) \in X$ *is the initial state of* $\mathcal{H}$;
- *for* $t \in [T_k, T_{k+1})$, $q(t) = q(T_k)$ *is constant and* $z(t)$ *is the solution of SDE*

$$dz(t) = b(q(T_k), z(t))dt + \sigma(q(T_k), z(t))dW_t;$$

- $T_{k+1} = T_k + t^*(q(T_k))$ *where* $t^*(q(T_k))$ *is the first exit time from the mode* $q(T_k)$ *as defined in* (2);
- *The probability distribution of* $q(T_{k+1})$ *is governed by the discrete kernel* $r((q(T_k), z(T_{k+1}^-)), \cdot)$ *and* $z(T_{k+1}) = z(T_{k+1}^-)$, *where* $z(T_{k+1}^-) := \lim_{t \uparrow T_{k+1}} z(t)$.

2.2   Example: Thermostatically Controlled Loads

Household appliances such as water boilers/heaters, air conditioners, and electric heaters -all referred to as *thermostatically controlled loads (TCLs)*- can store energy due to their thermal mass. TCLs have been extensively studied [12,21,25] for their role in energy management systems. TCLs generally operate within a dead-band around a temperature set-point and are naturally modeled using ct-SHS. The temperature evolution in a cooling TCL can be characterized by the following SDE:

$$d\theta(t) = \frac{1}{CR}(\theta_a - q(t)RP_{rate} - \theta(t))dt + \sigma(q(t))dW_t, \qquad (3)$$

---

[1] Solely for simplicity of exposition, we have put two restrictions on the ct-SHS model $\mathcal{H}$ in Definition 1. First, the model includes only forced jumps activated by reaching the boundaries of the invariant sets $\partial\mathcal{X}(q)$, $q \in Q$. Second, the state $z(t)$ remains continuous at the switching times as declared in Definition 2.

where θ_a is the ambient temperature, P_{rate} is the energy transfer rate of the TCL, and R and C are the thermal resistance and capacitance, respectively. The noise term W_t in (3) is a standard Wiener process. The model of the TCL has two discrete modes, $q(t) = 0$ for the OFF mode and $q(t) = 1$ for the ON mode. The temperature of the cooling TCL is regulated by a control signal $q(t^+) = f(q(t), \theta(t))$ based on discrete switching as

$$f(q, \theta) = \begin{cases} 0, \theta \leq \theta_s - \delta_d/2 =: \theta_- \\ 1, \theta \geq \theta_s + \delta_d/2 =: \theta_+ \\ q, \text{ else,} \end{cases} \tag{4}$$

where θ_s denotes a temperature set-point and δ_d a dead-band. Together, θ_s and δ_d characterize an operating temperature range. The model can be described by the ct-SHS $\mathcal{H}_{TCL} = (Q, \mathcal{X}, b, \sigma, x_0, r)$, where $Q = \{0, 1\}$ with the invariants $\mathcal{X}(0) = (-\infty, \theta_+)$ and $\mathcal{X}(1) = (\theta_-, +\infty)$, and $r(q^+ \mid q, \theta)$ is the Kronecker delta with $q^+ = f(q, \theta)$.

2.3 Problem Definition

For a given random variable defined on the executions of a ct-SHS, we study the problem of estimating its distribution function.

Problem 1. Let Y be a real-valued random variable defined on the executions of ct-SHS $\mathcal{H}$. Estimate $F_Y(s) := \mathbb{P}(Y \leq s)$, the distribution of Y for a given $s \in \mathbb{R}$.

Problem 2 (Probabilistic Safety). Compute the probability that an execution of a ct-SHS $\mathcal{H}$, with initial condition $x_0 \in X$, remains within a measurable set $A \subset X$ during the bounded time horizon $[0, s] \subset \mathbb{R}_{\geq 0}$:

$$\mathbb{P}(\mathcal{H} \text{ is safe over } [0, s]) = \mathbb{P}(Y > s) = 1 - F_Y(s) \tag{5}$$

where $Y := \min\{t \in \mathbb{R}_{\geq 0} \cup \{\infty\} \mid x(t) \notin A, x(0) = x_0\}$ and $F_Y(s) = \mathbb{P}(Y \leq s)$.

The random variable Y defined in Problem 2 is in fact the first exit time of the system $\mathcal{H}$ from the safe set A and its distribution can be represented as

$$F_Y(s) = \mathbb{E}\left(\mathbf{1}_{(-\infty, s]}(Y)\right). \tag{6}$$

Problem 3 (Specification of interest for TCL). Although the switching mechanism (4) is designed to keep the temperature inside the interval $[\theta_-, \theta_+]$, there is still a chance that the temperature goes out of this interval due to the Wiener process W_t. Define a random variable $Y = \max\{\theta_t, t \in [0, s]\}$. We aim to estimate the probability $\mathbb{P}(Y \leq \theta_+ + 0.1 \cdot \delta_d)$.

Analytic solution of Problems 1–3 is infeasible for the class of ct-SHS. In this work, we propose an approximate computation technique with a confidence bound. Our technique based on MLMC substantially improves the computational complexity of the standard Monte Carlo method. We first discuss standard Monte Carlo (SMC) method in Sect. 3 and then present the MLMC method in Sect. 4.

Algorithm 1. State update $(q_{k+1}, z_{k+1}) = \mathsf{Update}(\mathcal{H}, q_k, z_k, \Delta, W_k)$

Require: model $\mathcal{H} = (Q, \mathcal{X}, b, \sigma, x_0, r)$, current state (q_k, z_k), time step Δ, sampled noise W_k

1: compute z_{aux} according to the difference equation

$$z_{\mathsf{aux}} = z_k + b(q_k, z_k)\Delta + \sigma(q_k, z_k)\sqrt{\Delta}W_k \qquad (7)$$

2: **if** $z_{\mathsf{aux}} \in \mathcal{X}(q_k)$ **then**
3:    $z_{k+1} = z_{\mathsf{aux}}$ and $q_{k+1} = q_k$
4: **else**
5:    set z_{k+1} to be the normal projection of z_{aux} onto $\partial\mathcal{X}(q_k)$
6:    select q_{k+1} sampled from the distribution $r(q_k, z_{k+1})$
7: **end if**
Ensure: updated hybrid state $(q_{k+1}, z_{k+1}) = \mathsf{Update}(\mathcal{H}, q_k, z_k, \Delta, W_k)$

---

3   Standard Monte Carlo Method

In order to compute the quantities of interest in Problems 1–2 we need to estimate $\mathbb{E}P = \mathbb{E}g(Y)$, where Y is a function of the execution of ct-SHS $\mathcal{H}$, $g : \mathbb{R} \to \mathbb{R}$ is the indicator function over the interval $(-\infty, s]$ and $P := g(Y)$ is a one-dimensional random variable. The exact executions of $\mathcal{H}$ and thus exact samples of Y are not available is general but it is possible to construct approximate executions and approximate samples that converge to the exact ones.

Algorithm 1 presents a state update routine based on the *Euler-Maruyama* method that can be used to construct approximate executions. Given the model $\mathcal{H}$ and the current approximate state (q_k, z_k), this algorithm computes the approximate state (q_{k+1}, z_{k+1}) for the next time step of size Δ. Equation (7) in step 1 of the algorithm is the Euler-Maruyama approximation of the SDE (1). If z_{aux} is still inside the invariant of the current mode $\mathcal{X}(q_k)$, then the mode remains unchanged and z_{aux} will be the next state (steps 2–3). Otherwise, in steps 5–6 z_{aux} is projected onto the boundary $\partial\mathcal{X}(q_k)$ of the invariant and the mode is updated according to the discrete kernel $r(q_k, z_{k+1})$.

Algorithm 2 generates approximate executions of $\mathcal{H}$ and approximate samples of Y using Algorithm 1. The algorithm requires the model $\mathcal{H}$, the definition of Y as a function of the execution of $\mathcal{H}$, and the time interval $[0, s]$. The output of the algorithm θ^ℓ is an approximate sample of random variable Y. In steps 1–2 the number of time steps n is selected and the discretization time step Δ is computed. In order to highlight the dependency of the algorithm to the parameter n, we have opted to use ℓ in the representation $n = \kappa 2^\ell$ as the superscript of the variables. We call ℓ the *level* of approximation which is nicely connected to the MLMC terminology discussed in Sect. 4. Algorithm 2 initializes the approximate execution in step 3 as $x_0^\ell := (q_0^\ell, z_0^\ell)$ according to x_0 the initial state of $\mathcal{H}$. Then the algorithm iteratively computes the next approximate state $(q_{k+1}^\ell, z_{k+1}^\ell)$ by sampling from the m-dimensional standard normal distribution in step 5 and applying Algorithm 1 to $(\mathcal{H}, q_k^\ell, z_k^\ell, \Delta, W_k^\ell)$ in step 6. Finally, step 9 constructs the continuous-time approximate execution $(q^\ell(\cdot), z^\ell(\cdot))$ as the piecewise constant version of the discrete execution (q_k^ℓ, z_k^ℓ), which enables the computation of θ^ℓ by applying the definition of Y to $(q^\ell(\cdot), z^\ell(\cdot))$ (step 10).

Algorithm 2. Approximate sampling of random variable Y

Require: model $\mathcal{H} = (Q, \mathcal{X}, b, \sigma, x_0, r)$, Y a function of execution of $\mathcal{H}$, time interval $[0, s]$
1: select the number of time steps n and set $\kappa \geq 1, \ell \geq 0$ such that $n = \kappa 2^\ell$
2: compute the time step $\Delta := s/n$ and set $k := 0$
3: set the initial hybrid state $x_0^\ell := (q_0^\ell, z_0^\ell)$ according to $x_0 = (q_0, z_0) \in X$
4: **while** $k < n$ **do**
5:   sample W_k^ℓ from the standard m-dimensional normal distribution
6:   update the hybrid state $(q_{k+1}^\ell, z_{k+1}^\ell) = \mathsf{Update}(\mathcal{H}, q_k^\ell, z_k^\ell, \Delta, W_k^\ell)$ using Algorithm 1
7:   $k = k + 1$
8: **end while**
9: define for all $t \geq 0$, $z^\ell(t) = \sum_{k=0}^{n} z_k^\ell 1_{[k\Delta, (k+1)\Delta)}(t)$ and $q^\ell(t) = \sum_{k=0}^{n} q_k^\ell 1_{[k\Delta, (k+1)\Delta)}(t)$
10: compute θ^ℓ by applying the definition of Y to $(q^\ell(\cdot), z^\ell(\cdot))$
Ensure: θ^ℓ as approximate sample of Y

Algorithm 2 is parameterized by ℓ. Due to the nature of the Euler-Maruyama method in (7), we expect that the approximate samples θ^ℓ converge to Y as $\ell \to \infty$ in a suitable way. In fact, it is an unbiased estimator in the limit: $\lim_{\ell \to \infty} \mathbb{E} g\left(\theta^\ell\right) = \mathbb{E} g\left(Y\right)$. The idea behind standard Monte Carlo (SMC) method is to use the empirical mean of $g\left(\theta^\ell\right)$ as an approximation of $\mathbb{E} g\left(Y\right)$. The SMC estimator has the form

$$\hat{P} = \frac{1}{N} \sum_{i=1}^{N} g\left(\theta_i^\ell\right), \tag{8}$$

which is based on N replications of θ^ℓ. The replications $\{\theta_i^\ell, i = 1, \dots, N\}$ can be generated by running Algorithm 2 (with a fixed ℓ) N times, or running any other algorithm that generates such samples (cf. Algorithm 4 in Sect. 4). The SMC method is summarized in Algorithm 3, which approximates $\mathbb{E} g(Y)$ based on a general sampling algorithm $\mathcal{A}_\ell$. Note that Algorithm 3 can be used for estimating $\mathbb{E} g(Y)$ not only with $g(\cdot)$ being the indicator function but also any other functional that can be deterministically evaluated using the executions over the time interval $[0, s]$.

Owing to the randomized nature of algorithm $\mathcal{A}_\ell$ embedded in Algorithm 3, we quantify the quality of its outcome using *mean squared error*:[2]

$$MSE(\mathcal{A}_\ell) \equiv \mathbb{E}\left[\left(\hat{P} - \mathbb{E}P\right)^2\right] = \mathbb{E}\left[\left(\hat{P} - \mathbb{E}\hat{P}\right)^2\right] + \left[\mathbb{E}\hat{P} - \mathbb{E}P\right]^2. \quad (9)$$

The mean square error $MSE(\mathcal{A}_\ell)$ is decomposed into two parts: *Monte Carlo variance* and *squared bias error*. The latter is a systematic error arising from the fact that we might not sample our random variable exactly, but rather use a suitable approximation, while the former error comes from the randomized nature of the Monte Carlo algorithm. The Monte Carlo variance (first term in (9)) is proportional to N^{-1} as

$$\text{Var } \hat{P} = \text{Var}\left(\frac{1}{N}\sum_{i=1}^{N} g(\theta_i^\ell)\right) = \frac{1}{N^2}\text{Var}\left(\sum_{i=1}^{N} g(\theta_i^\ell)\right) = \frac{1}{N}\text{Var}\left(g(\theta^\ell)\right).$$

The cost of Algorithm 3 is typically taken to be the expected runtime in order to achieve a prescribed accuracy $MSE(\mathcal{A}_\ell) \leq \varepsilon$. A more convenient approach for theoretical comparison between different methods is to consider the cost associated to sampling algorithm $\mathcal{A}_\ell$,

$$C_\ell(\mathcal{A}_\ell) := \mathbb{E}\left[\#\text{operations and random number generations to calculate } g(\theta^\ell)\right],$$

which facilitates the definition of convergence rate of the algorithm.

Definition 3. *We say that Algorithm 3 based on sampling algorithm $\mathcal{A}_\ell$ converges with rate $\gamma > 0$ if $\lim_{\ell \to \infty} \sqrt{MSE(\mathcal{A}_\ell)} = 0$ and if there exist constants $c > 0, \eta \geq 0$ such that*

$$C_\ell(\mathcal{A}_\ell) \leq c \cdot \left(\sqrt{MSE(\mathcal{A}_\ell)}\right)^{-\gamma} \cdot \left(-\log \sqrt{MSE(\mathcal{A}_\ell)}\right)^{\eta}. \quad (10)$$

The definition of convergence rate in (10) indicates that for a desired accuracy $MSE(\mathcal{A}_\ell) \leq \varepsilon$ smaller convergence rate γ implies lower computational cost $C_\ell(\mathcal{A}_\ell)$.

The following theorem presents the convergence rate of the SMC method presented in Algorithm 3.

Theorem 1. *Let θ^ℓ denote the numerical approximation of the random variable Y according to an algorithm $\mathcal{A}_\ell$. Assume there exist positive constants α, ζ, c_1, c_2 such that for all $\ell \in \mathbb{N}_0$*

$$\left|\mathbb{E}[g(\theta^\ell) - g(Y)]\right| \leq c_1 2^{-\alpha \cdot \ell}, \quad \mathbb{E}[C_\ell] \leq c_2 2^{\zeta \cdot \ell}, \quad \text{and } \text{Var } g(\theta^\ell) < \infty. \quad (11)$$

Then the standard Monte Carlo method of Algorithm 3 based on sampling algorithm $\mathcal{A}_\ell$ converges with rate $\gamma = 2 + \dfrac{\zeta}{\alpha}$.

---

[2] We slightly abuse the notation and indicate by $MSE(\mathcal{A}_\ell)$ the mean square error of Algorithm 3 with the embedded sampling algorithm $\mathcal{A}_\ell$.

Algorithm 3. Standard Monte Carlo method to estimate $\mathbb{E}g(Y)$

Require: Sampling algorithm $\mathcal{A}_\ell$, number of samples N, functional $g(\cdot)$
1: $i := 1$
2: **while** $i < N$ **do**
3:     sample θ_i^ℓ using algorithm $\mathcal{A}_\ell$ (for example Algorithms 2 or 4)
4:     evaluate $g(\theta_i^\ell)$
5:     $i = i + 1$
6: **end while**
Ensure: $\hat{P} = \frac{1}{N}\sum_{i=1}^{N} g(\theta_i^\ell)$ as approximate estimate of $\mathbb{E}g(Y)$

Remark 1. Recall the role of ℓ in step 2 of Algorithm 2. Increasing ℓ results in an exponential increase in the number of time steps thus also in the number of samples. Therefore we have assumed in (11) an exponential bound on the increased cost and an exponential bound in the decreased bias as a function of ℓ.

The values of constants α, ζ, c_1, c_2 in Theorem 1 depend on the regularity of the functional g, sampling algorithm $\mathcal{A}_\ell$ and other parameters. In the next section we propose to use MLMC method that improves the convergence rate and substantially reduces the computational complexity of the estimation. We discuss a smoothing in Sect. 5 that replaces the indicator function $g(\cdot)$ with a smoothed function and discuss its effect on the algorithm's error.

4   Multilevel Monte Carlo Method

The multilevel Monte Carlo method (MLMC) relies on the simple observation of telescoping sum for expectation:

$$\mathbb{E}g\left(\theta^L\right) = \mathbb{E}g\left(\theta^0\right) + \sum_{l=1}^{L}\mathbb{E}\left[g\left(\theta^\ell\right) - g\left(\theta^{\ell-1}\right)\right]. \tag{12}$$

where θ^0 and θ^L correspond respectively to the coarsest and finest levels of numerical approximation. While any of the approximations $\{\theta^0, \theta^1, \ldots, \theta^L\}$ can be used individually in Algorithm 3 to approximate Y, instead, the MLMC method independently estimates each of the expectations on the right-hand side of (12) such that the overall variance is minimized for a given computational cost. The estimator $\hat{P}$ of $\mathbb{E}g\left(\theta^L\right)$ can be seen as a sum of independent estimators

$$\hat{P} = \sum_{\ell=0}^{L} P^\ell, \tag{13}$$

where P^0 is an estimator for $\mathbb{E}g\left(\theta^0\right)$ based on N_0 samples, and P^ℓ are estimates for $\mathbb{E}\left[g\left(\theta^\ell\right) - g\left(\theta^{\ell-1}\right)\right]$ based on N_ℓ samples. As we saw in the MSC method of Sect. 3, the simplest forms for P^0 and P^ℓ are the empirical means over all samples:

$$P^0 = \frac{1}{N_0} \sum_{i=1}^{N_0} g\left(\theta_i^0\right), \quad P^\ell = \frac{1}{N_\ell} \sum_{i=1}^{N_\ell} \left[g\left(\theta_i^\ell\right) - g\left(\theta_i^{\ell-1}\right)\right], \quad \ell = 1, \ldots, L. \quad (14)$$

Using the assumption of having independent estimators $\{P^0, P^1, P^2, \ldots, P^L\}$ and employing the telescoping sum (12) we can compute respectively the variance of $\hat{P}$ and bias as

$$\text{Var } \hat{P} = \text{Var}\left[\sum_{\ell=0}^{L} P^\ell\right] = \sum_{\ell=0}^{L} \text{Var } P^\ell, \quad \mathbb{E}P - \mathbb{E}\hat{P} = \mathbb{E}P - \mathbb{E}\left[\sum_{\ell=0}^{L} P^\ell\right] = \mathbb{E}P - \mathbb{E}g\left(\theta^L\right).$$

The computation of P^ℓ in (14) requires the samples $\theta_i^\ell, \theta_i^{\ell-1}$ to be generated from a common probability space. We utilize the fact that sum of normal random variables is still normally distributed. Algorithm 4 presents generation of approximate coupled samples $\theta_i^\ell, \theta_i^{\ell-1}$ for the random variable Y defined on the execution of a ct-SHS $\mathcal{H}$. As can be seen in steps 6–7 and 11, the approximate execution for the finer level ℓ is constructed exactly the same way as in Algorithm 2 with $n_f = \kappa 2^\ell$ time steps. The construction of approximate execution for the coarser level $(\ell - 1)$ with $n_c = \kappa 2^{\ell-1}$ is also similar except that the noise term in step 8 is obtained by taking the weighted sum of noise terms from the finer level $(W_{2k}^\ell + W_{2k+1}^\ell)/\sqrt{2}$. This choice preserves the properties of each approximation level while coupling the executions of levels $\ell-1, \ell$ thus also coupling approximate samples $\theta^{\ell-1}, \theta^\ell$.

Now we are ready to present the MLMC method in Algorithm 5. The method is parameterized by the number of levels L, number of samples for each level N_ℓ, $\ell = 0, 1, \ldots, L$ (which are gathered in $\mathfrak{S}$), and the initial number of time steps κ. Steps 2–3 performs the SMC method of Algorithm 3 with embedded sampling Algorithm 2 in order to estimate $\mathbb{E}g(\theta^0)$ with N_0 samples at the initial level $\ell = 0$. Then the algorithm iteratively estimate $\mathbb{E}[g(\theta^l) - g(\theta^{l-1})]$ in steps 6–7 using Algorithm 3 with number of samples $N = N_l$ and with the embedded coupled sampling Algorithm 4. The sum estimated quantity is reported in step 10 as the estimation of $\mathbb{E}g(Y)$.

Theorem 2. *Let θ^ℓ denote the level ℓ numerical approximation of the random variable Y. Assume the independent estimators P_ℓ used in Algorithm 5 satisfy*

$$\left|\mathbb{E}[g(\theta^\ell) - g(Y)]\right| \leq c_1 2^{-\alpha \ell} \quad \text{and} \quad \mathbb{E}[C_\ell] \leq c_2 2^{\varsigma \ell} \quad (15)$$

$$\mathbb{E}[P^\ell] = \begin{cases} \mathbb{E}[g(\theta^0)], & \ell = 0 \\ \mathbb{E}[g(\theta^\ell) - g(\theta^{\ell-1})], & \ell > 0 \end{cases} \quad \text{and} \quad \text{Var}[P^\ell] \leq c_3 N_\ell^{-1} 2^{-\beta \ell} \quad (16)$$

Algorithm 4. Approximate coupled samples $\theta^\ell, \theta^{\ell-1}$ of random variable Y

Require: model $\mathcal{H} = (Q, \mathcal{X}, b, \sigma, x_0, r)$, Y a function of execution of $\mathcal{H}$, time interval $[0, s]$, level ℓ

1: select the number of time steps $n_f = \kappa 2^\ell$ and $n_c = \kappa 2^{\ell-1}$ for some $\kappa \geq 1$
2: compute the time step $\Delta_c := s/n_c$, $\Delta_f := s/n_f$ and set $k := 0$
3: set the initial states $x_0^{\ell,c} := (q_0^{\ell,c}, z_0^{\ell,c})$ and $x_0^{\ell,f} := (q_0^{\ell,f}, z_0^{\ell,f})$ according to $x_0 = (q_0, z_0) \in X$
4: **while** $k < n_c$ **do**
5:    sample W_{2k}^ℓ, W_{2k+1}^ℓ independently from the standard m-dimensional normal distribution
6:    update hybrid state $(q_{2k+1}^{\ell,f}, z_{2k+1}^{\ell,f}) = \mathsf{Update}(\mathcal{H}, q_{2k}^{\ell,f}, z_{2k}^{\ell,f}, \Delta_f, W_{2k}^\ell)$ using Algorithm 1
7:    update hybrid state $(q_{2k+2}^{\ell,f}, z_{2k+2}^{\ell,f}) = \mathsf{Update}(\mathcal{H}, q_{2k+1}^{\ell,f}, z_{2k+1}^{\ell,f}, \Delta_f, W_{2k+1}^\ell)$ using Algorithm 1
8:    update hybrid state $(q_{k+1}^{\ell,c}, z_{k+1}^{\ell,c}) = \mathsf{Update}(\mathcal{H}, q_k^{\ell,c}, z_k^{\ell,c}, \Delta_c, (W_{2k}^\ell + W_{2k+1}^\ell)/\sqrt{2})$ using Algorithm 1
9:    $k = k + 1$
10: **end while**
11: define $z^{\ell,f}(t) = \sum_{k=0}^{n_f} z_k^{\ell,f} 1_{[k\Delta_f, (k+1)\Delta_f)}(t)$ and $q^{\ell,f}(t) = \sum_{k=0}^{n_f} q_k^{\ell,f} 1_{[k\Delta_f, (k+1)\Delta_f)}(t)$
12: define $z^{\ell,c}(t) = \sum_{k=0}^{n_c} z_k^{\ell,c} 1_{[k\Delta_c, (k+1)\Delta_c)}(t)$ and $q^{\ell,c}(t) = \sum_{k=0}^{n_c} q_k^{\ell,c} 1_{[k\Delta_c, (k+1)\Delta_c)}(t)$
13: compute θ^ℓ and $\theta^{\ell-1}$ by applying the definition of Y to $(q^{\ell,f}(\cdot), z^{\ell,f}(\cdot))$ and $(q^{\ell,c}(\cdot), z^{\ell,c}(\cdot))$ respectively
Ensure: $\theta^\ell, \theta^{\ell-1}$ as approximate sample of Y

---

for positive constants $\alpha, \beta, \zeta, c_1, c_2, c_3$ with $\alpha \geq \frac{1}{2}\min(\beta, \zeta)$. Then the MLMC method in Algorithm 5 converges with rate $2 + \dfrac{\max(\zeta - \beta, 0)}{\alpha}$.

Assumptions in (15) are exactly the same as the ones used in Theorem 1. Assumptions in (16) put restriction on the statistical properties of the estimators P^ℓ: they first enables us to use the telescoping property (12) and the second ensures the exponentially decaying variance as a function of level ℓ. In compare with the convergence rate of SMC method in Theorem 1, the improvement is due to the non-zero factor β which is the decaying rate of the variance of estimators.

Now that we have set up the MLMC method and the coupling technique that improves the convergence rate of the estimation, we focus on the following important problems associated with the approach: smoothing the discontinuous functional that leads to smaller values of α and β in Theorem 2, and defining the adaptive MLMC algorithm as the optimal values for parameters N_ℓ, L and the constants in Theorem 2 are in general not available.

Algorithm 5. MLMC method to estimate $\mathbb{E}g(Y)$

Require: model $\mathcal{H} = (Q, \mathcal{X}, b, \sigma, x_0, r)$, Y a function of execution of $\mathcal{H}$, time interval $[0, s]$, functional $g(Y)$
1: select the parameters: finest level of approximation L, number of samples for each level $\mathfrak{S} := (N_0, N_1, \ldots, N_L)$, initial number of time steps κ
2: define $\mathcal{A}_0$ to be Algorithm 2 with $\ell = 0$ and time step $n_0 = \kappa 2^0$ to generate samples θ^0
3: compute P^0 using Algorithm 3 with number of samples $N = N_0$ and functional $g(\theta^0)$ and with the embedded algorithm $\mathcal{A}_0$
4: $l = 1$
5: **while** $l < L$ **do**
6:     define $\mathcal{A}_l$ to be Algorithm 4 with time step $n_f = \kappa 2^\ell$ to generate samples θ^ℓ and $\theta^{\ell-1}$
7:     compute P^ℓ using Algorithm 3 with number of samples $N = N_l$ and functional $[g(\theta^l) - g(\theta^{l-1})]$ and with the embedded algorithm $\mathcal{A}_\ell$
8:     $l = l + 1$
9: **end while**
10: compute $\hat{P} = \sum_{\ell=0}^{L} P^\ell$ according to (13)
Ensure: $\hat{P}$ as approximate estimate of $\mathbb{E}g(Y)$

---

5   MLMC with Smoothed Indicator Function

The smoothing is based on the function $g^\delta : \mathbb{R} \to \mathbb{R}$, which is the rescaled translates of a function $g^0 : \mathbb{R} \to \mathbb{R}$ of the form

$$
g^0(x) = \begin{cases} 0, & x > 1 \\ \frac{1}{2} + \frac{1}{8}\left(5x^3 - 9x\right), & |x| \le 1 \\ 1, & x < -1, \end{cases} \quad \text{and } g^\delta(x) = g^0\left(\frac{x - s}{\delta}\right), \quad x \in \mathbb{R}. \quad (17)
$$

This is not the only possible choice for a smoothing function (see [19]), but is easy to implement and already provides significant gains in computational cost.

The new MLMC method that includes smoothing is defined by

$$
\mathcal{M}_{\mathfrak{S}}^{\delta, L} = \frac{1}{N_0} \cdot \sum_{i=1}^{N_0} g^\delta(\theta_i^0) + \sum_{\ell=1}^{L} \frac{1}{N_\ell} \cdot \sum_{i=1}^{N_\ell} \left(g^\delta(\theta_i^{\ell, f}) - g^\delta(\theta_i^{\ell, c})\right), \quad (18)
$$

with an independent family of $\mathbb{R}^2$-valued random variables $(\theta_i^{\ell, f}, \theta_i^{\ell, c})$ for $i = 1, \ldots, N_\ell$ and $\ell = 0, 1, \ldots, L$ such that equality in distribution holds for $(\theta_i^{\ell, f}, \theta_i^{\ell, c})$ and $(\theta^\ell, \theta^{\ell-1})$, where we used the notation $(\theta_i^{0, f}, \theta_i^{0, c}) = (\theta_i^0, 0)$ for the initial level $\ell = 0$. The next theorem gives the mean square error decomposition for (18).

Theorem 3. *For $\delta > 0$, the error of $\mathcal{M}_{\mathfrak{S}}^{\delta, L}$ in (18) with smoothing function (17) can be decomposed as*

$$MSE\left(\mathcal{M}_{\mathfrak{S}}^{\delta,L}\right) := \mathbb{E}\|\mathcal{M}_{\mathfrak{S}}^{\delta,L} - \mathbb{E}g(Y)\|^2$$

$$\leq \delta^4 + \left|\mathbb{E}(g^\delta(Y)) - \mathbb{E}(g^\delta(\theta^L))\right|^2 + \operatorname{Var}(\mathcal{M}_{\mathfrak{S}}^{\delta,L}) =: e_1^2 + e_2^2 + e_3. \qquad (19)$$

The error terms in (19) are related to smoothing, bias, and variance, respectively.

6   Adaptive MLMC Algorithm

In this section we present an adaptive algorithm to find the optimal parameters for the MLMC method. For a given $\varepsilon > 0$ we wish to select the parameters of the MLMC algorithm such that its error is at most ε and its cost is as small as possible. The adaptive algorithm assumes no prior knowledge on the smoothing parameter δ, along with bias and variance dependencies on it. The smoothing parameter δ is chosen from the discrete set of values $\delta_m = 1/2^m$, where $m \in \mathbb{N}$. With a slight abuse of notation we put $g^m = g^{\delta_m}$. We choose the parameters of our algorithm such that

$$e_1 \leq a_1 \varepsilon_*, \quad e_2 \leq a_2 \cdot \varepsilon_*, \quad e_3 \leq a_3^2 \cdot \varepsilon_*^2, \text{ where } \varepsilon_* := \frac{\varepsilon}{a_1 + a_2 + a_3}. \qquad (20)$$

The MLMC algorithm is parameterized by the value m for smoothing $\delta_m = 1/2^m$, the values of the maximal level L, and the replication numbers $\mathfrak{S} = (N_0, \dots, N_L)$. We use $y_{i,0}$ to denote *actual samples* of the random variable θ^0 and $(y_{i,\ell}, y_{i,\ell-1})$ to denote the actual samples of the random vector $(\theta^\ell, \theta^{\ell-1})$ for $\ell = 1, \dots, L$ as opposed to $\theta_i^{\ell,f}, \theta_i^{\ell,c}$ which were used previously for their respective *random variables*.

Theorem 2 relies on the assumption of exponential upper bounds in (15)–(16), which in general might be difficult to verify. Instead in this section we study asymptotic upper bounds. We replace assumptions (15)–(16) with the requirement that for every m there exists $c, \alpha > 0$ such that

$$\left|\mathbb{E}(g^m(\theta^\ell)) - \mathbb{E}(g^m(\theta^{\ell-1}))\right| \approx c \cdot 2^{-\ell \cdot \alpha} \quad \text{and} \quad \lim_{\ell \to \infty} \mathbb{E}g^m(\theta^\ell) = \mathbb{E}g^m(Y). \qquad (21)$$

We put $C_r = 2^{r+1}$ with $r = 3$, the degree of polynomial in (17), and suppose that there exists $c > 0$ such that $\left|\mathbb{E}(g^m(Y)) - \mathbb{E}(g^{m-1}(Y))\right| \approx c \cdot \delta_m^4$. This yields the asymptotic upper bound for the smoothing error with parameter δ_m,

$$\left|\mathbb{E}g(Y) - \mathbb{E}(g^m(Y))\right| \lesssim (C_r - 1)^{-1} \cdot \left|\mathbb{E}(g^m(Y)) - \mathbb{E}(g^{m-1}(Y))\right|. \qquad (22)$$

We estimate the expectations and variances with their versions:

$$\hat{b}_0 = \frac{1}{N_0} \cdot \sum_{i=1}^{N_0} g^m(y_{i,0}), \quad \hat{b}_\ell = \frac{1}{N_\ell} \cdot \sum_{i=1}^{N_\ell} (g^m(y_{i,\ell}) - g^m(y_{i,\ell-1})), \qquad (23)$$

$$\hat{v}_0 = \frac{1}{N_0} \cdot \sum_{i=1}^{N_0} |g^m(y_{i,0}) - \hat{b}_0|^2, \hat{v}_\ell = \frac{1}{N_\ell} \cdot \sum_{i=1}^{N_\ell} |g^m(y_{i,\ell}) - g^m(y_{i,\ell-1}) - \hat{b}_\ell|^2. \qquad (24)$$

We get that $\hat{v}(\mathfrak{S}) = \sum_{\ell=0}^{L} \frac{1}{N_\ell} \cdot \hat{v}_\ell$ serves as an empirical upper bound for the variance of the MLMC algorithm with any choice of replication numbers $\mathfrak{S} = (N_0, N_1, \ldots, N_L)$. If, for the present choice of replication numbers, this bound is too large compared to the upper bound for $\mathrm{Var}(\mathcal{M}_{\mathfrak{S}}^{\delta,L})$ in (20), i.e., if the variance constraint $\hat{v}(\mathfrak{S}) \leq a_3^2 \cdot \varepsilon_*^2$ is violated, we determine new values of $N_0', \ldots, N_L'$ by minimizing $c(N_0, \ldots, N_L)$ subject to the constraint $\hat{v}(\mathfrak{S}) \leq a_3^2 \cdot \varepsilon_*^2$, leading to

$$N_\ell' = \frac{\hat{v}_\ell^{1/2}}{(2^\ell + 1)^{1/2}} \cdot \sum_{\ell=0}^{L} \left(\hat{v}_\ell \cdot (2^\ell + 1)\right)^{1/2} \cdot \frac{\varepsilon_*^{-2}}{a_3^2}, \qquad \ell = 0, 1, \ldots, L, \qquad (25)$$

and extra samples of θ^0 and $(\theta^\ell, \theta^{\ell-1})$ have to be generated accordingly.

For estimating $|\mathbb{E}(g^m(\theta^\ell)) - \mathbb{E}(g^m(\theta^{\ell-1}))|$ we can use the values of $|\hat{b}_\ell|$ already available from (23) for the levels $\ell = 1, \ldots, L$. We estimate α and c in (21) by a least-squares fit, i.e., we take $\hat{\alpha}$ and $\hat{c}$ through least squares regression. This geometric upper bound can be used to set the stopping criterion of increasing the maximal level:

$$\hat{B}_L = \max\left(|\hat{b}_L|, |\hat{b}_{L-1}|/2^{\hat{\alpha}}, |\hat{b}_{L-2}|/2^{2\hat{\alpha}}\right) \leq a_2 \cdot (2^{\hat{\alpha}} - 1) \cdot \varepsilon_* \text{ for } L \geq 3. \qquad (26)$$

Similarly for the smoothing coefficient the present value m is accepted if

$$\hat{s} := \left| \frac{1}{N_L} \cdot \sum_{i=1}^{N_L} (g^m(y_{i,L}) - g^{m-1}(y_{i,L})) \right| \leq a_1 \cdot (C_r - 1) \cdot \varepsilon_*. \qquad (27)$$

We combine the above results in Algorithm 6.

Algorithm 6. Adaptive MLMC algorithm with smoothing

Require: sampling algorithm $\mathcal{A}_\ell$, functional $g(\cdot)$, target accuracy ε
1: initialize parameters $m = 2$; $L = 3$ $N_0 = N_1 = N_2 = 10^2$
2: generate N_0 samples of θ^0 and N_ℓ samples of $(\theta^\ell, \theta^{\ell-1})$ for $\ell = 1, 2$
3: compute $\hat{v}_0, \hat{v}_1, \hat{v}_2$, according to (24)
4: **repeat** {/* smoothing */}
5:     $m = m + 1$ and newlevel = false
6:     **repeat** {/* bias */}
7:         **if** newlevel **then**
8:             $L = L + 1$; $N_L = 100$
9:             generate N_L samples of (θ^L, θ^{L-1}) and compute $\hat{v}_L$ according to (24)
10:        **end if**
11:        **repeat**
12:            compute the replication numbers $N_0', \ldots, N_L'$ (see (25))
13:            $N_\ell = \max(N_\ell, N_\ell')$ for $\ell = 0, \ldots, L$
14:            generate new samples of θ^0 and $(\theta^\ell, \theta^{\ell-1})$ and compute $\hat{v}_0, \ldots, \hat{v}_L$ (see (24))
15:        **until** the variance constraint is satisfied
16:        estimate $\hat{\alpha}$, and $\hat{B}_L$; newlevel = true
17:    **until** the bias constraint (26) is satisfied
18:    compute $\hat{s}$ according to (27)
19: **until** the smoothing constraint (27) is satisfied
Ensure: $\mathcal{M}_{\mathfrak{S}}^{\delta,L}$ as an estimation of $\mathbb{E}g(Y)$

---

7   Simulation Results

Recall Problem 3 where the goal is to estimate the probability $\mathbb{P}(Y \leq \theta_+ + 0.1 \cdot \delta_d)$. The random variable Y is defined as $Y = \max\{\theta_t, t \in [0, s]\}$. We set the parameters of the TCL model (3)–(4) according to Table 1 and select the time horizon $s = 1\,\mathrm{h}$. We implement the MLMC Algorithm 6 for target accuracies $\varepsilon = 2^{-k}$, where $k \in \{3, \ldots, 8\}$. We set the parameters $a_1 = 4$, $a_2 = a_3 = 2$ in (20). With this choice we put less pressure on the smoothing error because the influence of the smoothing parameter δ on the variance and thus on the overal cost is severe. Due to the smoothing step we have to sample executions for the time duration of at least $(s + \delta)$ in order to evaluate the functional $g(Y)$, thus sampling executions for 1.5 h is sufficient.

Table 1. Parameters of a residential air conditioner as a TCL [13] modeled in (3)–(4).

Param.	Interpretation	Value	Param.	Interpretation	Value
θ_s	set-point	$20\,[^\circ C]$	R	thermal resistance	$1.5\,[^\circ C/kW]$
δ_d	dead-band width	$0.5\,[^\circ C]$	C	thermal capacitance	$10\,[kWh/^\circ C]$
θ_a	ambient temperature	$32\,[^\circ C]$	σ_0	standard deviation OFF mode	$0.2\,[^\circ C/\sqrt{hour}]$
P_{rate}	power	$14\,[kW]$	σ_1	standard deviation ON mode	$0.22\,[^\circ C/\sqrt{hour}]$

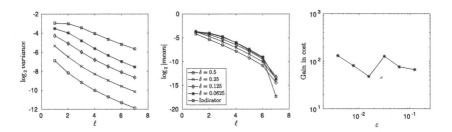

Fig. 1. Simulation results for Problem 3. Variance (left) and mean (center) of the estimation decay with respect to level ℓ for different smoothing coefficient. Computational gain (right) as ratio of the SMC and adaptive MLMC costs.

The left and center plots in Fig. 1 show the impact of the smoothing coefficient on the variance and mean decays respectively based on 10^6 runs of the algorithm. These plots indicate that the adaptive MLMC method is beneficial over SMC method (plots with $\ell = 1$ and with the indicator function) due to the strong variance and mean decay with respect to level ℓ as well as the use of smoothing function instead of the indicator function. The computational gain of the MLMC over SMC is presented on the right plot based on 100 runs. The plot compares the *expected* cost of the SMC method with the *estimated* cost of the adaptive MLMC method. The cost of SMC method is given by $\varepsilon^{-2-\frac{1}{\alpha}}$ (see Theorem 1), which bounds the cost of generating executions and evaluating functionals. The plot indicates larger computational gains for higher accuracies (smaller ε).

References

1. Abate, A., Prandini, M., Lygeros, J., Sastry, S.: Probabilistic reachability and safety for controlled discrete time stochastic hybrid systems. Automatica **44**(11), 2724–2734 (2008)
2. Baier, C., Haverkort, B., Hermanns, H., Katoen, J.-P.: Model-checking algorithms for continuous-time Markov chains. Trans. Softw. Eng. **29**(6), 524–541 (2003)
3. Baier, C., Katoen, J.-P.: Principles of Model Checking. MIT Press, Cambridge (2008)
4. Bujorianu, M.L., Lygeros, J.: Reachability questions in piecewise deterministic Markov processes. In: Maler, O., Pnueli, A. (eds.) HSCC 2003. LNCS, vol. 2623, pp. 126–140. Springer, Heidelberg (2003). doi:10.1007/3-540-36580-X_12
5. Bujorianu, M.L., Lygeros, J.: General stochastic hybrid systems: modelling and optimal control. In: Proceedings of 43rd IEEE Conference Decision Control, pp. 1872–1877 (2004)
6. Bulychev, P., David, A., Larsen, K.G., Legay, A., Mikučionis, M., Poulsen, D.B.: Checking and distributing statistical model checking. In: Goodloe, A.E., Person, S. (eds.) NFM 2012. LNCS, vol. 7226, pp. 449–463. Springer, Heidelberg (2012). doi:10.1007/978-3-642-28891-3_39
7. Cassandras, C.G., Lygeros, J. (eds.) Stochastic Hybrid Systems. Control Engineering, vol. 24. CRC Press, Boca Raton (2006)
8. Clarke, E.M., Zuliani, P.: Statistical model checking for cyber-physical systems. In: Bultan, T., Hsiung, P.-A. (eds.) ATVA 2011. LNCS, vol. 6996, pp. 1–12. Springer, Heidelberg (2011). doi:10.1007/978-3-642-24372-1_1
9. Daca, P., Henzinger, T.A., Křetínský, J., Petrov, T.: Faster statistical model checking for unbounded temporal properties. ACM Trans. Comput. Logic **18**(2), 12:1–12:25 (2017)
10. Esmaeil Zadeh Soudjani, S.: Formal abstractions for automated verification and synthesis of stochastic systems. Ph.D. thesis, Delft, NL, November 2014
11. Esmaeil Zadeh Soudjani, S., Abate, A.: Adaptive and sequential gridding procedures for the abstraction and verification of stochastic processes. SIAM J. Appl. Dyn. Syst. **12**(2), 921–956 (2013)
12. Esmaeil Zadeh Soudjani, S., Abate, A.: Aggregation and control of populations of thermostatically controlled loads by formal abstractions. IEEE Trans. Control Syst. Technol. **23**(3), 975–990 (2015)
13. Esmaeil Zadeh Soudjani, S., Gerwinn, S., Ellen, C., Fränzle, M., Abate, A.: Formal synthesis and validation of inhomogeneous thermostatically controlled loads. In: Norman, G., Sanders, W. (eds.) QEST 2014. LNCS, vol. 8657, pp. 57–73. Springer, Cham (2014). doi:10.1007/978-3-319-10696-0_6
14. Esmaeil Zadeh Soudjani, S., Majumdar, R., Abate, A.: Safety verification of continuous-space pure jump Markov processes. In: Chechik, M., Raskin, J.-F. (eds.) TACAS 2016. LNCS, vol. 9636, pp. 147–163. Springer, Heidelberg (2016). doi:10.1007/978-3-662-49674-9_9
15. Fränzle, M., Hermanns, H., Teige, T.: Stochastic satisfiability modulo theory: a novel technique for the analysis of probabilistic hybrid systems. In: Egerstedt, M., Mishra, B. (eds.) HSCC 2008. LNCS, vol. 4981, pp. 172–186. Springer, Heidelberg (2008). doi:10.1007/978-3-540-78929-1_13
16. Giles, M.B., Nagapetyan, T., Ritter, K.: Adaptive Multilevel Monte Carlo Approximation of Distribution Functions. ArXiv e-prints, June 2017

17. Giles, M.B.: Multilevel Monte Carlo path simulation. Oper. Res. **56**(3), 607–617 (2008)
18. Giles, M.B.: Multilevel Monte Carlo methods. Acta Numer. **24**, 259–328 (2015)
19. Giles, M.B., Nagapetyan, T., Ritter, K.: Multilevel Monte Carlo approximation of distribution functions and densities. SIAM/ASA J. Uncertain. Quantif. **3**(1), 267–295 (2015)
20. Heinrich, S.: Monte Carlo complexity of global solution of integral equations. J. Complexity **14**(2), 151–175 (1998)
21. Kamgarpour, M., Ellen, C., Esmaeil Zadeh Soudjani, S., Gerwinn, S., Mathieu, J.L., Mullner, N., Abate, A., Callaway, D.S., Fränzle, M., Lygeros, J.: Modeling options for demand side participation of thermostatically controlled loads. In: International Conference on Bulk Power System Dynamics and Control, pp. 1–15, August 2013
22. Křetínský, J.: Survey of statistical verification of linear unbounded properties: model checking and distances, pp. 27–45. Springer, Cham (2016)
23. Larsen, K.G., Legay, A.: Statistical model checking past, present, and future. In: Margaria, T., Steffen, B. (eds.) ISoLA 2014. LNCS, vol. 8803, pp. 135–142. Springer, Heidelberg (2014). doi:10.1007/978-3-662-45231-8_10
24. Legay, A., Delahaye, B., Bensalem, S.: Statistical model checking: an overview. In: Barringer, H., et al. (eds.) RV 2010. LNCS, vol. 6418, pp. 122–135. Springer, Heidelberg (2010). doi:10.1007/978-3-642-16612-9_11
25. Mathieu, J.L., Koch, S., Callaway, D.S.: State estimation and control of electric loads to manage real-time energy imbalance. IEEE Trans. Power Syst. **28**(1), 430–440 (2013)
26. Pola, G., Bujorianu, M.L., Lygeros, J., Di Benedetto, M.D.: Stochastic hybrid models: an overview. In: Analysis and Design of Hybrid Systems (2003)
27. Prajna, S., Jadbabaie, A., Pappas, G.J.: A framework for worst-case and stochastic safety verification using barrier certificates. IEEE Trans. Autom. Control **52**(8), 1415–1428 (2007)
28. Tkachev, I., Mereacre, A., Katoen, J.-P., Abate, A.: Quantitative automata-based controller synthesis for non-autonomous stochastic hybrid systems. In: Hybrid Systems: Computation and Control, pp. 293–302. ACM, New York (2013)
29. Wang, Y., Roohi, N., West, M., Viswanathan, M., Dullerud, G.E.: Statistical verification of dynamical systems using set oriented methods. In: Hybrid Systems: Computation and Control, pp. 169–178. ACM, New York (2015)

Author Index

Printed in the United States
By Bookmasters